Machine Learning in Radiation Oncology

Issam El Naqa • Ruijiang Li
Martin J. Murphy
Editors

Machine Learning in Radiation Oncology

Theory and Applications

Springer

Editors
Issam El Naqa
Department of Oncology
McGill University
Montreal
Canada

Ruijiang Li
Department of Radiation Oncology
Stanford University School of Medicine
Stanford, CA
USA

Department of Radiation Oncology
University of Michigan
Ann Arbor
USA

Martin J. Murphy
Department of Radiation Oncology
Virginia Commonwealth University
Richmond, VA
USA

ISBN 978-3-319-35464-4   ISBN 978-3-319-18305-3   (eBook)
DOI 10.1007/978-3-319-18305-3

Springer Cham Heidelberg New York Dordrecht London
© Springer International Publishing Switzerland 2015
Softcover reprint of the hardcover 1st edition 2015
This work is subject to copyright. All rights are reserved by the Publisher, whether the whole or part of the material is concerned, specifically the rights of translation, reprinting, reuse of illustrations, recitation, broadcasting, reproduction on microfilms or in any other physical way, and transmission or information storage and retrieval, electronic adaptation, computer software, or by similar or dissimilar methodology now known or hereafter developed.
The use of general descriptive names, registered names, trademarks, service marks, etc. in this publication does not imply, even in the absence of a specific statement, that such names are exempt from the relevant protective laws and regulations and therefore free for general use.
The publisher, the authors and the editors are safe to assume that the advice and information in this book are believed to be true and accurate at the date of publication. Neither the publisher nor the authors or the editors give a warranty, express or implied, with respect to the material contained herein or for any errors or omissions that may have been made.

Printed on acid-free paper

Springer International Publishing AG Switzerland is part of Springer Science+Business Media (www.springer.com)

Foreword

This book is based on one of the most consequential emergent results of the ongoing computer revolution, namely, that computers can be trained – under the right conditions – to reliably classify new data, such as patient data. This capability, called machine learning (or statistical learning), has been deployed in many areas of technology, commerce, and medicine. Data mining and statistical prediction models have already crept into many areas of modern life, including advertising, banking, sports, weather prediction, politics, science generally, and medicine in particular. The ability of computers to increasingly communicate with people in a natural way (understanding language and speaking to us), such as the famed IBM "Watson" appearance on Jeopardy, or "Siri" on iPhones, portends an accelerating role of sophisticated computer models that predict and respond to our requests. Fundamentally, these developments rely on the ability of statistical computer methods to pull (as Nate Silver puts it) "signal from the noise." While traditional statistical methods typically attempt to ascertain the role of particular variables in determining an outcome of interest (hence, needing many data points for every variable included in the prediction model), machine learning represents a different goal, to reliably predict an outcome, for example that an imaging abnormality is benign with a high degree of certainty. The statisticians and computer scientists working in this emerging area are often happy to use large numbers of variables (or previous data instances) that essentially vote together in a nonlinear fashion. Simplicity is happily traded for an improved ability to predict.

The chapters in this book comprehensively review machine learning and related modeling methods previously used in many areas of radiation oncology and diagnostic radiology. The editors and authors are explorers in this new territory, and have performed a great service by surveying and mapping the many achievements to date and outline many areas of potential application. Early chapters review the fundamental characteristics, and varieties, of machine learning methods, including difficult issues regarding evaluation of predictive model performance. The most well-developed use of machine learning reviewed is the creation of computer aided diagnosis (CAD) models to provide a reliable "second opinion" for radiologists reading mammograms to detect breast cancer. The increasing use of a wider range of imaging features referred to as "radiomics," in analogy to "genomics," presented in radiomics for disease detection, and radiomics for diagnosis, or "theragnostic" [1] chapters, which are devoted to details of image-based informatics formats and database systems, including tools to share and learn from institutional databases. Machine learning approaches to aid in the planning, delivery, and quality assurance of radiation therapy are reviewed. Efforts to predict response to radiation therapy are also reviewed in useful detail. Obtaining enough data of sufficient quality and diversity is the biggest challenge in predictive modeling. This is only possible if data are shared across institutional and national borders, both academic and community health-care systems [2].

Machine learning – coupled with computer vision and imaging processing techniques – has been demonstrated to be useful in diagnosis, treatment planning, and outcome prediction in radiation oncology and radiology. This is of particular importance since we know that doctors have increasing difficulties to predict the outcome of modernized complex patient treatments [3]. This book provides a wonderful summary of past achievements, current challenges, and emerging approaches in this important area of medicine. Unlike many other approaches to improving medicine, the use of improved and continuously updated prediction models put together in "Decision Support Systems" holds the potential of improved clinical decision making with minimal costs to patients [4]. An intuitively attractive characteristic of this approach is the user of *all* the data available (rather than using only one type of data such as dose or gene profile). We anticipate that predictive models-based Decisions Support Systems will ease the implementation of personalized (or precision) medicine.

Despite investment in efforts to improve the skills of clinicians, patients continue to report low levels of involvement [5]. There is indeed evidence level 1 from a Cochrane systematic review evaluating 86 studies involving 20,209 participants included in published randomized controlled trials demonstrating that decision aids increase people's involvement, support informed values-based choices in patient-practitioner communication, and improve knowledge and realistic perception of outcomes. We therefore believe the next step will be to integrate, whenever possible, Shared Decision Making approaches (see, e.g., www.treatmentchoice.info; www.optiongrid.org) to include the patient perspective on the best treatment of choice [6].

We are sincerely convinced that this book will continue to advance precision medicine in oncology.

References

1. Lambin P, Rios-Velazquez E, Leijenaar R et al. Radiomics: extracting more information from medical images using advanced feature analysis. Eur J Cancer. 2012;48:441–6.
2. Lambin P, Roelofs E, Reymen B, Velazquez ER, Buijsen J, Zegers CM, Carvalho S, Leijenaar RT, Nalbantov G, Oberije C, Scott Marshall M, Hoebers F, Troost EG, van Stiphout RG, van Elmpt W, van der Weijden T, Boersma L, Valentini V, Dekker A. 'Rapid Learning health care in oncology' – an approach towards decision support systems enabling customised radiotherapy'. Radiother Oncol. 2013;109(1):159–64.
3. Oberije C, Nalbantov G, Dekker A, Boersma L, Borger J, Reymen B, van Baardwijk A, Wanders R, De Ruysscher D, Steyerberg E, Dingemans AM, Lambin P. A prospective study comparing the predictions of doctors versus models for treatment outcome of lung cancer patients: a step toward individualized care and shared decision making. Radiother Oncol. 2014;112(1):37–43.

4. Lambin P, van Stiphout RG, Starmans MH et al. Predicting outcomes in radiation oncology--multifactorial decision support systems. Nature reviews. Clin Oncol. 2013;10:27–40.
5. Stacey D, Bennett CL, Barry MJ et al. Decision aids for people facing health treatment or screening decisions. Cochrane Database Syst Rev. 2011;(10):CD001431.
6. Stiggelbout AM, Van der Weijden T, De Wit MP et al. Shared decision making: really putting patients at the centre of healthcare. BMJ. 2012;344:e256.

Philippe Lambin, MD, PhD
Professor and Head, Department of Radiation Oncology,
Medical Director of the MAASTRO CLINIC
Division leader of the Research institute GROW
Maastricht University, The Netherlands

Joseph O. Deasy, PhD
Professor and Chair, Department of Medical Physics,
Enid A. Haupt Endowed Chair in Medical Physics,
Memorial Sloan Kettering Cancer Center, New York, NY, USA

Preface

Radiotherapy is a major treatment modality for cancer and is currently the main option for treating local disease at advanced stages. More than half of all cancer patients receive irradiation as part of their treatment, with curative or palliative intent to eradicate cancer or reduce pain, respectively, while sparing uninvolved normal tissue from detrimental side effects. Despite significant technological advances in treatment planning and delivery using image-guided techniques, the complex nature of radiotherapy processes and the massive amount of structured and unstructured heterogeneous data generated during radiotherapy from early patient consultation to patient simulation, to treatment planning and delivery, to monitoring response, to follow-up visits, invite the application of more advanced computational methods that can mimic human cognition and intelligent decision making to ensure safe and effective treatment. In addition, these computational methods need to compensate for human limitations in handling a large amount of flowing information in an efficient manner, in which simple errors can make the difference between life and death.

Machine learning is a technology that aims to develop computer algorithms that are able to emulate human intelligence by incorporating ideas from neuroscience, probability and statistics, computer science, information theory, psychology, control theory, and philosophy with successful applications in computer vision, robotics, entertainment, ecology, biology, and medicine. The essence of this technology is to *humanize computers* by learning from the surrounding environment and previous experiences, with or without a teacher. The development and application of machine learning has undergone a significant surge in recent years due to the exponential growth and availability of "big data" with machine learning techniques occupying the driver's seat to steer the understanding of such data in many fields, including radiation oncology.

The growing interest in applying machine learning algorithms to radiotherapy has been highlighted by special sessions at the annual meeting of the American Association of Physicists in Medicine (AAPM) and at the International Conference on Machine Learning and Applications (ICMLA). Ensuing discussions of compiling these disparate applications of machine learning in radiotherapy into a single succinct monograph led to the idea of this book. The goal is to provide interested readers with a comprehensive and accessible text on the subject to fill in an important existing void in radiotherapy and machine learning literature. Even as these discussions were taking place, the subject of machine learning in radiotherapy continued its growth from a peripheral subfield in radiotherapy into widespread applications that touch almost every area in radiotherapy from treatment planning, quality assurance, image guidance, and respiratory motion management to treatment response modeling and outcomes prediction. This rapid growth has driven the compilation of this textbook.

The textbook is intended to be an introductory learning guide for students and residents in medical physics and radiation oncology who are interested in exploring this new field of machine learning for their own curiosity or their research projects. In addition, the book is intended to be a useful and informative resource for more experienced practitioners, researchers, and members of both radiotherapy and applied machine learning as a two-way bridge between these communities. This is manifested by the fact that the book has been written by experts from both the radiotherapy and machine learning domains.

The book is structured into five sections:

- The first section provides an introduction to machine learning and is a must-read for individuals who are new to the field. It begins with a machine learning definition (Chap. 1), followed by discussion of the main computational learning principles using PAC or VC theories (Chap. 2), presentation of the most commonly used supervised and unsupervised learning algorithms with demonstrative applications drawn from the radiotherapy field (Chap. 3), and descriptions of different methods and techniques used for evaluating the performance of learning methods (Chap. 4). The ever-growing role of informatics infrastructure in radiotherapy and its application to machine learning are presented in Chap. 5. Finally, given the realistic challenges related to data sharing from a global radiotherapy network, this section concludes with a discussion of how machine learning could be extended to a distributed multicenter rapid learning framework.
- The second section summarizes years of successful application of machine learning in radiological sciences – a sister field to radiotherapy – as a computational tool for computer-aided detection (Chap. 7) and computer-aided diagnosis (Chap. 8).
- The third section presents applications of machine learning in radiotherapy treatment planning as a tool for image-guided radiotherapy (Chap. 9) and a computational vehicle for knowledge-based planning (Chap. 10).
- The fourth section demonstrates the application of machine learning to respiratory motion management – a rather challenging problem for accurate delivery of irradiation to a moving target – by discussing predictive respiratory models (Chap. 11) and image-based compensation techniques (Chap. 12).
- Quality assurance is at the heart of safe delivery of radiotherapy and is a major part of a medical physicist's job. Examples for application of machine learning to QA for detection and prediction of radiotherapy errors (Chap. 13), for treatment planning (Chap. 14), and for delivery (Chap. 15) validation are presented and discussed.

- In the era of personalized evidence-based medicine, machine learning predictive analytics can play an important role in the understanding of radiotherapy response (Chap. 16). Examples of successful machine learning applications to normal tissue complication probability (Chap. 17) and tumor control probability (Chap. 18) highlight the inherent power of this technology in deciphering complex radiobiological response.

This book is the product of a coordinated effort by the editors, authors, and publishing team to present the principles and applications of machine learning to a new generation of practitioners in radiation therapy and to present the present-day challenges of radiotherapy to the computer science community, with the hope of driving advancements in both fields.

Montreal, Canada	Issam El Naqa
Stanford, CA, USA	Ruijiang Li
Richmond, VA, USA	Martin J. Murphy

Contents

Part I Introduction

1. **What Is Machine Learning?** 3
    Issam El Naqa and Martin J. Murphy

2. **Computational Learning Theory** 13
    Issam El Naqa

3. **Machine Learning Methodology** 21
    Sangkyu Lee and Issam El Naqa

4. **Performance Evaluation in Machine Learning** 41
    Nathalie Japkowicz and Mohak Shah

5. **Informatics in Radiation Oncology** 57
    Paul Martin Putora, Samuel Peters, and Marc Bovet

6. **Application of Machine Learning for Multicenter Learning** 71
    Johan P.A. van Soest, Andre L.A.J. Dekker, Erik Roelofs,
    and Georgi Nalbantov

Part II Machine Learning for Computer-Aided Detection

7. **Computerized Detection of Lesions in Diagnostic Images** 101
    Kenji Suzuki

8. **Classification of Malignant and Benign Tumors** 133
    Juan Wang, Issam El Naqa, and Yongyi Yang

Part III Machine Learning for Treatment Planning

9. **Image-Guided Radiotherapy with Machine Learning** 157
    Yaozong Gao, Yanrong Guo, Yinghuan Shi, Shu Liao,
    Jun Lian, and Dinggang Shen

10. **Knowledge-Based Treatment Planning** 193
    Issam El Naqa

Part IV Machine Learning Delivery and Motion Management

11 **Artificial Neural Networks to Emulate and Compensate Breathing Motion During Radiation Therapy** 203
Martin J. Murphy

12 **Image-Based Motion Correction** 225
Ruijiang Li

Part V Machine Learning for Quality Assurance

13 **Detection and Prediction of Radiotherapy Errors** 237
Issam El Naqa

14 **Treatment Planning Validation** 243
Ruijiang Li and Steve B. Jiang

15 **Treatment Delivery Validation** 253
Ruijiang Li

Part VI Machine Learning for Outcomes Modeling

16 **Bioinformatics of Treatment Response** 263
Issam El Naqa

17 **Modelling of Normal Tissue Complication Probabilities (NTCP): Review of Application of Machine Learning in Predicting NTCP** 277
Sarah Gulliford

18 **Modeling of Tumor Control Probability (TCP)** 311
Issam El Naqa

Index 325

Part I
Introduction

What Is Machine Learning?

Issam El Naqa and Martin J. Murphy

Abstract
Machine learning is an evolving branch of computational algorithms that are designed to emulate human intelligence by learning from the surrounding environment. They are considered the working horse in the new era of the so-called big data. Techniques based on machine learning have been applied successfully in diverse fields ranging from pattern recognition, computer vision, spacecraft engineering, finance, entertainment, and computational biology to biomedical and medical applications. More than half of the patients with cancer receive ionizing radiation (radiotherapy) as part of their treatment, and it is the main treatment modality at advanced stages of local disease. Radiotherapy involves a large set of processes that not only span the period from consultation to treatment but also extend beyond that to ensure that the patients have received the prescribed radiation dose and are responding well. The degrees of the complexity of these processes can vary and may involve several stages of sophisticated human-machine interactions and decision making, which would naturally invite the use of machine learning algorithms into optimizing and automating these processes including but not limited to radiation physics quality assurance, contouring and treatment planning, image-guided radiotherapy, respiratory motion management, treatment response modeling, and outcomes prediction. The ability of machine learning algorithms to learn from current context and generalize into unseen tasks would allow improvements in both the safety and efficacy of radiotherapy practice leading to better outcomes.

I. El Naqa (✉)
Department of Oncology, McGill University, Montreal, QC, Canada

Department of Radiation Oncology, University of Michigan, Ann Arbor, USA
e-mail: issam.elnaqa@mcgill.ca; ielnaqa@med.umich.edu

M.J. Murphy
Department of Radiation Oncology, Virginia Commonwealth University, Richmond, VA, USA
e-mail: MMurphy@mcvh-vcu.edu

1.1 Overview

A machine learning algorithm is a computational process that uses input data to achieve a desired task without being literally programmed (i.e., "hard coded") to produce a particular outcome. These algorithms are in a sense "soft coded" in that they automatically alter or adapt their architecture through repetition (i.e., experience) so that they become better and better at achieving the desired task. The process of adaptation is called training, in which samples of input data are provided along with desired outcomes. The algorithm then optimally configures itself so that it can not only produce the desired outcome when presented with the training inputs, but can generalize to produce the desired outcome from new, previously unseen data. This training is the "learning" part of machine learning. The training does not have to be limited to an initial adaptation during a finite interval. As with humans, a good algorithm can practice "lifelong" learning as it processes new data and learns from its mistakes.

There are many ways that a computational algorithm can adapt itself in response to training. The input data can be selected and weighted to provide the most decisive outcomes. The algorithm can have variable numerical parameters that are adjusted through iterative optimization. It can have a network of possible computational pathways that it arranges for optimal results. It can determine probability distributions from the input data and use them to predict outcomes.

The ideal of machine learning is to emulate the way that human beings (and other sentient creatures) learn to process sensory (input) signals in order to accomplish a goal. This goal could be a task in pattern recognition, in which the learner wants to distinguish apples from oranges. Every apple and orange is unique, but we are still able (usually) to tell one from the other. Rather than hard code a machine with many, many exact representations of apples and oranges, it can be programmed to learn to distinguish them through repeated experience with actual apples and oranges. This is a good example of *supervised learning*, in which each training example of input data (color, shape, odor, etc.) is paired with its known classification label (apple or orange). It allows the learner to deal with similarities and differences when the objects to be classified have many variable properties within their own classes but still have fundamental qualities that identify them. Most importantly, the successful learner should be able to recognize an apple or an orange that it has never seen before.

A second type of machine learning is the so-called *unsupervised algorithm*. This might have the objective of trying to throw a dart at a bull's-eye. The device (or human) has a variety of degrees of freedom in the mechanism that controls the path of the dart. Rather than try to exactly program the kinematics a priori, the learner practices throwing the dart. For each trial, the kinematic degrees of freedom are adjusted so that the dart gets closer and closer to the bull's-eye. This is unsupervised in the sense that the training doesn't associate a particular kinematic input configuration with a particular outcome. The algorithm finds its own way from the training input data. Ideally, the trained dart thrower will be able to adjust the learned kinematics to accommodate, for instance, a change in the position of the target.

A third type of machine learning is *semi-supervised learning*, where part of the data is labeled and other parts are unlabeled. In such a scenario, the labeled part can be used to aid the learning of the unlabeled part. This kind of scenario lends itself to most processes in nature and more closely emulates how humans develop their skills.

There are two particularly important advantages to a successful algorithm. First, it can substitute for laborious and repetitive human effort. Second, and more significantly, it can potentially learn more complicated and subtle patterns in the input data than the average human observer is able to do. Both of these advantages are important to radiation therapy. For example, the daily contouring of tumors and organs at risk during treatment planning is a time-consuming process of pattern recognition that is based on the observer's familiarity and experience with the appearance of anatomy in diagnostic images. That familiarity, though, has its limits, and consequently, there are uncertainty and interobserver variability in the resulting contours. It is possible that an algorithm for contouring can pick up subtleties of texture or shape in one image or simultaneously incorporate data from multiple sources or blend the experience of numerous observers and thus reduce the uncertainty in the contour.

More than half of the patients with cancer receive ionizing radiation (radiotherapy) as part of their treatment, and it is the main treatment modality at advanced stages of disease. Radiotherapy involves a large set of processes that not only span the period from consultation to treatment but also extend beyond, to ensure that the patients have received the prescribed radiation dose and are responding well. The complexity of these processes can vary and may involve several stages of sophisticated human-machine interactions and decision making, which would naturally invite the use of machine learning algorithms to optimize and automate these processes, including but not limited to radiation physics quality assurance, contouring and treatment planning, image-guided radiotherapy, respiratory motion management, treatment response modeling, and outcomes prediction.

1.2 Background

Machine learning is the technology of developing computer algorithms that are able to emulate human intelligence. It draws on ideas from different disciplines such as artificial intelligence, probability and statistics, computer science, information theory, psychology, control theory, and philosophy [1–3]. This technology has been applied in such diverse fields as pattern recognition [3], computer vision [4], spacecraft engineering [5], finance [6], entertainment [7, 8], ecology [9], computational biology [10, 11], and biomedical and medical applications [12, 13]. The most important property of these algorithms is their distinctive ability to learn the surrounding environment from input data with or without a teacher [1, 2].

Historically, the inception of machine learning can be traced to the seventeenth century and the development of machines that can emulate human ability to add and subtract by Pascal and Leibniz [14]. In modern history, Arthur Samuel from IBM coined the term "machine learning" and demonstrated that computers could be programmed to learn to play checkers [15]. This was followed by the development of the perceptron by Rosenblatt as one of the early neural network architectures in 1958 [16]. However, early enthusiasm about the perceptron was dampened by the observation made by Minsky that the perceptron classification ability is limited to linearly separable problems and not common nonlinear problems such as a simple XOR logic [17]. A breakthrough was achieved in 1975 by the development of the multilayer perceptron (MLP) by Werbos [18]. This was followed by the development of decision trees by Quinlan in 1986 [19] and support vector machines by Cortes and Vapnik [20]. Ensemble machine learning algorithms, which combine multiple learners, were subsequently proposed, including Adaboost [21] and random forests [22]. More recently, distributed multilayered learning algorithms have emerged under the notion of deep learning [23]. These algorithms are able to learn good representations of the data that make it easier to extract useful information when building classifiers or other predictors [24].

1.3 Machine Learning Definition

The field of machine learning has received several formal definitions in the literature. Arthur Samuel in his seminal work defined machine learning as "a field of study that gives computers the ability to learn without being explicitly programmed" [15]. Using a computer science lexicon, Tom Mitchell presented it as "A computer program is said to learn from experience (E) with respect to some class of tasks (T) and performance measure (P), if its performance at tasks in T, as measured by P, improves with experience E" [1]. Ethem Alpaydin in his textbook defined machine learning as the field of "Programming computers to optimize a performance criterion using example data or past experience" [2]. These various definitions share the notion of coaching computers to intelligently perform tasks beyond traditional number crunching by learning the surrounding environment through repeated examples.

1.4 Learning from Data

The ability to learn through input from the surrounding environment, whether it is playing checkers or chess games, or recognizing written patterns, or solving the daunting problems in radiation oncology, is the main key to developing a successful machine learning application. Learning is defined in this context as estimating dependencies from data [25].

The fields of data mining and machine learning are intertwined. Data mining utilizes machine learning algorithms to interrogate large databases and discover hidden knowledge in the data, while many machine learning algorithms employ data mining methods to preprocess the data before learning the desired tasks [26]. However, it should be noted that machine learning is not limited to solving database-like problems but also extends into solving complex artificial intelligence challenges by learning and adapting to a dynamically changing situation, as is encountered in a busy radiation oncology practice, for instance.

Machine learning has both engineering science aspects such as data structures, algorithms, probability and statistics, and information and control theory and social science aspects by drawing in ideas from psychology and philosophy.

1.5 Overview of Machine Learning Approaches

Machine learning can be divided according to the nature of the data labeling into supervised, unsupervised, and semi-supervised as shown in Fig. 1.1. Supervised learning is used to estimate an unknown (input, output) mapping from known (input, output) samples, where the output is labeled (e.g., classification and regression). In unsupervised learning, only input samples are given to the learning system (e.g., clustering and estimation of probability density function). Semi-supervised learning is a combination of both supervised and unsupervised where part of the data is partially labeled and the labeled part is used to infer the unlabeled portion (e.g., text/image retrieval systems).

From a concept learning perspective, machine learning can be categorized into transductive and inductive learning [27]. Transductive learning involves the inference from specific training cases to specific testing cases using discrete labels as in clustering or using continuous labels as in manifold learning. On the other hand, inductive learning aims to predict outputs from inputs that the learner has not encountered before. Along these lines, Mitchell argues for the necessity of an

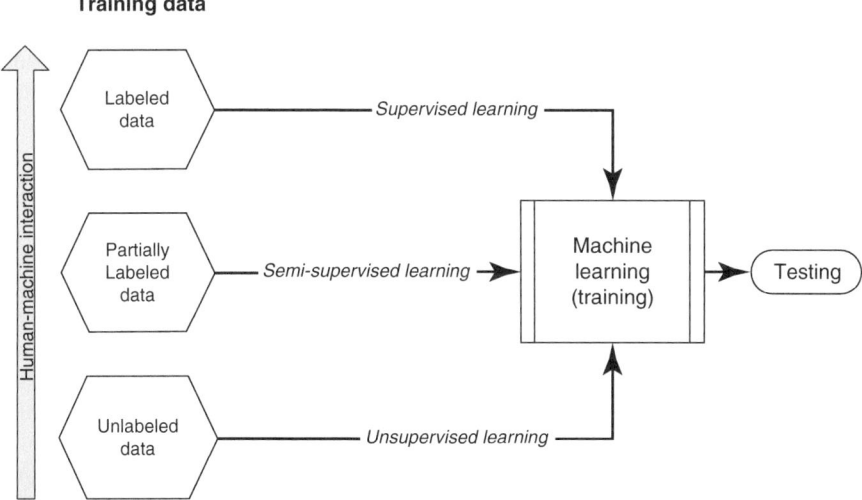

Fig. 1.1 Categories of machine learning algorithms according to training data nature

inductive bias in the training process to allow for a machine learning algorithm to generalize beyond unseen observation [28].

From a probabilistic perspective, machine learning algorithms can be divided into discriminant or generative models. A discriminant model measures the conditional probability of an output given typically deterministic inputs, such as neural networks or a support vector machine. A generative model is fully probabilistic whether it is using a graph modeling technique such as Bayesian networks, or not as in the case of naïve Bayes.

Another interesting class of machine learning algorithms that attempts to control learning by accommodating a feedback system is reinforcement learning, in which an agent attempts to take a sequence of actions that may maximize a cumulative reward such as winning a game of checkers, for instance [29]. This kind of approach is particularly useful for online learning applications.

1.6 Application in Biomedicine

Machine learning algorithms have witnessed increased use in biomedicine, starting naturally in neuroscience and cognitive psychology through the seminal work of Donald Hebb in his 1949 book [30] developing the principles of associative or Hebbian learning as a mechanism of neuron adaptation and the work of Frank Rosenblatt developing the perceptron in 1958 as an intelligent agent [16]. More recently, machine learning algorithms have been widely applied in breast cancer detection and diagnosis [31–33]. Reviews of the application of machine learning in biomedicine and medicine can be found in [12, 13].

1.7 Application in Medical Physics and Radiation Oncology

Early applications of machine learning in radiation oncology focused on predicting normal tissue toxicity [34–36], but its application has since branched into almost every part of the field, including tumor response modeling, radiation physics quality assurance, contouring and treatment planning, image-guided radiotherapy, respiratory motion management, as seen from the examples presented in this book.

1.8 Steps to Machine Learning Heaven

For the successful application of machine learning in general and in medical physics and radiation oncology in particular, one first needs to properly characterize the nature of problem, in terms of the input data and the desired outputs. Secondly, despite the robustness of machine learning to noise, a good model cannot substitute for bad data, keeping in mind that models are primarily built on approximations, and it has been stated that "All models are wrong; some models are useful (George

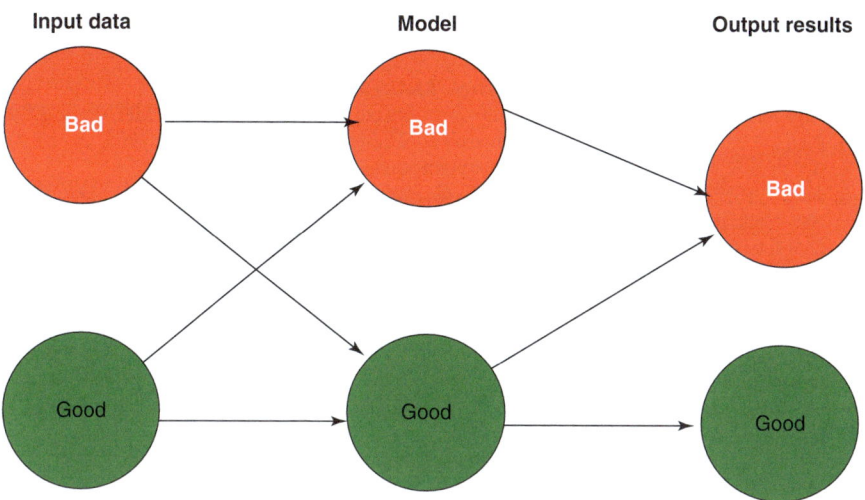

Fig.1.2 GIGO paradigm. Learners cannot be better than the data

Box)." Additionally, this has been stated as the GIGO principle, garbage in garbage out as shown in Fig. 1.2 [37].

Thirdly, the model needs to generalize beyond the observed data into unseen data, as indicated by the inductive bias mentioned earlier. To achieve this goal, the model needs to be kept as simple as possible but not simpler, a property known as parsimony, which follows from Occam's razor that "Among competing hypotheses, the hypothesis with the fewest assumptions should be selected." Analytically, the complexity of a model could be derived using different metrics such as Vapnik–Chervonenkis (VC) dimension discussed in chapter 2. for instance [25]. Finally, a major limitation in the acceptance of machine learning by the larger medical community is the "black box" stigma and the inability to provide an intuitive interpretation of the learned process that could help clinical practitioners better understand their data and trust the model predictions. This is an active and necessary area of research that requires special attention from the machine learning community working in biomedicine.

> **Conclusions**
>
> Machine learning presents computer algorithms that are able to learn from the surrounding environment to optimize the solution for the task at hand. It builds on expertise from diverse fields such as artificial intelligence, probability and statistics, computer science, information theory, and cognitive neuropsychology. Machine learning algorithms can be categorized into different classes according to the nature of the data, the learning process, and the model type. Machine learning has a long history in biomedicine, but its application in medical physics and radiation oncology is in its infancy, with high potential and promising future to improve the safety and efficacy of radiotherapy practice.

References

1. Mitchell TM. Machine learning. New York: McGraw-Hill; 1997.
2. Alpaydin E. Introduction to machine learning. 3rd ed. Cambridge, MA: The MIT Press; 2014.
3. Bishop CM. Pattern recognition and machine learning. New York: Springer; 2006.
4. Apolloni B. Machine learning and robot perception. Berlin: Springer; 2005.
5. Ao S-I, Rieger BB, Amouzegar MA. Machine learning and systems engineering. Dordrecht/New York: Springer; 2010.
6. Györfi L, Ottucsák G, Walk H. Machine learning for financial engineering. Singapore/London: World Scientific; 2012.
7. Gong Y, Xu W. Machine learning for multimedia content analysis. New York/London: Springer; 2007.
8. Yu J, Tao D. Modern machine learning techniques and their applications in cartoon animation research. 1st ed. Hoboken: Wiley; 2013.
9. Fielding A. Machine learning methods for ecological applications. Boston: Kluwer Academic Publishers; 1999.
10. Mitra S. Introduction to machine learning and bioinformatics. Boca Raton: CRC Press; 2008.
11. Yang ZR. Machine learning approaches to bioinformatics. Hackensack: World Scientific; 2010.
12. Cleophas TJ. Machine learning in medicine. New York: Springer; 2013.
13. Malley JD, Malley KG, Pajevic S. Statistical learning for biomedical data. Cambridge: Cambridge University Press; 2011.
14. Ifrah G. The universal history of computing: from the abacus to the quantum computer. New York: John Wiley; 2001.
15. Samuel AL. Some studies in machine learning using the game of checkers. IBM: J Res Dev. 1959;3:210–29.
16. Rosenblatt F. The perceptron: a probabilistic model for information storage and organization in the brain. Psychol Rev. 1958;65:386–408.
17. Minsky ML, Papert S. Perceptrons; an introduction to computational geometry. Cambridge, MA: MIT Press; 1969.
18. Werbos PJ. Beyond regression: new tools for prediction and analysis in the behavioral sciences; PhD thesis, Harvard University, 1974.
19. Quinlan JR. Induction of decision trees. Mach Learn. 1986;1:81–106.
20. Cortes C, Vapnik V. Support-vector networks. Mach Learn. 1995;20:273–97.
21. Schapire RE. A brief introduction to boosting. In: Proceedings of the 16th international joint conference on artificial intelligence, vol. 2. Stockholm: Morgan Kaufmann Publishers Inc; 1999. p. 1401–6.
22. Breiman L. Random forests. Mach Learn. 2001;45:5–32.
23. Hinton GE. Learning multiple layers of representation. Trends Cogn Sci. 2007;11:428–34.
24. Bengio Y, Courville A, Vincent P. Representation learning: a review and new perspectives. IEEE Trans Pattern Anal Mach Intell. 2013;35:1798–828.
25. Cherkassky VS, Mulier F. Learning from data: concepts, theory, and methods. 2nd ed. Hoboken: IEEE Press/Wiley-Interscience; 2007.
26. Kargupta H. Next generation of data mining. Boca Raton: CRC Press; 2009.
27. Vapnik VN. Statistical learning theory. New York: Wiley; 1998.
28. Mitchell TM. The need for biases in learning generalizations. New Brunswick: Rutgers University; 1980.
29. Sutton RS, Barto AG. Reinforcement learning: an introduction. Cambridge, MA: MIT Press; 1998.
30. Hebb DO. The organization of behavior; a neuropsychological theory. New York: Wiley; 1949.
31. El-Naqa I, Yang Y, Wernick MN, Galatsanos NP, Nishikawa RM. A support vector machine approach for detection of microcalcifications. IEEE Trans Med Imaging. 2002;21:1552–63.

32. Gurcan MN, Chan HP, Sahiner B, Hadjiiski L, Petrick N, Helvie MA. Optimal neural network architecture selection: improvement in computerized detection of microcalcifications. Acad Radiol. 2002;9:420–9.
33. El-Naqa I, Yang Y, Galatsanos NP, Nishikawa RM, Wernick MN. A similarity learning approach to content-based image retrieval: application to digital mammography. IEEE Trans Med Imaging. 2004;23:1233–44.
34. Gulliford SL, Webb S, Rowbottom CG, Corne DW, Dearnaley DP. Use of artificial neural networks to predict biological outcomes for patients receiving radical radiotherapy of the prostate. Radiother Oncol. 2004;71:3–12.
35. Munley MT, Lo JY, Sibley GS, Bentel GC, Anscher MS, Marks LB. A neural network to predict symptomatic lung injury. Phys Med Biol. 1999;44:2241–9.
36. Su M, Miften M, Whiddon C, Sun X, Light K, Marks L. An artificial neural network for predicting the incidence of radiation pneumonitis. Med Phys. 2005;32:318–25.
37. Tweedie R, Mengersen K, Eccleston J. Garbage in, garbage out: can statisticians quantify the effects of poor data? Chance. 1994;7:20–7.

Computational Learning Theory

Issam El Naqa

Abstract

The conditions for learnability of a task by a computer algorithm provide guidance for understanding their performance and guidance for selecting the appropriate learning algorithm for a particular task. In this chapter, we present the two main theoretical frameworks—probably approximately correct (PAC) and Vapnik–Chervonenkis (VC) dimension—which allow us to answer questions such as which learning process we should select, what is the learning capacity of the algorithm selected, and under which conditions is successful learning possible or impossible. Practical methods for selecting proper model complexity are presented using techniques based on information theory and statistical resampling.

2.1 Introduction

In many computational learning problems, we are given a relatively small number of observed data samples from the general population and asked to understand the functional dependencies and make decisions or perform tasks based on the data accordingly. In standard statistics introduced by Ronald Fisher in the 1920–1930s in his classical textbooks [1, 2], learning dependencies are based on the concepts of sufficiency and ancillary statistics, which requires representing dependencies by a finite set of parameters and then estimating these using maximum likelihood or Bayesian techniques. However, a paradigm shift in learning theory was introduced in the 1960s by Vladimir Vapnik and colleagues in which the parameter estimation restrictions imposed by Fisher's paradigm are replaced by knowledge of some

---

I. El Naqa
Department of Oncology, McGill University, Montreal, QC, Canada

Department of Radiation Oncology, University of Michigan, Ann Arbor, USA
e-mail: issam.elnaqa@mcgill.ca; ielnaqa@med.umich.edu

general properties of the set of functions to which the unknown dependencies belong. The determination of the general conditions for estimating the unknown data dependency, description of the inductive learning of relationships, and the development of algorithms to implement these principles are the subjects of the modern computational learning theory [3].

In this framework of learning theory, the focus is on small sample size statistics, in which a machine-learning algorithm is trained on a subset of the data (training data) that is used to identify the learning function to achieve the desired response of the task at hand and is built with the goal of predicting response to unseen data (out-of-sample or testing data). This is a challenging task that poses several questions regarding which learning process we should select, what is the learning capacity of the algorithm selected, what are the expected errors or their bounds, under what conditions is successful learning possible and impossible, and under what conditions is a particular learning algorithm assured of learning successfully [3, 4]. In this chapter, we will start by highlighting the differences between statistical analysis and statistical modeling. We will present the theoretical background for computational learning. Specifically, two specific frameworks for analyzing learning algorithms, namely, the probably approximately correct (PAC) and Vapnik–Chervonenkis (VC) theory, will be discussed. Finally, practical methods for estimating learning generalization ability and model complexity will be presented.

2.2 Computational Modeling Versus Statistics

There is a common mix-up between statistical analysis and computational modeling of data. The objective of statistical analysis is to use statistics to describe data and make inferences on the population for *hypothesis testing* purposes; for instance, variable x is significant while variable y is not in explaining the observed clinical endpoint of interest. In the case of computational modeling, the objective is to provide an adequate description of data dependencies and summarize its features for *hypothesis generation* as summarized in Fig. 2.1 [5].

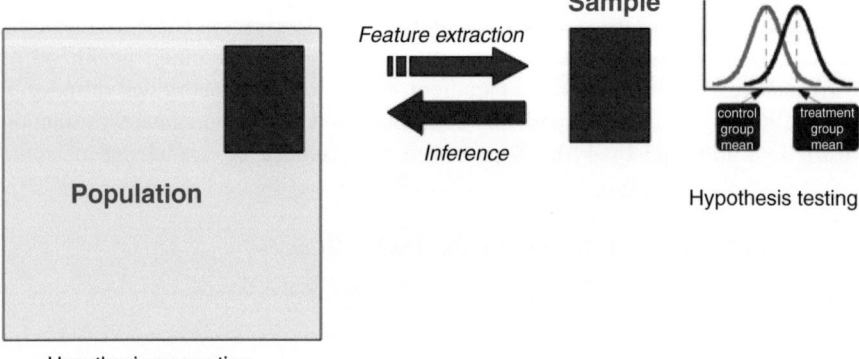

Fig. 2.1 Computational modeling vs. statistical analysis (Adapted from [5])

Machine learning is a branch of computational modeling that inherited many of its properties and utilizes statistical modeling techniques as part of its arsenal. For instance, machine-learning models of quality assurance (QA) in radiotherapy can capture many salient features in the data that may impact quality of delivered treatment and their possible interdependencies, which could be further tested for varying hypotheses for their severity and possible action levels to mitigate their effect. However, development of computational modeling techniques could be achieved using both deterministic and statistical methodologies.

2.3 Learning Capacity

Learning capacity or "learnability" defines the ability of a machine-learning algorithm to learn the task at hand in terms of model complexity and the number of training samples required to optimize a performance criteria. Using formal statistical learning taxonomy [6], assuming a training set Ξ of n-dimensional vectors, $x_i^n, i = 1:m$, each labeled (by 1 or 0) according to a target function, f, which is unknown to the learner and called the target concept and is denoted by c, which belongs to the set of functions, C, the space of target functions as illustrated in Fig. 2.2. The probability of any given vector X being presented in Ξ is $P(X)$. The goal of the training is to guess a function, $h(X)$ based on the labeled samples in Ξ, called the hypothesis. We assume that the target function is an element of a set, H, the space of hypotheses. For instance, in our QA example, if we are interested in developing a treatment plan quality metric, we would have a list of input features X (e.g., energies, beam arrangements, monitor units, etc.) that is governed in our pool of treatment plans with a certain joint probability density function P. Based on clinical experience, a set of these plans are considered to be good while others are bad, which would constitute the target concept (c) of interest with an unknown functional form f that we aim to estimate. During the training process, we attempt to identify a hypothesis function $h(X)$ that would approximate the mapping to c using varying possible machine-learning algorithms, and the higher the overlap between our hypothesized mapping function and the target quality

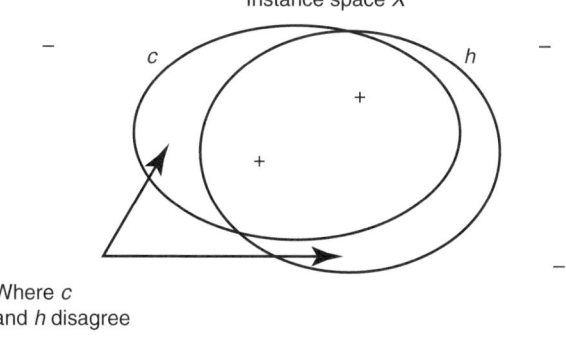

Fig. 2.2 Illustration of learning concepts (From Nilsson and Nilsson [6])

metric concept, the more successful the learning process is as indicated in the Venn diagram of Fig. 2.2.

There are two main theories that attempt to characterize the learnability of machine-learning algorithms: the PAC and the VC theories as discussed below.

2.4 PAC Learning

One method to characterize the learnability of a machine-learning algorithm is by number of training examples needed to learn a hypothesis $h(X)$ as mentioned earlier. This could be measured by the probability of learning a hypothesis that is approximately correct (PAC). Formally, this could be defined as follows. Consider the concept class C defined over a set of instances X of length m and a learner L using hypothesis space H. C is PAC learnable by L using H.

If for all c ∈ C, distributions D over X, ε such that $0 < \varepsilon < 1/2$ and δ such that $0 < \delta < 1/2$, there is a learner L with probability at least $(1-\delta)$ that will output a hypothesis h ∈ H such that error $D(h) \leq \varepsilon$, in time that is polynomial in $1/\varepsilon$, $1/\delta$, n, and size (c) [4]. For a finite hypothesis space H, the number of training examples (m) required to reduce the probability of error below a desired level δ is given by assuming a zero training error:

$$m \geq \frac{1}{\varepsilon}\left(\ln|H| + \ln(1/\delta)\right) \qquad (2.1)$$

This estimated number of training examples is sufficient to ensure that any consistent hypothesis will be probably (with probability $(1-\delta)$) approximately (within error ε) correct. In the case the training error is not necessarily zero, the number of required training examples becomes:

$$m \geq \frac{1}{2\varepsilon^2}\left(\ln|H| + \ln(1/\delta)\right) \qquad (2.2)$$

It is recognized that such estimate could be in practice an overestimate [4]. Another problem in PAC is that it includes the size of the hypothesis space H, which in practice could be infinite.

2.5 VC Dimension

An alternative approach to measure learnability that overcomes the limitations of PAC is to use Vapnik–Chervonenkis (VC) dimension. The VC dimension measures the complexity of the hypothesis space H, not by the number of distinct hypotheses H as in PAC but rather by the number of distinct instances from X that can be completely discriminated using H. VC(H), of hypothesis space H defined over instance space X, is the size of the largest finite subset of X shattered by H. If arbitrarily large finite sets of X can be shattered by H, then $VC(H) = \infty$. This is noted that for any finite H, $VC(H) \leq \log 2|H|$. To see this, suppose that $VC(H) = d$. Then, H will require $2d$ distinct hypotheses to shatter d instances. Hence, $2^d \leq \log 2|H|$, and $d = \leq \log 2|H|$. This is illustrated in Fig. 2.2.

2.6 Model Complexity Analysis in Practice

Any multivariate analysis often involves a large number of variables or features in the data samples [7]. The complexity of a learning model increases with the number of input features (i.e., the dimensionality of the input feature vector); therefore, it is desirable to focus on the most important features that characterize the observations. These are usually unknown. Therefore, practical dimensionality reduction or subset selection aims to find the "significant" set of features. Finding the best subset of features is definitely challenging, especially in the case of nonlinear models. The objective is to reduce the model complexity, decrease the computational burden, and improve the generalizability on unseen data as explained earlier. A straightforward approach is to make an educated guess based on experience and domain knowledge and then apply feature transformation (e.g., principal component analysis (PCA)) [8–10] or sensitivity analysis by using organized searches such as sequential forward selection or sequential backward selection or combination of both [9]. A recursive elimination technique that is based on machine learning has been also suggested [11]. In this technique, the data set is initialized to contain the whole set, train the predictor (e.g., SVM classifier) on the data, rank the features according to a certain criteria (e.g., $\|W\|$), and keep iterating by eliminating the lowest ranked one. It should be noted that the specific definition of model order changes depending on the functional form. It could be identified by the number of parameters in logistic regression, or by the number of neurons and layers in the case of neural networks (cf. Fig. 2.3), etc. However, in any of these forms, the model order creates a balance between complexity (increased model order) and the model ability to generalize to unseen data. Finding this balance is referred to in statistical learning theory as the bias–variance dilemma (see Fig. 2.4), in which an oversimple model is expected to underfit the data (large bias and small variance), whereas a too complex model is expected to overfit data (small bias and large variance) [12]. Hence, the objective is to achieve an optimally parsimonious model, i.e., a model with the correct degree of complexity to fit the data and also a maximum ability to generalize to new, unseen, data sets, in other words to derive its VC dimension from the data itself. Practical approaches utilize information theoretic methods or statistical resampling as discussed below.

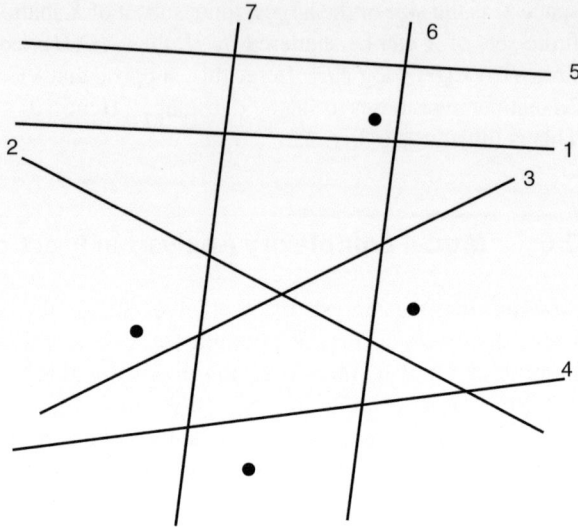

Fig. 2.3 An example of 14 dichotomies shattering 4 points in 2D (From Nilsson and Nilsson [6])

14 dichotomies of 4 points in 2 dimensions

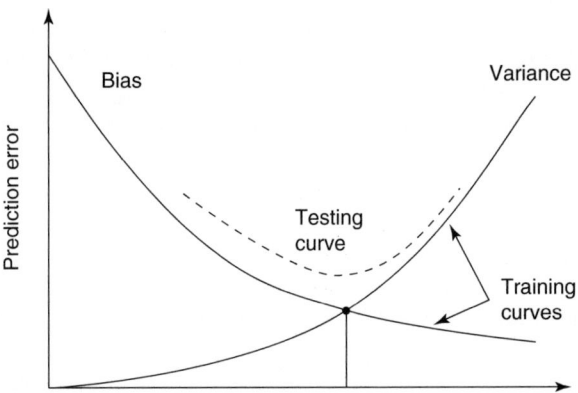

Fig. 2.4 This figure illustrates a common trade-off in model predictive power between prediction bias (average error) and prediction variance (square error). As model complexity increases, the average prediction error (bias) tends to decrease while the average square error tends to decrease. The point of optimal complexity tends to be near the point when average and square errors are of similar magnitude (Reproduced with permission from Deasy and El Naqa [13])

2.6.1 Model Order Based on Information Theory

Information theory provides intuitive measures of model order optimality; among the most commonly used are Akaike information criteria (AIC) and the Bayesian information criteria (BIC) [14]. AIC is an estimate of predictive power of a model, which includes both the maximum likelihood principle and a model complexity term that penalizes models with an increasing number of parameters (to avoid overfitting the data). BIC is derived from Bayesian theory, which results in a penalty term that increases linearly with the number of parameters.

2.6.2 Model Order Based on Resampling Methods

Resampling techniques are used for model selection and performance comparison purposes to provide statistically sound results when the available data set is limited (which is almost always the case in radiotherapy). We use two types of fit-then-validate methods: cross-validation methods and bootstrap resampling techniques. *Cross-validation* [9] uses some of the data to train the model and some of the data to test the model validity. The type we most often use is the "leave-one-out" cross-validation (LOO-CV) procedure (also known as the "jackknife"). In each LOO-CV iteration, all the data are used for training/fitting except for one data point left out for testing, and this is repeated so that each data point is left out exactly once. The overall success of predicting the left-out data is a quantitative estimate of model performance on new data sets. *Bootstrapping* [15] is an inherently computationally intensive procedure but generates more realistic results. Typically, a bootstrap pseudo-data set is generated by making copies of original data points and randomly selected with a probability of inclusion of 63 %. The bootstrap often works acceptably well even when data sets are small or unevenly distributed. To achieve valid results, this process must be repeated many times, typically several hundred or thousand times. Examples of applying these methods to outcomes modeling in radiotherapy could be found in our previous work [16] and are discussed in details in [13].

> **Conclusions**
>
> In this chapter, we discussed some of the guiding principles of computational learning. Within the probably approximately correct (PAC) framework, we identify classes of hypotheses that can and cannot be learned from a polynomial number of training examples and we define a natural measure of complexity for hypothesis spaces that allows bounding the number of training examples required for inductive learning. Within the mistake-bound framework, we examine the number of training errors that will be made by a learner before it determines the correct hypothesis [4]. The VC dimension offers an alternative approach for measuring learnability by estimating the number of instances necessarily to discriminate among hypotheses. Beside the theoretical approaches, we also presented practical methods based on information theory and statistical resampling for estimating model complexity. Resampling techniques such as cross-validation and bootstrapping arm among the most used methods in the literature and will be further discussed in the context of performance evaluation in chapter 4.

References

1. Fisher RA. Statistical methods for research workers. Edinburgh: Oliver and Boyd; 1925.
2. Fisher RA. The design of experiments. Edinburgh: Oliver and Boyde; 1935.
3. Vapnik VN. The nature of statistical learning theory. New York: Springer; 1995.
4. Mitchell TM. Machine learning. New York: McGraw-Hill; 1997.
5. Berry MJA, Linoff G. Data mining techniques: for marketing, sales, and customer relationship management. 2nd ed. Indianapolis: Wiley Pub; 2004.
6. Nilsson NJ, Nilsson NJ. The mathematical foundations of learning machines. San Mateo: Morgan Kaufmann; 1990.
7. Guyon I, Elissee A. An introduction to variable and feature selection. J Mach Learn Res. 2003;3:1157–82.
8. Dawson LA, Biersack M, Lockwood G, Eisbruch A, Lawrence TS, Ten Haken RK. Use of principal component analysis to evaluate the partial organ tolerance of normal tissues to radiation. Int J Radiat Oncol Biol Phys. 2005;62:829–37.
9. Kennedy R, Lee Y, Van Roy B, Reed CD, Lippman RP. Solving data mining problems through pattern recognition. Upper Saddle River, NJ: Prentice Hall; 1998.
10. Härdle W, Simar L. Applied multivariate statistical analysis. Berlin/New York: Springer; 2003.
11. Guyon I, Weston J, Barnhill S, Vapnik V. Gene selection for cancer classification using support vector machines. Mach Learn. 2002;46:389–422.
12. Hastie T, Tibshirani R, Friedman JH. The elements of statistical learning: data mining, inference, and prediction: with 200 full-color illustrations. New York: Springer; 2001.
13. Deasy JO, El Naqa I. Image-based modeling of normal tissue complication probability for radiation therapy. Cancer Treat Res. 2008;139:215–56.
14. Burnham KP, Anderson DR. Model selection and multimodal inference: a practical information-theoretic approach. 2nd ed. New York: Springer; 2002.
15. Efron B, Tibshirani R. An introduction to the bootstrap. New York: Chapman & Hall; 1993.
16. El Naqa I, Bradley JD, Lindsay PE, Blanco AI, Vicic M, Hope AJ, et al. Multi-variable modeling of radiotherapy outcomes including dose-volume and clinical factors. Int J Radiat Oncol Biol Phys. 2006;64:1275–86.

Machine Learning Methodology

3

Sangkyu Lee and Issam El Naqa

Abstract
There is a variety of patterns we desire to learn from radiation oncologic data. The previous chapters described how these various learning objectives can commonly be formulated in theoretical nomenclatures. This chapter introduces different machine learning algorithms that could cater to readers' specific learning goals. We intend to provide conceptual outlines of some of the widely used algorithms with minimal mathematical conundrum and examples drawn from the radiotherapy literature. In this chapter we classify the algorithms into three types, based on the availability of information: unsupervised, supervised, and reinforcement learning. The methods illustrated in this chapter include principal component analysis and clustering (unsupervised), logistic regression, neural network, support vector machine, decision tree, Bayesian networks, and naive Bayes (supervised) in addition to reinforcement learning.

3.1 Introduction

Learning is defined in this context as estimating dependencies from data. There are two common types of learning: supervised and unsupervised. Supervised learning is used to estimate an unknown (input, output) mapping from known (input, output)

S. Lee (✉)
Medical Physics Unit, Montreal General Hospital, McGill University,
L5-211, 1650 Cedar Ave, Montreal, QC H3G 1A4, Canada
e-mail: sangkyu.lee@mail.mcgill.ca

I. El Naqa
Department of Oncology, McGill University, Montreal, QC, Canada

Department of Radiation Oncology, University of Michigan, Ann Arbor, USA
e-mail: issam.elnaqa@mcgill.ca; ielnaqa@med.umich.edu

samples (e.g., classification and regression). In unsupervised learning, only input samples are given to the learning system (e.g., clustering and estimation of probability density function) [19]. In this study, we focus mainly on supervised learning where a teacher provides the output samples.

3.2 Unsupervised Learning

3.2.1 Linear Principal Component Analysis

Suppose we have treatment data for a group of patients, some of whom developed a late complication and others who did not. The data might include patient age and weight, diagnostic factors such as Gleason score, dose delivered to the PTV, dose to one or more critical structures, etc. We would like to know if there are patterns in these data that can predict for the complication. The first step is to reduce the data set to its most informative elements. Often there will be more than one datum that measures more or less the same thing. We would like to reduce the data vector to a smaller dimension containing only components that are clearly distinctive (i.e., uncorrelated with one another). To do this, we arrange the patient data in a matrix X so that each row and column represent one patient and a variable, respectively. As a pre-processing step, each column in the matrix X is normalized to zero mean and unity variance (z-score). Principal component analysis (PCA) is then applied to the normalized X to identify a set of principal components (PCs) which are given by:

$$\text{PC} = U^\text{T}\mathbf{X} = \Sigma V^\text{T} \tag{3.1}$$

where $U\Sigma V^\text{T}$ is the singular value decomposition of X. This is equivalent to transformation into a new coordinate system such that the greatest variance by any projection of the data would lie on the first coordinate (first PC), the second greatest variance on the second coordinate (second PC), and so on. For visualization purposes with the PCA, the heterogeneous variables are typically normalized using z-scoring (zero mean and unity variance). The term variance explained, used in PCA plots (Fig. 3.1), refers to the variance of the data model about the mean prognostic input factor values. The data model is formed as a linear combination of its principal components. Thus, if the PC representation of the data explains the spread (variance) of the data about the full data mean, it would be expected that this PC representation capture enough information for modeling. Moreover, PCA analysis can provide an indication about class separability; however, it should be cautioned that PCA is an indicator and is not necessarily optimized for this purpose as supervised linear discriminant analysis, for instance [16].

Figure 3.1 shows three examples of PCA applied to patient data for three different prognostic challenges: prediction of xerostomia, esophagitis, and pneumonitis. The main purpose of PCA in this case is to visualize a degree of separation between patients with and without complications. For the case of xerostomia, PCA revealed several significant principal modes in the prognostic data, the first two of which accounted for only 60 % of the total varianc among the data components. However, the first two principal components already show a fairly clear distinction between

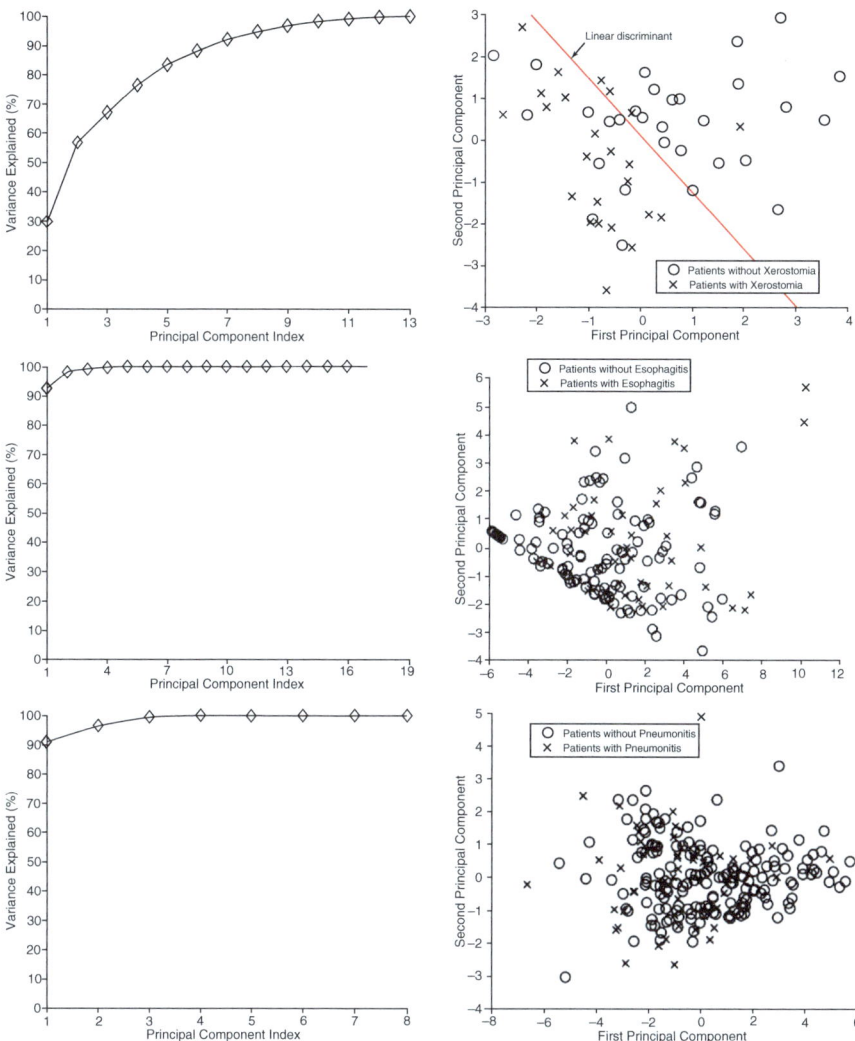

Fig. 3.1 Projection of prognostic factors for xerostomia (*top*), esophagitis (*middle*), and radiation pneumonitis (*bottom*) into a two-dimensional space consisting of the first and second principal components (*the right column*). The *left column* shows variation explanation versus principle component index. Linear separation in the xerostomia dataset is well demonstrated but not as much for the pneumonitis case (as seen from a wide class overlap) (Reproduced from El Naqa et al. [16])

the cases with and without xerostomia, meaning that the rest of the variance may not be relevant to the complication. In contrast, PCA reveals only one strong principal mode for esophagitis and pneumonitis, i.e., the original data components are so highly correlated that PCA reduces them to a single principal component. The projected data do not demonstrate clear separation among cases, which calls for a nonlinear modeling approach such as kernel methods (see Sect. 3.3.4).

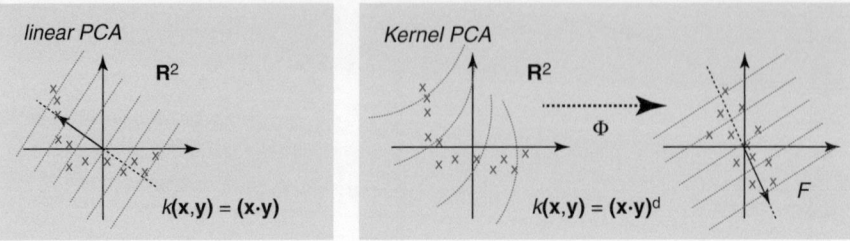

Fig. 3.2 A cartoon describing the utility of kernel PCA in linearizing a nonlinear pattern by feature transformation Φ via a polynomial kernel. The *dotted lines* are contour lines of the same value of projection to the first principal component (Reproduced from Scholkopf et al. [42])

3.2.2 Kernel Principal Component Analysis

Kernel PCA is a nonlinear form of the principal component analysis by use of a kernel technique (see the upcoming section on support vector machine). It is useful for detecting nonlinear behaviors in data that cannot be represented in terms of linear combination of the existing variables. The kernel trick effectively transforms an input space into a higher-dimensional feature space in which nonlinear patterns can be discoverable in a linear fashion (Fig. 3.2). However, the input space transformation does not need to be defined explicitly, as the PCA only requires the knowledge of a covariance matrix in the transformed space. The (i,j)-th component of a covariance matrix for the data $x_1, x_2, ..., x_n$ can be computed directly from the kernel function $k(\cdot, \cdot)$:

$$K_{ij} = \langle \Phi(x_i), \Phi(x_j) \rangle = k(x_i, x_j) \tag{3.2}$$

where $\Phi(x)$ denotes the input space transformation. K is then diagonalized to extract a set of principal components and corresponding eigenvalues. Kernel PCA becomes more computationally expensive than linear PCA when the number of samples exceeds input dimension. Nevertheless, when applied to problems containing nonlinear patterns (e.g., handwriting), a nonlinear PCA could be more suitable than the linear one for reducing data dimension prior to a classification task [42].

3.2.3 Clustering

Cluster analysis refers to detection of collective patterns in data based on similarity criteria. It can be performed either in a supervised or unsupervised fashion. Grouping data points into clusters is useful in several ways. First, it can provide intuitive and succinct representation of the nature of data prior to major investigation. Secondly, clustering can be applied to compressing complex data distribution into a group of vectors corresponding to cluster centroids (vector quantization).

The *K*-means clustering is one of the most popular clustering methods. It begins with randomized partitions with the given number (*K*) of clusters. The partitions are then iteratively refined by the following steps: (1) an assignment step (reassignment of the cluster membership of each data point based on a distance to cluster centroids) and (2) an update step (recalculation of cluster centroids as a geometric mean of the updated membership). The Minkowski distance between the d-dimensional vectors **a** and **b**, also known as a L_p norm, is used as a measure of proximity:

$$L_k(\mathbf{a},\mathbf{b}) = \left(\sum_{i=1}^{d} |\mathbf{a_i} - \mathbf{b_i}|^k \right)^{1/k} \tag{3.3}$$

The widely used Euclidean and Manhattan distance refer to the Minkowski distance at $p=2$ and $p=1$, respectively.

The *K*-means gained popularity thanks to its simplicity and fast convergence [25]. One of the drawbacks of this algorithm is its tendency to converge to local minima when initial partitions are not carefully chosen [25]. This can be partially overcome by introducing seeding heuristis such as the *K*++ means algorithm by Arthur and Vassilvitskii [1]. Furthermore, the original *K*-means requires the number of clusters (*K*) to be given a priori. The choice can either be made based on domain knowledge or optimized in data in a cross-validated fashion. The optimization method employs for an objective function the Bayesian information criteria (BIC) [38] or the minimum description length (MDL) [2] that penalizes the larger number of clusters.

Another clustering algorithm gaining popularity is a neural network-derived method called a self-organizing map (SOM) or a Kohonen map [30]. In a SOM, distinct patterns in input data are represented by nodes which are typically arranged in a two-dimensional hexagonal or rectangular grid for better visualization. Each node is assigned with its location in the grid and a vector of weights on input variables. The learning algorithm begins with randomizing node weights. Then, one training example is sampled from the training set and the node at the closest distance from it (Minkowski metrics can be used) is identified as a best matching unit (BMU). The weight vectors for the BMU and the nodes in its vicinity are adjusted to decrease the distance to the training example according to the following update formula:

$$\mathbf{w}_v(t+1) = \mathbf{w}_v(t) + \alpha(t)\Lambda\left(\|\mathbf{w}_{BMU}(t) - \mathbf{w}_v(t)\|\right)\left(\mathbf{x}_i - \mathbf{w}_v(t)\right) \tag{3.4}$$

where $\mathbf{w}_v(t)$ is a weight vector for a node *v* at iteration *t* and **x***i* is the *i*-th input sample. The magnitude of the update is determined by the factors that depend on the distance from the BMU $|\mathbf{w}_{BMU}(t) - \mathbf{w}_v(t)|$ and the number of iteration (*t*). A window function (Λ) is the highest when $v = BMU$ and tapers off to zero as a node goes farther away from the BMU. It ensures the nodes will be topologically ordered (neighboring nodes have similar weight patterns). The learning rate, $\alpha(t)$, typically decreases with iterations to ensure convergence. After the learning is repeated through all the training samples, the nodes tend to clump toward the weights that

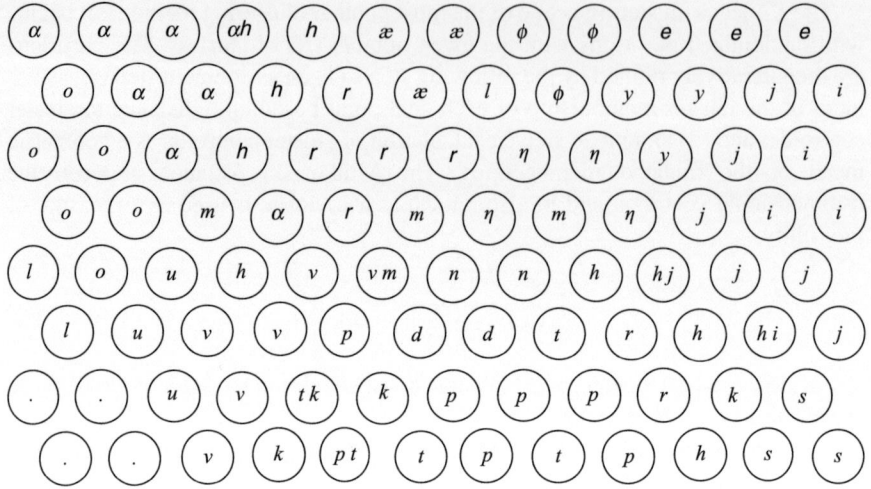

Fig. 3.3 Self-organizing map learned from natural Finnish speech analysis by Kohonen [30]. Each node represents one acoustic unit of speech called a phoneme

appears in input patterns frequently (topological ordering). A SOM has been shown useful in some areas such as speech recognition, linguistics, and robot control (Fig. 3.3).

Many challenges in bioinformatics are framed as a clustering problem, such as identifying a group of genes showing similar patterns of expression under certain conditions or diseases. A work by Svensson et al. [48] is a good example from radiotherapy toxicity modeling. They grouped 1,182 candidate late toxicity marker genes into two groups using their expression patterns in lymphocytes after radiation, although the grouping did not correlate with toxicity status. In contrast, a SOM of radiation pneumonitis risk factors built by Chen et al. [8] showed that grouping patterns among the factors can be exploited for predicting the toxicity with decent accuracy (AUC=0.73).

3.3 Supervised Learning

3.3.1 Logistic Regression

In radiation outcomes modeling, the response will usually follow an S-shaped curve. This suggests that models with sigmoidal shape are more appropriate to use [3–5, 21, 23, 24, 34, 49]. A commonly used sigmoidal form is the logistic model, which also has nice numerical stability properties. The logistic model is given by [22, 51]:

$$f(\mathbf{x}_j) = \frac{e^{g(x_j)}}{1+e^{g(x_j)}}, \quad i=1,2,\ldots,n \qquad (3.5)$$

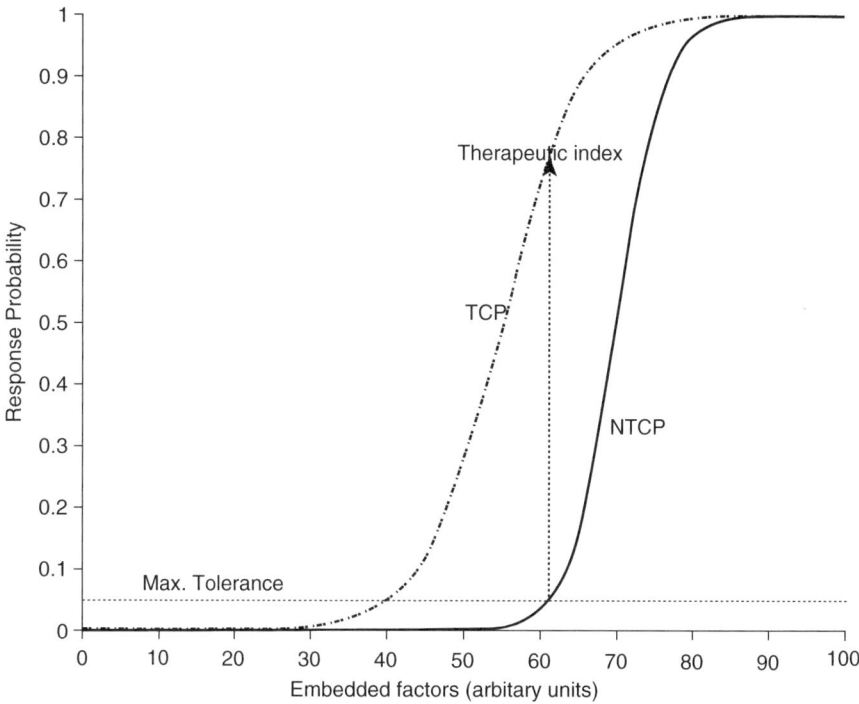

Fig. 3.4 Sigmoidally shaped response curves (for tumor control probability of normal tissue complication probability) are constructed as a function of a linear weighting of various factors, for a given dose distribution, which may include multiple dose-volume metrics as well as clinical factors. The units of the x-axis may be thought of as equivalent dose units (Reproduced from El Naqa et al. [14])

where n is the number of cases (patients) and $\mathbf{x}$ is a vector of the input variable values used to predict $f(\mathbf{x}_i)$ for the outcome y_i of the i_{th} patient. The $f(\cdot)$ is referred to as the logic transformation. The "x-axis" summation $g(\mathbf{x}_i)$ is given by:

$$g(\mathbf{x}_i) = \beta_0 + \sum_{j=1}^{S} \beta_j x_{ij}, \quad i = 1,...,n, \quad j = 1,...,s \quad (3.6)$$

where s is the number of model variables and the β's are the set of model coefficients that are determined by maximizing the probability that the data gave rise to the observations (i.e., the likelihood function). Many commercially available software packages, such as SAS, SPSS, and Stata, provide estimates of the logistic regression model coefficients and their statistical significance. The results of this type of approach are not expressed in closed form as above, but instead, the model parameters are chosen in a stepwise fashion to define the abscissa of a regression model as shown in Fig. 3.4 However, it is the analyst's responsibility to test for interaction effects on the estimated response, which can potentially be corrected by adding cross terms to Eq. 3.6. However, this transformation suffers from limited

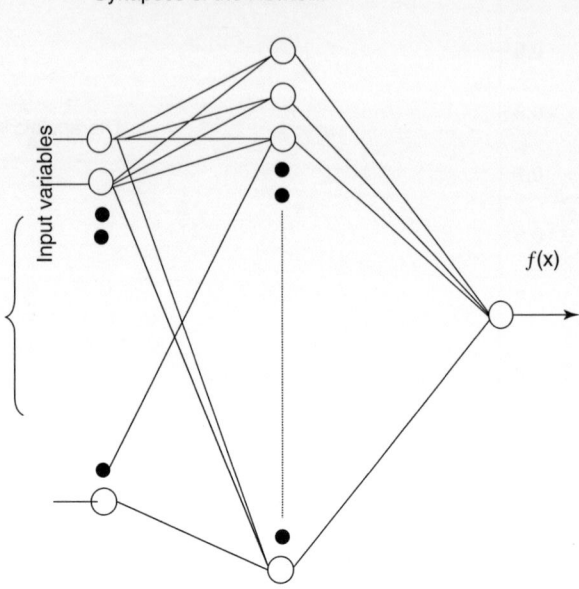

Fig. 3.5 Neural network architecture consisting of an input layer, middle (hidden) layer(s), and an output layer. The synapses of the network consist of neurons that fire depending on their chosen activation functions

learning capacity. In such a model, it is the user's responsibility to determine whether interaction terms or higher order terms should be added. A solution to ameliorate this problem is offered by applying artificial intelligence methods.

3.3.2 Feed-Forward Neural Networks (FFNN)

Neural networks are described as adaptive massively parallel-distributed computational models that consist of many nonlinear elements arranged in patterns similar to a simplistic biological neuron network. A typical neural network architecture is shown in Fig. 3.5.

Neural networks have been applied successfully to model many different types of complicated nonlinear processes, including many pattern recognition problems [41]. A three-layer FFNN network would have the following model for the approximated functional:

$$f(\mathbf{x}) = \mathbf{y}^T \mathbf{w}^{(2)} + b^{(2)} \qquad (3.7)$$

where **v** is a vector, the elements of which are the output of the hidden neurons, i.e.,

$$v = s\left(\mathbf{x}^T \mathbf{w}_i^{(1)} + b^{(1)}\right) \qquad (3.8)$$

where **x** is the input vector and $\mathbf{w}^{(j)}$ and $\mathbf{b}^{(j)}$ are the interconnect weight vector and the bias of layer j, respectively, $j = 1,2$. In the FFNN, the activation function $s(\cdot)$ is usually a sigmoid, but radial basis functions were also used [46]. The FFNN could be trained in two ways: batch mode or sequential mode. In the batch mode, all the training examples are used at once; in sequential mode, the training examples are presented on a pattern basis, in the order that is randomized from one epoch (cycle) to another. The number of neurons is a user-defined parameter that determines the complexity of the network; the larger the number of neurons, the more complex the network would be. The number is determined during the training phase.

3.3.3 General Regression Neural Networks (GRNN)

The GRNN [44] is a probabilistic regression model based on neural network architecture. It is characterized as non-parametric, which means that it does not require any pre-determined functional form (e.g., polynomials). Instead, it estimates the joint density of input variables **x** and a target y from training data. The regression output using the GRNN is obtained by taking the expectation value of y for a given observation **X** and the joint density $g(\mathbf{x}, y)$:

$$\hat{y}(\mathbf{X}) = E(y \mid \mathbf{X}) = \frac{\int_{-\infty}^{\infty} y g(\mathbf{X}, y) dy}{\int_{-\infty}^{\infty} g(\mathbf{X}, y) dy} \quad (3.9)$$

The joint density $g(\mathbf{x}, y)$ is estimated from training examples $\mathbf{X}_i$ and y_i via the Parzen estimator where the density is regarded as the superposition of Gaussian kernels centered at the observation points with a spread σ. The resulting form of the regression function is:

$$\hat{y}(\mathbf{X}) = \frac{\sum_{i=1}^{n} y_i \exp\left(\frac{-D_i^2}{2\sigma^2}\right)}{\sum_{i=1}^{n} \exp\left(\frac{-D_i^2}{2\sigma^2}\right)} \quad (3.10)$$

where $D_i^2 = (\mathbf{X} - \mathbf{X}_i)^T (\mathbf{X} - \mathbf{X}_i)$, denoting the Euclidean distance between the testing data **X** and the i-th training data **Xi**.

The GRNN is fairly simple to train, with only the Gaussian width σ to be tuned. Thus, implementation of the GRNN does not require an optimization solver to obtain the weights, as in the case of FFNN. However, the output is obtained as a weighted sum of all the training samples, which could make it less efficient during running time. This could be improved by performing cluster analysis on training data (see Sect. 3.2.3) to compress it into a few cluster centers so that the metric D_i can be computed only between those center points and a testing example. The computational speed can also benefit from parallelized neural network implementation since each summation can be performed independently using synapses and an exponential activation function. In our previous work, we demonstrated that GRNN can outperform traditional FFNN in radiotherapy outcomes prediction [15].

3.3.4 Kernel-Based Methods

Kernel-based methods and its most prominent member, support vector machines (SVMs), are universal constructive learning procedures based on the statistical learning theory [50]. For discrimination between patients who are at low risk versus patients who are at high risk of radiation therapy, the main idea of the kernel-based technique would be to separate these two classes with hyper-planes that maximizes the margin between them in the nonlinear feature space defined by implicit kernel mapping. The optimization problem is formulated as minimizing the following cost function:

$$L(\mathbf{x}, \xi) = \frac{1}{2}\mathbf{w}^T\mathbf{w} + C\sum_{i=1}^{n}\xi_i \quad (3.11)$$

subject to the constraints:

$$y_i\left(\mathbf{w}^T\Phi(\mathbf{x}_i) + b\right) \geq 1 - \zeta_i$$

$$3\zeta_i \geq 0 \text{ for all } i$$

where $\mathbf{w}$ is a weighting vector and $\Phi(\cdot)$ is a nonlinear mapping function. The ζ_i represents the tolerance error allowed for each sample being on the wrong side of the margin. Note that minimization of the first term in Eq. 3.11 increases the separation (improves generalizability) between the two classes, whereas minimization of the second term (penalty term) improves fitting accuracy. The trade-off between complexity and fitting error is controlled by the regularization parameter C. However, such nonlinear formulation would suffer from the curse of dimensionality (i.e., the dimension of the problem becomes too large to solve) [19, 20]. However, computational efficiency is achieved from solving the dual optimization problem instead of the equation which is convex with a complexity that is dependent only on the number of samples [50]. The prediction function in this case is characterized only by a subset of the training data known as support vectors s_i:

$$C^+\sum_{i \in Z^+}\xi_i + C^-\sum_{i \in Z^-}\xi_i \quad (3.12)$$

where n_s is the number of support vectors, α s are the dual coefficients determined by quadratic programming, and $K(\cdot,\cdot)$ is the kernel function as discussed next. Typically, used nonlinear kernels include:

$$\text{Polynomials}: K(\mathbf{x}, \mathbf{x}') = \left(\mathbf{x}^T\mathbf{x}' + c\right)$$

$$\text{Radial basis function (RBF)}: K(\mathbf{x}, \mathbf{x}') = exp\left(\frac{\|\mathbf{x} - \mathbf{x}'\|}{2\sigma^2}\right)$$

where c is a constant, q is the order of the polynomial, and σ is the width of the radial basis functions. The kernel-based approach is very flexible, which allows for constructing a neural network by using combination of sigmoidal kernels or chooses

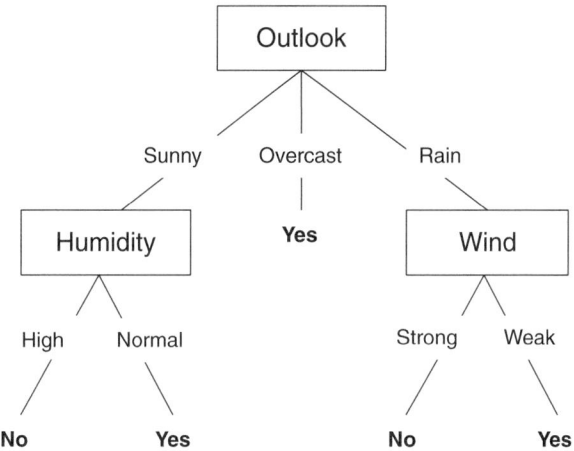

Fig. 3.6 An example decision tree that classifies whether to go play tennis or not (written in *bold*) based on three attributes (outlook, humidity, wind) shown in *box nodes*. Values of the three attributes are written on the corresponding branches (Reproduced from Mitchell [35])

a logistic regression equivalent kernel by replacing the hinge loss with a binomial deviance [19].

SVM has been widely used for many radiotherapy outcome prediction cases where complex relationships between risk factors are expected. Examples include lung cancer prognosis [12, 26, 29], radiation pneumonitis [7, 45], and GI/genitourinary toxicity [37].

3.3.5 Decision Tree

A decision tree is suitable for generating the hypotheses that consist of multiple Boolean conditions on attributes (disjunctive hypotheses). Although it can also perform regression, we will limit the discussion to its application to classification. A decision tree divides an input space into several disjoint subregions. A testing instance falls into one of the subregions after successive tests on its attribute values. Then, the instance is given for its classification result the value that is assigned to the subregion. The tests are organized in the order specified by a tree structure (Fig. 3.6). A tree consists of nodes, branches, and leaves, each representing the following:

- Node: the attribute to be tested
- Branch: the outcome of the test, for example, is the body temperature of a patient higher than 37° (continuous attribute) or is the patient taking aspirin (categorical attribute)?
- Leaf node: the node located at the terminus of a tree representing a subset of data and a class label assigned to the subset

The tree and its parameters are learned from training data in a supervised fashion. The learning process can be thought of as dividing training instances into subgroups (corresponding to nodes) in a way that class labels in the subgroups are made as homogeneous as possible. The major questions in decision tree learning are (1) in which order the attributes be tested, (2) what level of purity the partition class labels is desired as a result of a single test, and (3) how many nodes are needed.

The ID3 (iterative dichotomizer 3) algorithm is a primitive form of the decision tree learning algorithm that aims to arrive at an optimal decision tree via a greedy search [39]. The ID3 algorithm is initiated by identifying the first attribute (root node) to create the first set of partitions, and the tree is further branched by applying the same procedure to the resulting subsets and the remaining attributes. At each round of partitioning, the attribute to split is chosen based on how well it can predict a target class by itself. In the context of decision tree learning, the predictive value of an attribute A with respect to a class C is measured by its information gain, which is defined as:

$$\text{gain}(A) = H(A) - H(A|C) \tag{3.13}$$

where H, entropy, is a measure of information conveyed by a probability distribution. For a variable A with the distribution of c discrete states and corresponding probabilities $p_1, p_2, ..., p_c$, the entropy is:

$$H(A) = \sum_{i=1}^{c} -p_i \log_2 p_i. \tag{3.14}$$

In the case of continuous attributes, a threshold (A_{th}) is set to split the data into two subsets with proportions p_1 and p_2 where $p_1 = p(A < A_{th})$ and $p_2 = p(A > A_{th})$. The value of A_{th} is chosen so that the resulting information gain is the largest.

A branch of the tree stops growing when all the attributes have been used or all the partitions of the branch are purified to one class. However, when no regulatory measures are taken, a tree can easily overfit the data by adding more branches until every training instance is correctly classified. A number of preventive methods have been proposed to improve the generalizability of a tree. Reduced-error pruning [40] reduces the size of a tree after it was learned by applying iterative pruning to branches. The branches closer to leaves are removed first and the pruning propagates upstream until the validation performance of the pruned tree begins to decrease.

Overfitting can also be alleviated by a meta-algorithm called *ensemble learning*. The idea is to train a group of classifiers with a given dataset and combine their output in order to compensate for the high variance of an individual model. Breiman [6] applied this concept to tree learning, which is dubbed as the *random forest*. In creating a bag of models, the random forest algorithm introduces two levels of randomization: First, it randomizes training samples by resampling with replacements (bootstrapping). Second, at each branching step it chooses an attribute to split among a randomly selected subset of attributes. After a bag of trees is trained, prediction is made for all the individual trees and the most frequent class selected by the trees is taken as a final result. *Boosting* is another ensemble meta-algorithm that is often used in conjunction with decision tree. In this setting, trees are learned

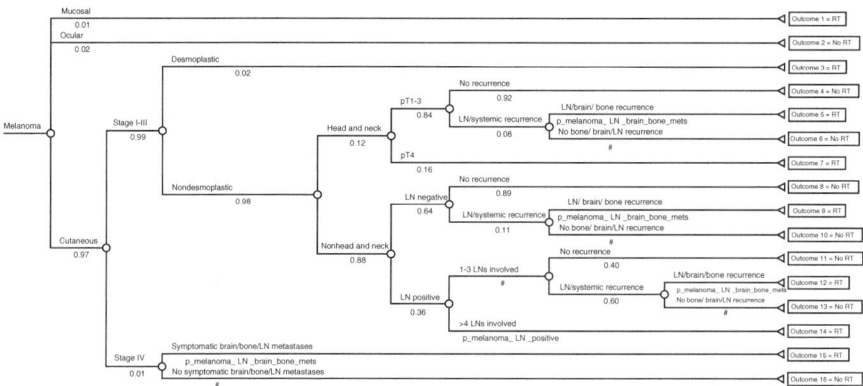

Fig. 3.7 A tree representing decision rules that determine whether radiotherapy should be used for melanoma patients based on characteristics of the disease. *RT* radiotherapy, *LN* lymph node (Reproduced from Delaney et al. [13])

sequentially in the following way: after a tree is learned, the incorrectly classified training examples are assigned with larger weights and the subsequent tree is learned with the reweighed training set. The final classification result is taken as an average output of the group of trees. Detailed algorithm can be consulted in a paper by Freund and Schapire [17]

Decision trees have been a popular choice for many decision support systems, especially in the field of medicine, because their representation of hypotheses as sequential "if-then" clauses is easy to interpret and somewhat resembles human reasoning. For example, Delaney et al. [13] conducted a literature survey to construct a tree to determine recommendation for radiotherapy to melanoma patients based on several clinical attributes (Fig. 3.7). Das et al. [11] trained an ensemble of trees that combined dosimetric and non-dosimetric risk factors for radiation pneumonitis and showed that the prediction can be improved by combining a larger number of trees.

3.3.6 Bayesian Network

Bayesian belief network, or Bayesian network, is designed to model probabilistic relationships among a set of random variables. A key feature of Bayesian network is graphical representation of the relationships via a directed acyclic graph (DAG) which encodes the presence and direction of influence between variables. In a DAG, each variable is assigned to a node and connected to each other via an edge (vertex) which originates from a variable (parent) that influences the probability of the variable it is connected to (child). Thus, probability of a random variable is set to be conditional upon its parent variable(s). The connectivity information in a DAG derives conditional independence relationships that can be stated as random

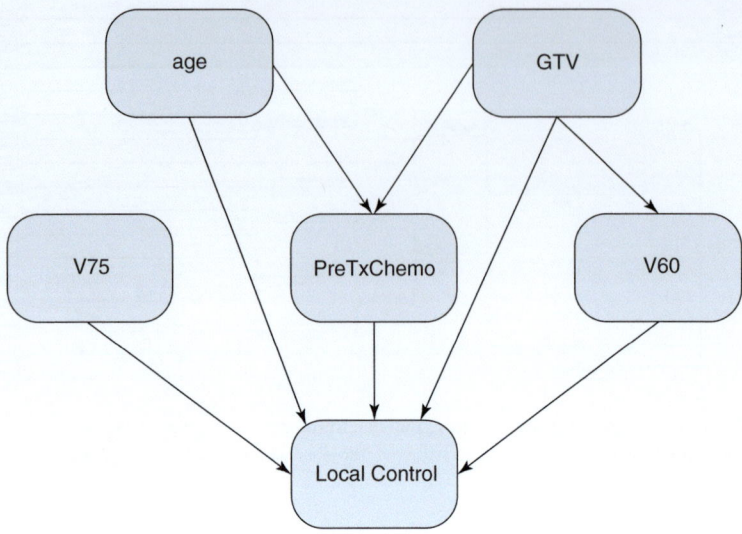

Fig. 3.8 A Bayesian network DAG for predicting local control of NSCLC using radiotherapy and clinical variables. The DAG was trained from clinical data by Oh et al. [36]

variables X and Y are conditionally independent given another variable set $Z_1, Z_2,..., Z_n$ if and only if:

$$P(X|Y, Z_1, Z_2,..., Z_n) = P(X|Z_1, Z_2,..., Z_n) \qquad (3.15)$$

A set of conditional independence relationships specified in a DAG greatly simplifies computation of probability distributions by use of this convenient property: joint probability distribution between the entire variable set, $X = X_1, X_2,..., X_n$, can be obtained by taking the product of all the conditional probabilities for each parents-child set (*the chain rule for Bayesian networks* [31]). Figure 3.8 demonstrates a network of local control of non-small-cell lung cancer (LC) in relation to the following clinical and dosimetric variables: age (*A*), GTV volume (*G*), PTV coverage (V75, V60), and pre-treatment chemo (*P*) [36]. Using the chain rule, a joint probability can be factorized into:

$$P(LC, A, G, V75, V60, C)$$
$$= P(A)P(G)P(V75)P(C|A,G)P(V60|G)P(LC|A,G,C,V75,V60)$$

Conditional probability values are often referred to as the "parameters" of Bayesian network. The parameters can be trained from data as a maximum likelihood estimate or maximum a posteriori (MAP) which incorporates a prior probability with the likelihood obtained from observations.

A DAG can be constructed using prior knowledge on the study domain. When the domain knowledge is not sufficient, observational data can be used to search for the DAG that can best describe the data. DAG searching can be solved as an optimization problem where a predefined a scoring function is maximized over a space of possible DAG configurations. Searching algorithms can vary according to a choice of the scoring function and searching procedures. Widely used scoring functions include a marginal likelihood (Bayesian) score and a Bayesian information criteria (BIC) score. Both scores aim at achieving a balance between the fitness to data (an edge is more likely to be formed between the variables with stronger correlation in data) and complexity of a graph (quantified by the number of edges or parameters), although difference exists in a degree to which complexity is penalized. Mathematical details can be consulted in a primer by Koller and Friedman [31].

Since the number of possible DAGs grows super-exponentially with the number of variables, it is impractical to search exhaustively over the entire graph space for the highest-scoring DAG. Various heuristic approaches have been suggested to reduce a computational cost. For example, a greedy search algorithm begins with the empty graph and keeps adding on edges only when it leads to a higher graph score. Also, constraints on graph topology can be imposed to the search algorithm in order to confine a search domain. For example, the search can be restricted to treelike structures (Chow-Liu trees) [9] or a certain variable ordering that permits only the edges between the variables in descending order ($K2$ algorithm) [10]. High-scoring DAGs can be discovered by a sampling method such as the Markov Chain Monte Carlo (MCMC) [33]. The MCMC algorithm generates samples of DAGs encountered during a random walk over the graph space (Markov chain), which can be approximated as a posterior distribution of DAGs upon convergence of a chain.

The probabilistic approach of BN makes it suitable for handling uncertainties. Especially in a medical domain, missing records or test results could have a negative impact on prediction performance. Bayesian network does not require the full observation on its features for prediction, as it is capable of building and marginalizing joint probability using the conditional dependence relationships between the features. This advantage, in comparison to non-probabilistic classifiers such as SVM, was shown in survival prediction of lung cancer patients by Jayasurya et al. [26]. Other applications of the BN in radiation oncology include a prognostic network for prostate cancer [43] and lung cancer [36].

3.3.7 Naive Bayes

Naive Bayes is a simplified derivative of Bayesian network that is used solely for classification. This method makes an assumption that feature variables are considered independent given a class variable. This so-called naive independence

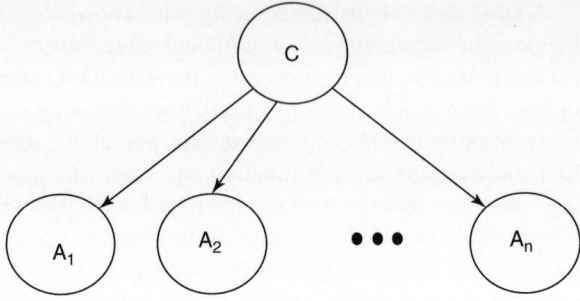

Fig. 3.9 Directed graph representation of the naive Bayes model for a class C and features $A_1, A_2,, An$

assumption can be graphically represented by the Bayesian DAG as shown in Fig. 3.9. Inference of the most probable state for a class, C_{MAP}, is derived from the maximum a posteriori (MAP) rule, using the independence assumption:

$$C_{MAP} = \arg\max_C P(C|A_1, A_2, ..., A_n):$$

$$= \arg\max_C \frac{P(A_1, A_2, ..., A_N|C)P(C)}{P(A_1, A_2, ..., A_n)}$$

$$= \arg\max_C \frac{P(C)\prod_{i=1}^{N} P(A_i|C)}{P(A_1, A_2, ..., A_n)}$$

$$\propto \arg\max_C P(C)\prod_{i=1}^{N} P(A_i|C)$$

Naive Bayes is effective for classification in a high-dimensional space where estimating joint probability of a full variable set is challenging. Its theoretical property is shown to be less sensitive to noisy variation in input, which contributes to its robust performance [18]. However, naive Bayes is not suitable for direct estimation of class posterior as the unrealistic independence assumption results in inaccurate probability estimate. Nevertheless, it has been applied to many medical prognostic problems where classification of a disease state is the only interest. For example, Kazmierska and Malicki predicted brain tumor relapse from a set of 96 features with naive Bayes which accuracy surpassed Bayesian network and decision tree algorithms [27].

3.4 Other Methods

3.4.1 Reinforcement Learning

Reinforcement learning (RL) is a class of machine learning algorithms in which a learner or software agent attempts to take a sequence of actions that would maximize a cumulative reward such as winning a game of checker or chess, for instance [47].

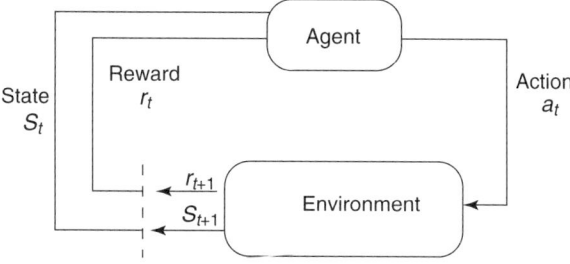

Fig. 3.10 Reinforcement learning system

To an extent RL mimics the way human learns by combining the fields of Markov decision processes (e.g., dynamic programming) with supervised learning. An RL could be depicted as shown in Fig. 3.10 [32], in which at any time point (*t*) actions (a_t) taken by the agent lead to rewards (r_{t+1}) from the current environment state (s_t). The objective is to maximize expected discounted returns value (*V*) at particular state to a given policy (π):

$$V^\pi(s_t) = E\{R/s_t, \pi\} \quad (3.16)$$

where *R* is the return function $R = E\{\gamma r_{t+1}\} = \sum_{t=0}^{\infty} \gamma^t r_{t+1}$ and $0 \le \gamma \le 1$ are discounted return rates. An example of applying RL to radiotherapy is presented by Kim et al. [28], where they showed numerical examples of modifying dose fractionation schedules using a Markov decision process for adaptive radiotherapy applications.

References

1. Arthur D, Vassilvitskii S. K-means++: the advantages of careful seeding. In: Proceedings of the eighteenth annual ACM-SIAM symposium on discrete algorithms, SODA'07. Society for Industrial and Applied Mathematics; Philadelphia:2007. p. 1027–35.
2. Bischof H, Leonardis A, Selb A. Minimum description length principle for robust vector quantisation. Pattern Anal Appl. 1999;2(1):59–72.
3. Blanco AI, Chao KSC, El Naqa I, Franklin GE, Zakarian K, Vicic M, Deasy JO. Dose-volume modeling of salivary function in patients with head-and-neck cancer receiving radiotherapy. Int J Radiat Oncol Biol Phys. 2005;62(4):1055–69.
4. Bradley J, Deasy JO, Bentzen S, El-Naqa I. Dosimetric correlates for acute esophagitis in patients treated with radiotherapy for lung carcinoma. Int J Radiat Oncol Biol Phys. 2004;58(4):1106–13.
5. Bradley JD, Hope A, El Naqa I, Apte A, Lindsay PE, Bosch W, Matthews J, Sause W, Graham MV, Deasy JO. A nomogram to predict radiation pneumonitis, derived from a combined analysis of rtog 9311 and institutional data. Int J Radiat Oncol Biol Phys. 2007;69(4):985–92.
6. Breiman L. Random forests. Mach Learn. 2001;45(1):5–32.
7. Chen S, Zhou S, Yin F-F, Marks LB, Das SK. Investigation of the support vector machine algorithm to predict lung radiation-induced pneumonitis. Med Phys. 2007;34(10):3808–14.

8. Chen S, Zhou S, Yin FF, Marks LB, Das SK. Using patient data similarities to predict radiation pneumonitis via a self-organizing map. Phys Med Biol. 2008;53(1):203.
9. Chow C, Liu C. Approximating discrete probability distributions with dependence trees. IEEE Trans Inf Theor. 2006;14(3):462–7.
10. Cooper GF, Herskovits E. A Bayesian method for the induction of probabilistic networks from data. Mach Learn. 1992;9(4):309–47.
11. Das SK, Zhou S, Zhang J, Yin F-F, Dewhirst MW, Marks LB. Predicting lung radiotherapy-induced pneumonitis using a model combining parametric Lyman probit with nonparametric decision trees. Int J Radiat Oncol Biol Phys. 2007;68(4):1212–21.
12. Dehing-Oberije C, Yu S, Ruysscher DD, Meersschout S, Beek KV, Lievens Y, Meerbeeck JV, Neve WD, Rao B, van der Weide H, Lambin P. Development and external validation of prognostic model for 2-year survival of non small cell lung cancer patients treated with chemoradiotherapy. Int J Radiat Oncol Biol Phys. 2009;74(2):355–62.
13. Delaney G, Barton M, Jacob S. Estimation of an optimal radiotherapy utilization rate for melanoma. Cancer. 2004;100(6):1293–301.
14. El Naqa I, Bradley J, Blanco AI, Lindsay PE, Vicic M, Hope A, Deasy JO. Multivariable modeling of radiotherapy outcomes, including dose-volume and clinical factors. Int J Radiat Oncol Biol Phys. 2006;64(4):1275–86.
15. El Naqa I, Bradley J, Deasy J. Machine learning methods for radiobiological outcome modeling. In: Mehta M, Paliwal B, Bentzen S, editors. Physical, chemical, and biological targeting in radiation oncology. Madison: Medical Physics Pub.; 2005. p. 150–9.
16. El Naqa I, Bradley JD, PE L, Hope AJ, Deasy JO. Predicting radiotherapy outcomes using statistical learning techniques. Phys Med Biol. 2009;54(18):S9.
17. Freund Y, Schapire RE. A brief introduction to boosting. In: Proceedings of the sixteenth international joint conference on artificial intelligence. San Francisco: Morgan Kaufmann; 1999. p. 1401–6.
18. Friedman JH. On bias, variance, 0/1ñloss, and the curse-of-dimensionality. Data Min Knowl Discov. 1997;1(1):55–77.
19. Hastie T, Tibshirani R, Friedman J. The elements of statistical learning: data mining, inference, and prediction. New York: Springer; 2009.
20. Haykin S. Neural networks: a comprehensive foundation. Upper Saddle River: Prentice Hall PTR; 1998.
21. Hope AJ, Lindsay PE, Naqa IE, Alaly JR, Vicic M, Bradley JD, Deasy JO. Modeling radiation pneumonitis risk with clinical, dosimetric, and spatial parameters. Int J Radiat Oncol Biol Phys. 2006;65(1):112–24.
22. Hosmer D, Lemeshow S. Applied logistic regression. New York: John Wiley; 2000.
23. Huang EX, Bradley JD, El Naqa I, Hope AJ, Lindsay PE, Bosch WR, Matthews JW, Sause WT, Graham MV, Deasy JO. Modeling the risk of radiation-induced acute esophagitis for combined Washington University and rtog trial 93-11 lung cancer patients. Int J Radiat Oncol Biol Phys. 2012;82(5):1674–9.
24. Huang EX, Hope AJ, Lindsay PE, Trovo M, El Naqa I, Deasy JO, Bradley JD. Heart irradiation as a risk factor for radiation pneumonitis. Acta Oncol. 2011;50(1):51–60.
25. Jain AK, Murty MN, Flynn PJ. Data clustering: a review. ACM Comput Surv. 1999;31(3):264–323.
26. Jayasurya K, Fung G, Yu S, Dehing-Oberije C, De Ruysscher D, Hope A, De Neve W, Lievens Y, Lambin P, Dekker ALAJ. Comparison of Bayesian network and support vector machine models for two-year survival prediction in lung cancer patients treated with radiotherapy. Med Phys. 2010;37(4):1401–7.
27. Kazmierska J, Malicki J. Application of the nave Bayesian classifier to optimize treatment decisions. Radiother Oncol. 2008;86(2):211–6.
28. Kim M, Ghate A, Phillips MH. A Markov decision process approach to temporal modulation of dose fractions in radiation therapy planning. Phys Med Biol. 2009;54(14):4455.
29. Klement R, Allgauer M, Appold S, Dieckmann K, Ernst I, Ganswindt U, Holy R, Nestle U, Nevinny-Stickel M, Semrau S, Sterzing F, Wittig A, Andratschke N, Guckenberger M. Support

vector machine-based prediction of local tumor control after stereotactic body radiation therapy for early-stage non-small cell lung cancer. Int J Radiat Oncol Biol Phys. 2014;88(3):732–8.
30. Kohonen T. The self-organizing map. Proc IEEE. 1990;78(9):1464–80.
31. Koller D, Friedman N. Probabilistic graphical models: principles and techniques – adaptive computation and machine learning. Cambridge: The MIT Press; 2009.
32. Kulkarni P. Reinforcement and systemic machine learning for decision making. Hoboken: Wiley-IEEE Press; 2012.
33. Madigan D, York J, Allard D. Bayesian graphical models for discrete data. Int Stat Rev. 1995;63(2):215–32.
34. Marks LB. Dosimetric predictors of radiation-induced lung injury. Int J Radiat Oncol Biol Phys. 2002;54(2):313–6.
35. Mitchell TM. Machine learning. 1st ed. New York: McGraw-Hill, Inc.; 1997.
36. Oh JH, Craft JM, Townsend R, Deasy JO, Bradley JD, El Naqa I. A bioinformatics approach for biomarker identification in radiation-induced lung inflammation from limited proteomics data. J Proteome Res. 2011;10(3):1406–15.
37. Pella A, Cambria R, Riboldi M, Jereczek-Fossa BA, Fodor C, Zerini D, Torshabi AE, Cattani F, Garibaldi C, Pedroli G, Baroni G, Orecchia R. Use of machine learning methods for prediction of acute toxicity in organs at risk following prostate radiotherapy. Med Phys. 2011;38(6):2859–67.
38. Pelleg D, Moore A. X-means: extending k-means with efficient estimation of the number of clusters. In: Proceedings of the 17th international conference on machine learning. San Francisco: Morgan Kaufmann; 2000. p. 727–34.
39. Quinlan JR. Induction of decision trees. Mach Learn. 1986;1:81–106.
40. Quinlan JR. Simplifying decision trees. Int J Man Mach Stud. 1987;27(3):221–34.
41. Ripley BD. Pattern recognition and neural networks. Cambridge/New York: Cambridge University Press; 1996.
42. Scholkopf A, Smola J, Muller KR. Kernel principal component analysis. Cambridge: MIT Press; 1999. p. 327–52.
43. Smith WP, Doctor J, Meyer J, Kalet IJ, Phillips MH. A decision aid for intensity-modulated radiation-therapy plan selection in prostate cancer based on a prognostic Bayesian network and a Markov model. Artif Intell Med. 2009;46(2):119–30.
44. Specht DF. A general regression neural network. IEEE Trans Neural Netw. 1991;2(6):568–76.
45. Spencer SJ, Bonnin DA, Deasy JO, Bradley JD, El Naqa I. Bioinformatics methods for learning radiation-induced lung inflammation from heterogeneous retrospective and prospective data. J Biomed Biotechnol. 2009(2009), 892863. doi:10.1155/2009/892863.
46. Su M, Miften M, Whiddon C, Sun X, Light K, Marks L. An artificial neural network for predicting the incidence of radiation pneumonitis. Med Phys. 2005;32(2):318–25.
47. Sutton RS, Barto AG. Introduction to reinforcement learning. Cambridge: MIT Press; 1998.
48. Svensson JP, Stalpers LJA, Lange REEE, Franken NAP, Haveman J, Klein B, Turesson I, Vrieling H, Giphart-Gassler M. Analysis of gene expression using gene sets discriminates cancer patients with and without late radiation toxicity. PLoS Med. 2006;3(10):e422.
49. Tucker SL, Cheung R, Dong L, Liu HH, Thames HD, Huang EH, Kuban D, Mohan R. Dose-volume response analyses of late rectal bleeding after radiotherapy for prostate cancer. Int J Radiat Oncol Biol Phys. 2004;59(2):353–65.
50. Vapnik V. Statistical learning theory. New York: Wiley; 1998.
51. Vittinghoff E, Glidden D, Shiboski S, McCulloch C. Regression methods in biostatistics: linear, logistic, survival, and repeated measures models. New York: Springer; 2006.

Performance Evaluation in Machine Learning

Nathalie Japkowicz and Mohak Shah

Abstract
Performance evaluation is an important aspect of the machine learning process. However, it is a complex task. It, therefore, needs to be conducted carefully in order for the application of machine learning to radiation oncology or other domains to be reliable. This chapter introduces the issue and discusses some of the most commonly used techniques that have been applied to it. The focus is on the three main subtasks of evaluation: measuring performance, resampling the data, and assessing the statistical significance of the results. In the context of the first subtask, the chapter discusses some of the confusion matrix-based measures (accuracy, precision, recall or sensitivity, and false alarm rate) as well as receiver operating characteristic (ROC) analysis; several error estimation or resampling techniques belonging to the cross-validation family as well as bootstrapping are involved in the context of the second subtask. Finally, a number of nonparametric statistical tests including McNemar's test, Wilcoxon's signed-rank test, and Friedman's test are covered in the context of the third subtask. The chapter concludes with a discussion of the limitations of the evaluation process.

N. Japkowicz, PhD (✉)
School of Information Technology and Engineering, University of Ottawa,
Ottawa, ON, Canada
e-mail: nat@site.uottawa.ca; http://www.site.uottawa.ca/~nat

M. Shah, PhD
Research and Technology Center - North America, Robert Bosch LLC,
Palo Alto, CA, USA
e-mail: mohak@mohakshah.com

4.1 Introduction

While developing and applying machine learning tools to problems in radiation oncology or other domains are what will allow new advances to be made in these domains, it is important to realize that without proper means of evaluating the new methods, there is no way to know whether or not they are effective. While researchers and practitioners of machine learning have long known that and used general evaluation methods to judge the effectiveness of their algorithms, until recently, very little attention has been paid to the details of how this evaluation should be carried out. Instead, a uniform methodology was consistently applied without any concern regarding the appropriateness of that methodology for the particular cases considered. In this chapter, we begin by giving an overview of machine learning evaluation, pointing to some issues that may creep in if it is not conducted appropriately. We then follow with a discussion of evaluation metrics, resampling methods, and statistical testing. We conclude the chapter with a consideration of the limitations of the evaluation process. The discussion in this chapter is based on [1], which gives much more detail about the issue and its solution.

4.2 An Overview of Machine Learning Evaluation

While not as exciting a process as the design of machine learning algorithms or its application to difficult problems, the issue of machine learning evaluation needs to be considered very carefully. Indeed, there exist many approaches to evaluation and it remains unclear when or why certain approaches are more appropriate than others. In this chapter, we clarify these questions in at least a few cases that may be of interest to the radiation oncology research community.

To begin with, Fig. 4.1 presents the various steps of classifier evaluation along with the interaction between these steps. At each of these steps, choices must be made. In particular, the researcher must decide on which algorithms will be used in the study, which data sets they will be applied, and what performance measure, resampling technique, and statistical tests will be used. Each of these questions is quite complex because the choices made at one step may impact on the other steps. For example, if the data set on which the evaluation will be based contains very few instances of X-rays with malignant tumors and many instances of X-rays with benign tumors, then the performance measure (or metric) to be used cannot be the same as if there were as many instances of each cases. In addition, the choices to be made at each step depend on the purpose of the evaluation. Here are four common scenarios:

- Comparison of a *new algorithm* to other (may be generic or application-specific) classifiers on a *specific domain* (e.g., when proposing a novel learning algorithm)
- Comparison of a *new generic algorithm* to other generic ones on a set of *benchmark domains* (e.g., to demonstrate general effectiveness of the new approach against other approaches)

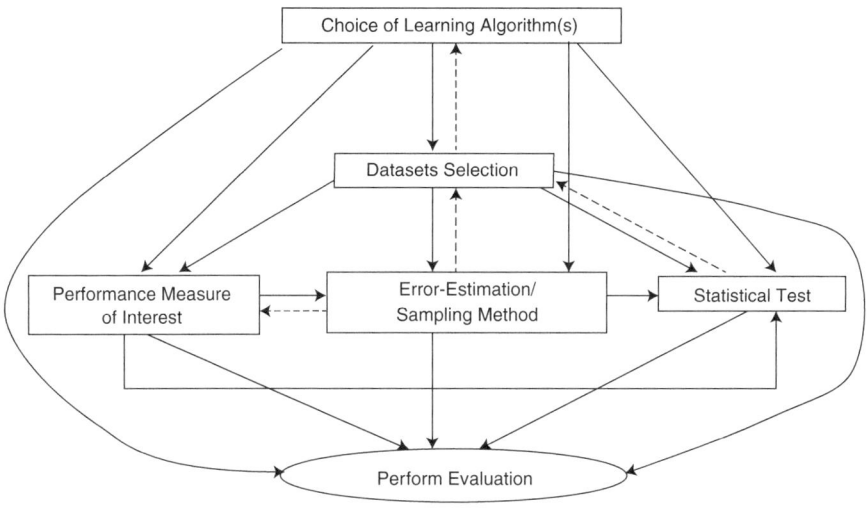

Fig. 4.1 The main steps of evaluation

Table 4.1 The performance of eight different classifiers according to nine different performance measures. There is clear disagreement among the evaluation measures

Algo	Acc	RMSE	TPR	FPR	Prec	Rec	F	AUC	Info S
NB	71.7	.4534	.44	.16	.53	.44	.48	.7	48.11
C4.5	75.5	.4324	.27	.04	.74	.27	.4	.59	34.28
3NN	72.4	.5101	.32	.1	.56	.32	.41	.63	43.37
Ripp	71	.4494	.37	.14	.52	.37	.43	.6	22.34
SVM	69.6	.5515	.33	.15	.48	.33	.39	.59	54.89
Bagg	67.8	.4518	.17	.1	.4	.17	.23	.63	11.30
Boost	70.3	.4329	.42	.18	.5	.42	.46	.7	34.48
RanF	69.23	.47	.33	.15	.48	.33	.39	.63	20.78

- Characterization of *generic classifiers* on *benchmarks domains* (e.g., to study the algorithms' behavior on general domains for subsequent use)
- Comparison of *multiple classifiers* on a *specific domain* (e.g., to find the best algorithm for a given application task)

To better illustrate the difficulties of making the appropriate choices at each step, we look at an example involving the choice of an appropriate performance measure. Table 4.1 shows the performance obtained by eight different classifiers (naive Bayes [NB], C4.5, three-nearest neighbor [3NN], ripper [Rip], support vector machines

[SVM], bagging [Bagg], boosting [Boost], random forest [RF]) on a given data set (the UCI breast cancer data set [2]) using nine different performance measures (accuracy [Acc], root-mean-square error [RMSE], true positive rate [TPR], false positive rate [FPR], precision [Prec], recall [Rec], F-measure [F], area under the ROC curve [AUC], information score [Info S]). As can be seen from the table, each measure tells a different story. For example, accuracy ranks C4.5 as the best classifier for this domain, while according to the AUC, C4.5 is the worst classifier (along with SVM, which accuracy did not rank highly either). Similarly, the F-measure ranks naive Bayes in the first place, whereas it only reaches the 5th place as far as RMSE is concerned. This suggests that one may obtain very different conclusions depending on what performance measure is used. Generally speaking, this example points to the fact that classifier evaluation is not an easy task and that not taking it seriously may yield grave consequences.

The next section looks at performance measures in more detail, while the next two sections will discuss resampling and statistical testing.

4.3 Performance Measures

Figure 4.2 presents an overview of the various performance measures commonly used in machine learning. This overview is not comprehensive, but touches upon the main measures. In the figure, the first line, below the "all measures" box indicates the kind of information used by the performance measure to calculate the value. All measures use the confusion matrix, which will be presented next, but some add additional information such as the classifier's uncertainty or the cost ratio of the data set, while others also use other information such as how comprehensible the result of the classifier is or how generalizable it is, and so on. The next line in the figure indicates what kind of classifier the measure applies to deterministic classifiers, scoring classifiers, or continuous and probabilistic classifiers. Below this line comes information about the focus (e.g., multiclass with chance correction), format (e.g., summary statistics), and methodological basis (e.g., information theory) of the measures. The leaves of the tree list the measures themselves.

As just mentioned, all the measures of Fig. 4.2 are based on the confusion matrix. The template for a confusion matrix is given in Table 4.2:

TP, FP, FN, and TN stand for true positive, false positive, false negative, and true negative, respectively. Some common performance measures calculated directly from the confusion matrix are:

- Accuracy $=(TP+TN)/(P+N)$
- Precision $=TP/(TP+FP)$
- Recall, sensitivity, or true positive rate $=TP/P$
- False alarm rate or false positive rate $=FP/N$

For a more comprehensive list of measures including sensitivity, specificity, likelihood ratios, positive and negative predictive values, and so on, please refer to [1].

4 Performance Evaluation in Machine Learning

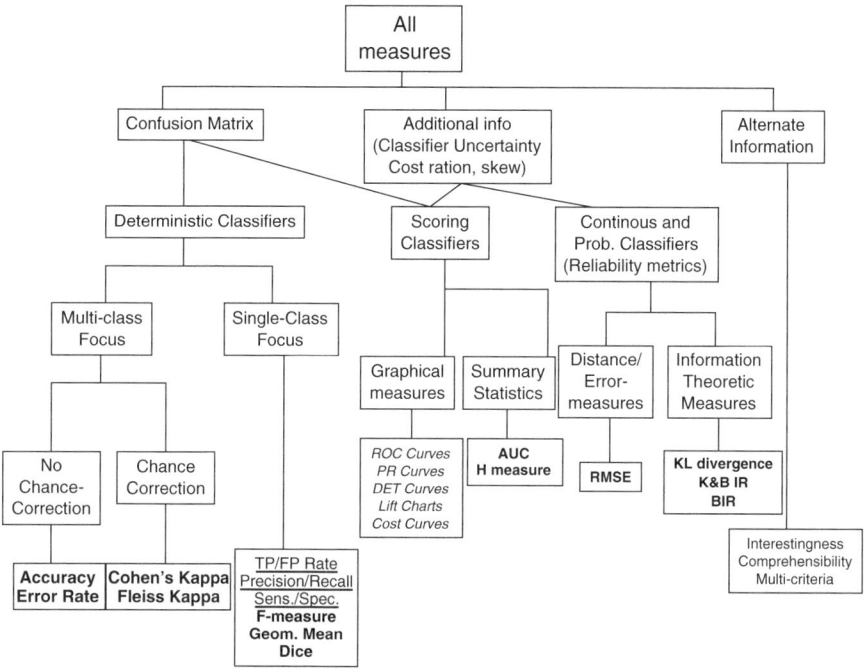

Fig. 4.2 An overview of performance measures

Table 4.2 A generic confusion matrix

True class → hypothesized|class V	Pos	Neg
Yes	TP	FP
No	FN	TN
	P=TP+FN	N=FP+TN

Table 4.3 The confusion matrices of two very different classifiers with the same accuracy

True class →	Pos	Neg
Yes	200	100
No	300	400
	P=500	N=500

True class →	Pos	Neg
Yes	400	300
No	100	200
	P=500	N=500

We now illustrate some of the problems encountered with accuracy, precision, and recall since they represent important problems in evaluation. Consider the confusion matrices of Table 4.3. The accuracy for both matrices is 60 %. However, the two matrices t classifiers with the same accuracy account for two very different classifier behaviors. On the left, the classifier exhibits a weak positive recognition

Table 4.4 The confusion matrices of two very different classifiers with the same precision and recall

True class →	Pos	Neg
Yes	200	100
No	300	400
	P=500	N=500

True class →	Pos	Neg
Yes	200	100
No	300	0
	P=500	N=100

rate and a strong negative recognition rate. On the right, the classifier exhibits a strong positive recognition rate and a weak negative recognition rate. In fact, accuracy, while generally a good and robust measure, is extremely inappropriate in the case of class imbalance data, such as the example, mentioned in Sect. 4.2 where there were only very few instances of X-rays containing malignant tumors and many instances containing benign ones. For example, in the extreme case where, say, 99.9 % of all the images would not contain any malignant tumors and only 0.1 % would, the rough classifier consisting of predicting "benign" in all cases would produce an excellent accuracy rate of 99.9 %. Obviously, this is not representative of what the classifier is really doing because, as suggested by its 0 % recall, it is not an effective classifier at all, specifically if what it is trying to achieve is the recognition of rare, but potentially important, events. The problem of classifier evaluation in the case of class imbalance data is discussed in [3].

Table 4.4 illustrates the problem with precision and recall. Both classifiers represented by the table on the left and the table on the right obtain the same precision and recall values of 66.7 and 40 %. Yet, they exhibit very different behaviors: while they do show the same positive recognition rate, they show extremely different negative recognition rates; in the left confusion matrix, the negative recognition rate is strong, while in the right confusion matrix, it is nil! This certainly is information that is important to convey to a user, yet, precision and recall do not focus on this kind of information. Note, by the way, that accuracy which has a multiclass rather than a single-class focus has no problem catching this kind of behavior: the accuracy of the confusion matrix on the left is 60 %, while that of the confusion matrix on the right is 33 %!

Because the class imbalance problem is very pervasive in machine learning, ROC analysis and its summary measure and the area under the ROC curve (AUC), which do not suffer from the problems encountered by accuracy, have become central to the issue of classifier evaluation. We now give a brief description of that approach. In the context of the class imbalance problem, the concept of ROC analysis can be interpreted as follows. Imagine that instead of training a classifier f only at a given class imbalance level, that classifier is trained at all possible imbalance levels. For each of these levels, two measurements are taken as a pair, the true positive rate (or sensitivity) and the false positive rate (FPR) (or false alarm rate). Many situations may yield the same measurement pairs, but that does not matter since

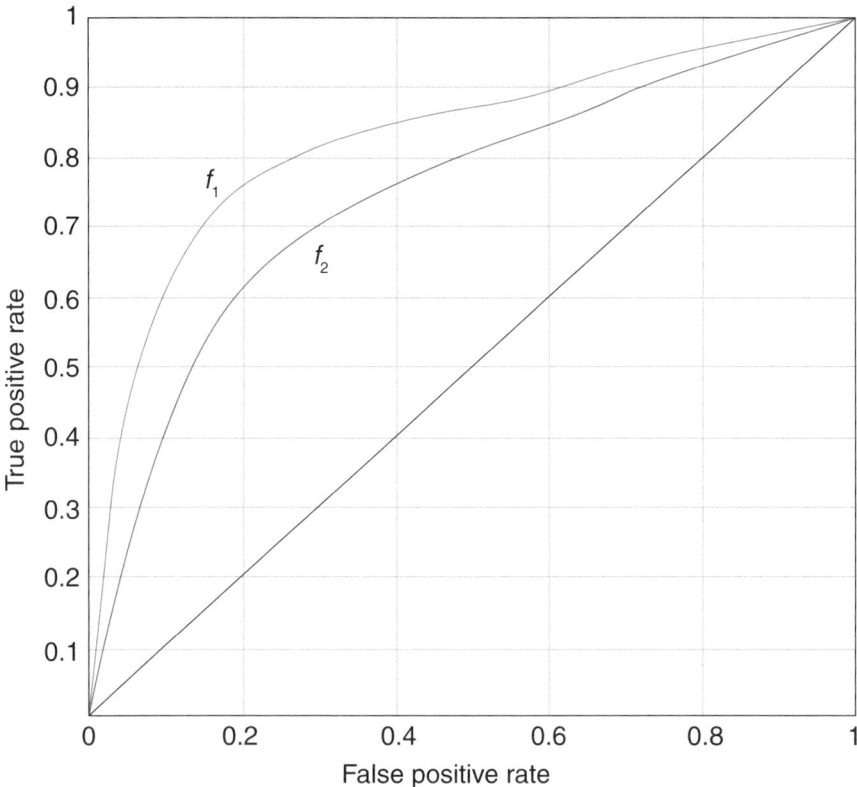

Fig. 4.3 The ROC curves of two classifiers f1 and f2. f1 performs better than f2 in all parts of the ROC space

duplicates are ignored. Once all the measurements have been made, the points represented by all the obtained pairs are plotted in what is called the *ROC space*, a graph that plots the true positive rate as a function of the false positive rate. The points are then joined in a smooth curve, which represents the ROC curve for that classifier. Figure 4.3 shows two ROC curves representing the performance of two classifiers *f1* and *f2* across all possible operating ranges.

The closer a curve representing a classifier *f* is from the top-left corner of the ROC space (small false positive rate, large true positive rate), the better the performance of that classifier. For example, *f*1 performs better than *f*2 in the graph of Fig. 4.3. However, the ideal situation of Fig. 4.3 rarely occurs in practice. More often than not, one is faced with a situation such as that of Fig. 4.4, where one classifier dominates the other in some parts of the ROC space, but not in others.

The reason why ROC analysis is well suited to the study of class imbalance domains is twofold. First, as in the case of the single-class focus metrics of the previous section, rather than being combined together into a single multiclass focus

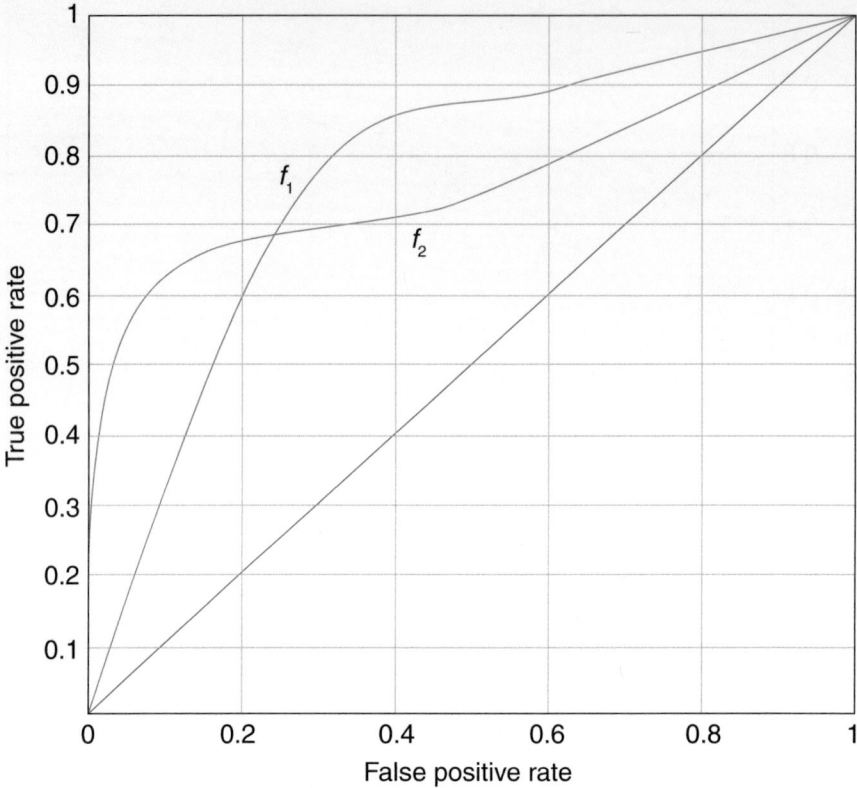

Fig. 4.4 The ROC curves of two classifiers f1 and f2. f2 performs better than f1 on the left side of the ROC space. After the two curves, cross f1 performs better than f1

metric, performance on each class is decomposed into two distinct measures. Second, the imbalance ratio that truly applies in a domain is rarely precisely known. ROC analysis gives an evaluation of what may happen in diverse situations.

We now move on to discussing the question of data resampling.

4.3.1 Resampling

What is the purpose of resampling? Ideally, we would have access to the entire population or a lot of representative data from it. This, unfortunately, is usually not the case, and the limited data available has to be reused in clever ways in order to be able to estimate the error of our classifiers as reliably as possible. Resampling is divided into two categories: *simple resampling* (where each data point is used for testing only once) and *multiple resampling* (which allows the use of the same data point more than once for testing). In addition to discussing a few resampling approaches, this section will underline the issues that may arise when applying

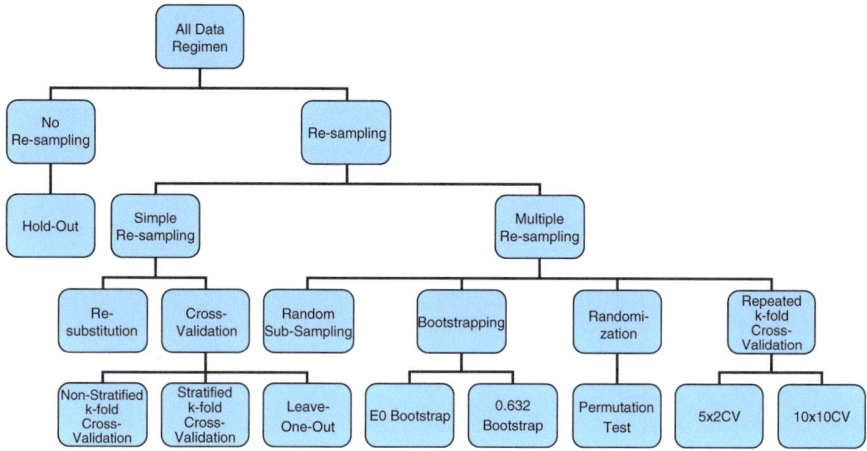

Fig. 4.5 Overview of resampling methods

them. Figure 4.5 gives an overview of various resampling regimens. We will discuss a few of them. For a more detailed presentation, please see [1].

When the data set is very large and all cases are well represented, then no resampling method is needed, and it is possible to use the holdout method where a portion of the data set is reserved for training while the rest of the data set is used for testing. Please note that the practice of training and testing on the same data set (re-substitution) is unacceptable when the goal of the study is to test the predictive capability of the learning tool. Such a practice gives an optimistic assessment of the tool's capability. In general, the classifier will overfit the data it was trained on, which means that it will perform very well on that data and obtain much poorer results on data it has never seen before. To a certain extent and for many algorithms, the better the classifier performs on the known data, the worse it will perform on unknown data.

In most cases, there is not enough data to use the holdout method. The most commonly used resampling method then is k-fold cross validation and its variants, stratified k-fold cross validation, and leave-one-out, also known as the jackknife. 10×10-fold cross validation has also become quite common and the 0.632 bootstrap is sometimes used as well. We will present each of these schemes in turn and discuss the situations in which each scheme is believed to be most appropriate.

Figure 4.6 illustrates the k-fold cross-validation process in the following ways: each line of the graph symbolizes the entire data set. It is randomly divided into k subsets (on the graph, $k=10$) as symbolized by the $k=10$ rectangles that compose each line. The first line corresponds to Fold 1, the second to Fold 2, and so on. In Fold 1, the first rectangle is shaded differently from the others. This signifies that in this fold, the data represented by the first rectangle will be used as testing data while the data represented by the other $k-1$ rectangles will be used as training data. In Fold 2, it is the data of the second rectangle that is used as the testing set, while the data

Fig. 4.6 The k-fold cross-validation process

represented by the other rectangles are used as the training set. This goes on k times so that each of the rectangles is used as a testing set. This is an interesting scheme which guarantees that (1) at each fold, the training and the testing set are separate; (2) once the entire scheme has been executed, every data point has been used as a testing point; (3) no data point has been used more than once as a testing point; and (4) every data point has been used k-1 times as a training point. So in summary, there is no overlap in the testing sets, but there is overlap in the training set. The facts that there is no overlap in the testing set, that this scheme is very simple to implement, and that it is not very computer intensive make it a very popular approach believed to yield a good error estimate. Because of the high overlap in the training set, however, the method can yield a bias in the error estimate, but this is mitigated in the case of moderate to large data sets.

When the data set is imbalanced, k-fold cross validation as just described can yield problems. In particular, the random division of the data into k subsets may yield situations where the data of the minority class is not at all represented in the subset. The performance of the classifier on such a data set would be misleading as it would be overly optimistic. Similarly, if the training data contained an even smaller proportion of minority examples than the actual data set, the classifier's performance would be overly pessimistic. In order to avoid both problems, a process called stratified k-fold cross validation is used to ensure that the distribution is respected in the training and testing sets created at every fold. This would not necessarily be the case if a pure random process were used.

Another issue arises when the data set is quite small. In such cases, k-fold cross validation may cause the training portion of the data at each fold to be too small for effective learning to take place. In such cases, it is common to set k to the size of the data set, meaning that (1) there are as many folds as there are data points and, at each fold, (2) the testing set includes a single data point and (3) the classifier is trained on all the data but this particular point. This process is commonly called leave-one-out or the jackknife. It has the advantage of yielding a relatively unbiased classifier (since virtually all the data is used for training at each fold, although since the data set is small to begin with, the classifier is probably not unbiased); however, the error estimate is likely to show high variance since only one example is tested at every fold, resulting in a 0 or 100 % accuracy rate for each fold. In addition, it is a very time-consuming process since the number of folds equals the size of the data set.

A further issue with the family of k-fold cross-validation approaches just discussed concerns the stability of the estimate it produces. In order to improve the stability of that estimate, it has become commonplace to run the k-fold cross-validation process multiple times, each with different random partitions of the data into k-folds. The most common combination is the 10×10-fold cross validation [4], though 5×2-fold cross validation [5] had also been proposed early on as an alternative to tenfold cross validation.

We conclude this discussion with a presentation of bootstrapping, an alternative to the k-fold cross-validation schemes. Bootstrapping assumes that the available sample is representative and creates a large number of new samples by drawing from replacement from the available sample. Bootstrapping is useful in practice when the sample is too small for cross-validation or leave-one-out approaches to yield a good estimate. There are two bootstrap estimates that are useful in the context of classification: the €0 and the e632 bootstraps. The €0 bootstrap tends to be pessimistic because it is only trained on 63.2 % of the data in each run. The e632 attempts to correct for this. The listing below is an informal description of the algorithms for the €0 and e632 bootstraps.

- Given a data set D of size m, we create k bootstrap samples Bi of size m, by sampling from D with replacement (k is typically ≥ 200).
- At each run, each of the k bootstraps represent the training set while the testing set is made up of a single copy of the examples from D that did not make it to Bi.
- At each run, a classifier is trained and tested and $€o_i$ represents the performance of the classifier at that run.
- €o represents the average of all the $€o_i$'s.

$$e632 = 0.632 \times €o + 0.368 \times err(f)$$

Where err(f) is the optimistically biased re-substitution error (error rate obtained when training and testing on D)

As previously mentioned, bootstrapping is a good estimator when the data set is too small to run k-fold cross validation or leave-one-out. In particular, it was shown to have low variance in such cases. On the other hand, bootstrapping is not a useful estimator in the case of classifiers that do not benefit from the presence of duplicate instances such as k-nearest neighbors.

4.4 Significance Testing

The performance metrics discussed in Sect. 4.2 allow us to make observations about different classifiers, and the resampling approaches discussed in Sect. 4.3 allow us to reuse the available data in order to obtain results believed to be more reliable. The question we ask in this section is related to the issue raised by resampling in Sect. 4.3. In particular, we ask to what extent the observed results are, indeed, reliable. More specifically, can the observed results be attributed to the real characteristics of the classifiers under scrutiny or are they observed by chance? The purpose of statistical significance testing is to help us gather evidence of the extent to which the results returned by an evaluation metric on the resampled data sets are representative of the general behavior of our classifiers.

Although some researchers have argued against the use of statistical tests mainly because it is often difficult to perform properly and its results are often overvalued and limit the search for new ideas [6, 7], statistical testing remains the norm in most experimental settings. Nonetheless, in line with the critics, it is important to conduct and interpret such tests properly. We will discuss basic aspects of the practice in what follows. In particular two issues arise:

1. Do we have enough information about the underlying distributions of the classifiers' results to apply a parametric test?
2. What kind of problem are we considering?
   - The comparison of two algorithms on a single domain
   - The comparison of two algorithms on several domains
   - The comparison of multiple algorithms on multiple domains

Figure 4.7 overviews the various statistical tests in relation to these two problems. The first line in the figure differentiates between the kinds of problems considered. The next line lists the different statistical tests available in each situation. The tests in red boxes are parametric tests while those in green boxes represent nonparametric tests. Parametric tests have the advantage of being more powerful than nonparametric ones, but they apply in a more limited number of situations than the nonparametric ones since they require knowledge of the underlying distribution. Nonparametric tests are more flexible than the parametric ones since they do not take into account the underlying distribution. Instead, they use ranking information.

A comprehensive discussion of all these tests can be found in [1]. In this chapter, we will focus on three versatile nonparametric tests: McNemar's test, Wilcoxon's signed-rank test for matched pairs, and Friedman's test (followed by Nemenyi's test). McNemar's test applies in the case of two algorithms and one domain; Wilcoxon's test applies in the case of two algorithms tested on multiple domains and Friedman's test applies to the case of multiple algorithms executed over multiple domains.

McNemar's test calculates four variables:

- The number of instances misclassified by both classifiers (C_{00})
- The number of instances misclassified by the first classifier but correctly classified by the second (C_{01})

4 Performance Evaluation in Machine Learning

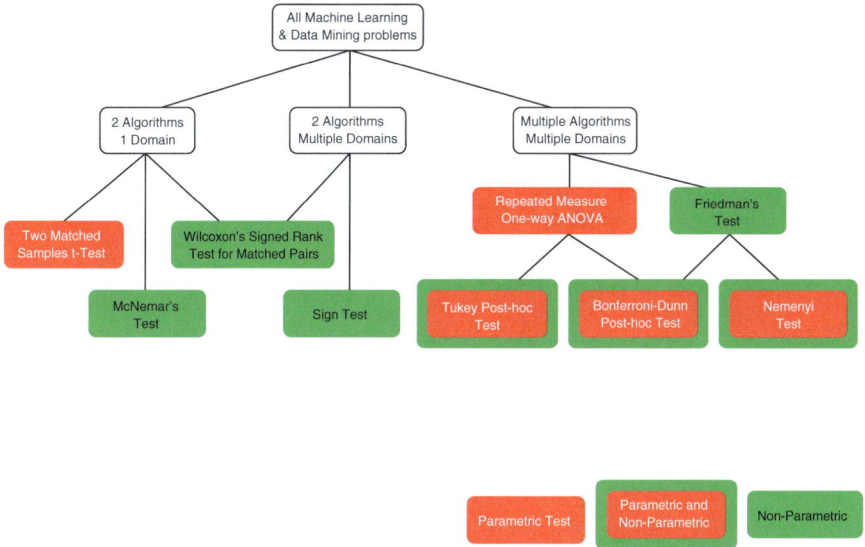

Fig. 4.7 Overview of statistical tests

- The number of instances misclassified by the second classifier but correctly classified by the first (C_{10})
- The number of instances correctly classified by both classifiers (C_{11})

McNemar's χ^2 statistics is given by

$$\chi^2_{MC} = \left(\left|C_{01} - C_{10}\right| - 1\right)^2 / \left(C_{01} + C_{10}\right)$$

If $C_{01} + C_{10} < 20$, then the test cannot be used.

Otherwise, the χ^2_{MC} statistics is compared to the χ^2 statistics. If χ^2_{MC} exceeds the $\chi^2_{1,\,1-\alpha}$ statistic, then we can reject the null hypothesis that assumes that the first and second classifiers perform equally well with $1-\alpha$ confidence.

Wilcoxon's signed-rank test deals with two classifiers on multiple domains. It is also nonparametric. Here is its description:

- For each domain, we calculate the difference in the performance of the two classifiers.
- We rank the absolute values of these differences and graft the signs in front of the ranks.
- We calculate the sum of positive and negative ranks, respectively (W_{S1} and W_{S2}).
- We compute T_{Wilcox} such that $T_{Wilcox} = \min(W_{S1}, W_{S2})$.
- We compare T_{Wilcox} to critical value V_α. If $V_\alpha \geq T_{Wilcox}$, we reject the null hypothesis that the performance of the two classifiers is the same at the α confidence level.

Wilcoxon's signed-rank test is illustrated in Table 4.5 and in the discussion below the table. In this example, NB and SVM are compared on ten different domains.

Table 4.5 Wilcoxon's signed-rank test for NB and SVM on 10 different domains

Data	NB	SVM	NB-SVM	|NB-SVM|	Ranks	+/− ranks
1	.9643	.9944	−0.0301	0.0301	3	−3
2	.7342	.8134	−0.0792	0.0792	6	−6
3	.7230	.9151	−0.1921	0.1921	8	−8
4	.7170	.6616	+0.0554	0.0554	5	+5
5	.7167	.7167	0	0	Remove	Remove
6	.7436	.7708	−0.0272	0.0272	2	−2
7	.7063	.6221	+0.0842	0.0842	7	+7
8	.8321	.8063	+0.0258	0.0258	1	+1
9	.9822	.9358	+0.0464	0.0464	4	+4
10	.6962	.9990	−0.3028	0.3028	9	−9

From the table, we find that $W_{S1}=17$ and $W_{S2}=28$, which means that $T_{\text{Wilcox}}=\min(17, 28)=17$. For $n=10-1$ degrees of freedom and $\alpha=0.005$, $V=8$ (see Table 4.5 in Appendix A of [1]) for the 1-sided test. V must be larger than T_{Wilcox} in order to reject the hypothesis. Since $17>8$, we cannot reject the hypothesis that NB's performance is equal to that of SVM at the 0.005 level.

In the case where multiple algorithms are to be compared on multiple domains, Friedman's test is a simple and good alternative. It is conducted as follows:

- All the classifiers are ranked on each domain separately. Ties are broken by adding the ranks of the tied algorithms and dividing them by the number of algorithms involved in the tie. The result is assigned to each of the algorithms involved in the tie.
- For each classifier, the sum of ranks obtained on all domains is calculated and labeled $R_{.j}^2$ where j symbolizes the classifier.
- Friedman's statistics is then calculated as follows:

$$\chi_F^2 = \left[12 \Big/ n.k.(k+1). \sum_{j=1}^{k} R_{.j}^2 \right] - 3.n.(k+1)$$

where n represents the number of domains and k the number of classifiers

Table 4.6 illustrates Friedman's test on a synthetic example. The table on the left lists the accuracies obtained by classifiers fA, fB, and fC on domains 1, 2, … 10. The table on the left calculates the rank of each classifier on each domain. These ranks in each column are then added yielding the $R_{.j}$'s. Applying the formula, we find that $\chi_F^2=15.05$. From Table 7 in Appendix A of [1], we find that for a 2-tailed test at the 0.05 level of significance, the critical value is 7.8. Since $\chi_F^2>7.8$, we can reject the null hypothesis that all three algorithms perform equally well.

Note that while Friedman's test shows that there is a significant difference among the algorithms being tested, it does not say where that difference is. In such cases,

4 Performance Evaluation in Machine Learning

Table 4.6 Friedman's test applied to three classifiers fA, fB, and fC on ten different domains

Domain	fA	fB	fC
1	85.83	75.86	84.19
2	85.91	73.18	85.90
3	86.12	69.08	83.83
4	85.82	74.05	85.11
5	86.28	74.71	86.38
6	86.42	65.90	81.20
7	85.91	76.25	86.38
8	86.10	75.10	86.75
9	85.95	70.50	88.03
10	86.12	73.95	87.18

Domain	fA	fB	fC
1	1	3	2
2	1.5	3	1.5
3	1	3	2
4	1	3	2
5	2	3	1
6	1	3	2
7	2	3	1
8	2	3	1
9	2	3	1
10	2	3	1
R.j	15.5	30	14.5

Nemenyi's test (or other post hoc tests) can be used to pinpoint where that difference lies. Here is how Nemenyi's test works.

- Let R_{ij} be the rank of classifier f_j on data set S_i; we compute the mean rank of classifier f_j on all data sets as

$$\overline{R._j} = \frac{1}{n}\sum_{i=1}^{n} R_{ij}$$

- Let q_{yz} be the statistic between classifier f_y and f_z. The formula is

$$q_{yz} = \frac{\overline{R._y} - \overline{R._z}}{\sqrt{\frac{k(k+1)}{6n}}}$$

(n is the number of domains and k the number of classifiers).
- Nemenyi's test proceeds by calculating all the q_{yz} statistics. Then, those that exceed a critical value q_α are said to indicate a significant difference between classifiers f_y and f_z at the α significance level.

To illustrate Nemenyi's test, we calculate the following values from Friedman's test we just ran[1]:

$$\overline{R._A} = 1.55, \overline{R._B} = 3, \text{ and } \overline{R._C} = 1.45$$

---

[1] Please note that there is an error in the textbook. We present, herein, the corrected solution.

- Replacing $R_{.y}$ and $R_{.z}$ by the above values in

$$q_{yz} = \frac{\overline{R.y} - \overline{R.z}}{\sqrt{\dfrac{k(k+1)}{6n}}}$$

we obtain $q_{AB} = -3.22$, $q_{AC} = .222$, and $q_{BC} = 3.44$.

- $q_\alpha = 2.55$ for $\alpha = 0.05$ [see [1]] (q_α must be larger than q_{yz} for the hypothesis that y and z perform equally to be rejected).
- Therefore, we reject the null hypothesis in the case of classifiers A and B and B and C (please note that we consider the absolute value of the q_{xy} quantity), but not in the case of A and C.

Conclusion

This chapter presented the most common methods of evaluating the performance of classifiers on applied domains. Unfortunately, there is no preexisting recipe that satisfies every situation. In most cases, the user must reflect about what he or she is trying to verify, understand the restrictions of the experimental setting (e.g., too little data, data skews (or imbalances), and so on), and apply the best combination of evaluation methods that is available in these conditions. It is important to note that due to the fact that there is a lot of unknown in the data, certain assumptions about the data may end up being violated. It remains unknown to what extent this will invalidate the results. Last but not least, it is important to understand how to interpret the results one observes. These results should be thought of as support for a hypothesis or evidence about certain effects. They do not prove that a hypothesis is correct. Classifier evaluation thus remains an art rather than a perfect science.

Bibliography

1. Japkowicz N, Shah M. Evaluating learning algorithms: a classification perspective. Cambridge/New York: Cambridge University Press; 2011.
2. Lichman M. UCI machine learning repository [http://archive.ics.uci.edu/ml]. Irvine: University of California, School of Information and Computer Science; 2013.
3. Japkowicz N. Assessment metrics for imbalanced learning. In: Haibo He, Yunqian Ma, editors. Imbalanced learning: foundations, algorithms, and applications. 1st ed. Hoboken: Wiley; 2013.
4. Bouckaert R. Choosing between two learning algorithms based on calibrated tests. In: Proceedings of the 20th international conference on machine learning (ICML-03). Washington, DC; 2003. p. 51–58.
5. Thomas D. Approximate statistical tests for comparing supervised classification learning algorithms. Neural Comput. 1998;10(7):1895–923.
6. Drummond C. Machine learning as an experimental science (revisited). In: Proceedings of the twenty-first national conference on artificial intelligence: workshop on evaluation methods for machine learning. AAAI Press technical report WS-06-06. 2006. p. 1–5.
7. Demšar J. On the appropriateness of statistical tests in machine learning. In: Proceedings of the 25th international conference on machine learning: workshop on evaluation methods for machine learning. Helsinki, Finland; 2008.

Informatics in Radiation Oncology

5

Paul Martin Putora, Samuel Peters, and Marc Bovet

Abstract

Radiation oncology informatics includes informatics from the perspectives of every discipline involved in radiation oncology. As there are many open questions and an abundance of data, machine learning technologies can be valuable. Available data includes handwritten notes on paper, imaging data available in digital formats, radiation treatment plan details, financial data, and multilevel multicenter databases, to name a few. Tools of various complexity for various goals are available. The following chapter aims to portray this domain and present a selection of available tools.

5.1 Introduction

Radiation oncology (RO) is the discipline dealing with the treatment of cancer with ionizing radiation (radiation therapy) for cure or palliation. Radiation therapy is often applied in combination with other treatment modalities (chemotherapy, surgery, hormonal therapy, etc.). Radiation therapy is typically delivered with linear accelerators (linacs); although other modalities and devices exist, linac-based treatments represent the vast majority of radiotherapy treatments in modern radiation oncology units.

P.M. Putora, MD, PhD, MA (✉) • S. Peters
Department of Radiation Oncology, Kantonsspital St. Gallen,
St. Gallen, Switzerland
e-mail: paul.putora@kssg.ch

M. Bovet, PhD
Direktion ICT, Radiotherapy Applications, Zürich University Hospital, Zürich University,
Zürich, Switzerland
e-mail: marc.bovet@usz.ch

Radiation oncology informatics includes informatics from the perspectives of every discipline involved in radiation oncology. This may range from data administration to radiobiology, from clinical to dosimetry issues.

5.2 Radiation Oncology Process: How Data Surrounds the Patient

Before treatment can be delivered, several steps take place in preparation. Initially, diagnosis and staging of the cancer take place. This includes the definition of the tumor type and extent of disease (tumor staging) [1]. Multidisciplinary cancer conferences, often including radiation oncologists, medical oncologists, surgeons, and other disciplines, lead to a treatment recommendation [2]. When radiation therapy is indicated and the patient has given his/her consent, the patient will enter into the process of radiotherapy treatment preparation.

The patient is accompanied by information regarding his/her disease, history, and general information (age, sex, address) when entering a radiation oncology department. Often, standardized questionnaires may be used to quantify symptoms, side effects [3], or general well-being [4]. Any healthcare process can produce vast quantities of data [5]; this is even more so in radiation oncology with large amounts of imaging and radiotherapy planning data being generated. Somewhat like a snowball rolling down a hill, more and more information is associated with the patient as he proceeds along this process (see Fig. 5.1). Planning computed tomography (CT) will be performed to provide the basis for further planning. Depending on the proposed treatment, the patient will be required to be in a specific position (e.g., arms over head, supine, prone). This imaging modality produces images which provide the basis for dose calculations in the correct treatment position.

The imaging data, accompanied by identifiers, is then transferred to a treatment planning system (TPS) in which the treatment target and normal organs – organs at risk (OARs) – are defined. Based on the prescribed dose and fractionation (e.g. 50 Gy (Gray) in 2 Gy fractions five times per week for 5 weeks), a treatment plan is generated. After verification of the dose, which is a double check of the performed calculations, the treatment plan is transferred to the linac where the patient is treated, typically by delivering daily doses (fractionated treatment) over several days to weeks. During treatment, to verify correct positioning, it is common practice to obtain 2D, 3D, or even 4D (e.g., breathing-dependent 3D) imaging, adding to the already significant amount of data associated with the patient. Additional information on side effects, symptoms, and further imaging is collected during follow-up.

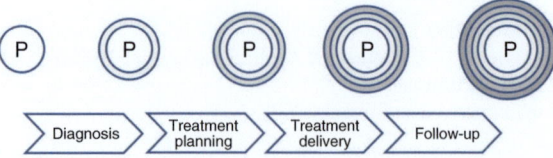

Fig. 5.1 With each step along the process, more information is created and collected which is associated with the patient

Due to the typical processes involved, a patient undergoing radiotherapy will have collected a significant amount of information associated with him. This information is present in multiple formats from handwritten notes on paper and filled-out questionnaires to three-dimensional imaging data and complex treatment plan information.

5.3   Where Is This Data?

In radiation oncology, probably more than in other medical specialties, the design and implementation of electronic information systems occur due to the need to reduce and eliminate human error in treatment – such as during the transfer of information from treatment planning to treatment delivery [6]. Although there is certainly a trend toward paperless or paper-light departmental organization, this is still work in progress in most units: digitalization touches several issues from legal, practical to know-how. While it has been demonstrated that a purely digital workflow is possible [7], this is not the case in many centers. The level of digital information available varies, heavily dependent on department-specific configurations. Costs and logistical and technical issues have been perceived as barriers in the implementation of a purely digital workflow [8]. Simply reproducing previous paper-based workflows can lead to unsafe processes and inefficient information flow [9]. On the other hand, key data in radiation oncology (e.g., treatment plans, 3D imaging) is not replaceable with paper so a transfer from a hybrid to a paperless system seems intuitive.

Similar to the level of digitalization, integration can be implemented to various extents. The purpose has been clear for decades now, to capture and store data consistently in order to rearrange and display the data as needed [10]. From a machine learning point of view, a strong fragmentation with redundancies may be present within an RO department or hospital. Heterogeneity of computer systems is ubiquitous and has multiple healthcare domain-specific causes [11]: departmental organization within a hospital, medical devices with inflexible built-in software, either general or niche vendors, selecting the "best" software for each specific purpose ignoring the big picture, as well as legacy applications [12, 13]. Often, third-party providers supply picture archiving and communication systems (PACS) or systems specific for laboratories or, for example, pathology reports.

These issues have been recognized a long time ago and have not been resolved to date. Radiation oncology units are typically embedded in, or at least associated with a broader healthcare system. Often, some information is associated and stored in a hospital information system (HIS) before the patient reaches a radiation oncology unit. This would include general demographics and patient identifiers (date of birth, gender, etc.), whose uniqueness is decisive for assuring the consistency between databases.

Integration is not only an issue between a radiation oncology unit and the hospital. It is nontrivial among different suppliers within a radiation oncology unit. IHE-RO stands for *I*ntegrating the *H*ealthcare *E*nterprise-*R*adiation *O*ncology

(IHE-RO) and is an ASTRO-sponsored initiative for improving the functionality of the radiation oncology clinic. The IHE-RO task force develops IHE integration profiles, which specify how industry standards are to be used to address specific clinical problems and ambiguities [14]. This process has succeeded in defining standards that have helped advance radiation oncology-specific integration [15].

Problems with integration are complicated by patient data confidentiality and security [16, 17]. A universal database where all relevant information is well structured and available is hardly to be expected. There are however attempts to centralize and integrate most relevant information by several vendors. Patient management systems, designed around the requirements of a radiation oncology unit, are available such as ARIA [18] by Varian and MOSAIQ [19] by Elekta. In theory, integration with third-party applications is not a problem; in practice, we see this is not always the case. Unfortunately, a vendor-independent radiation oncology-specific data model, i.e., xml standard or similar, is still not established.

Some departments are involved in developing their own models and systems, these with various capabilities, ranging from patient management to radiation delivery.

5.4 Accessing and Analyzing

Innumerable levels and forms of information are present in radiation oncology and may be approachable with data mining, operations research, and machine learning [20]. These range from nonstructured free-text in writing (physician notes) to highly structured digital information (e.g., imaging and radiation dose distribution) (Fig. 5.2) [15].

A modern radiation oncology unit has been described to have three distinct computer systems: clinical medical electronic record, a computerized treatment planning system, and a record and verify system (R&V) [21]. Possibly to a lesser extent, but one should mention the quality assurance applications too. The electronic

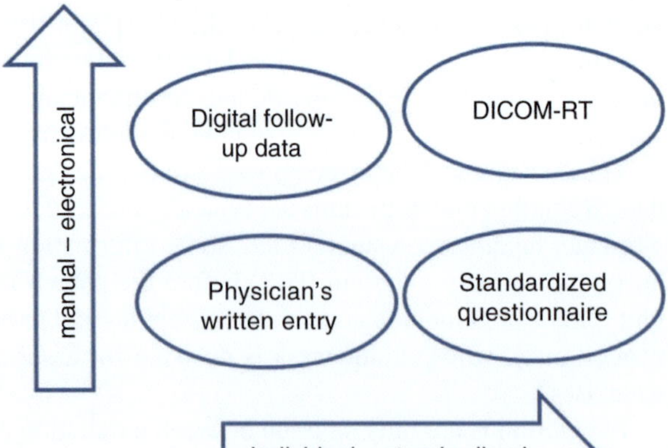

Fig. 5.2 Data and information are available in various forms in radiation oncology ranging from unformatted free-text manual entries (individual, manual) to highly standardized electronically available data sets such as a DICOM-RT treatment plan (standardized, electronic)

medical record may be integrated into a broader patient management system (PMS). Most of the time, the clinical medical records are distributed in the hospital PMS and the RO expert system, i.e., the R&V system. The information structure may vary from one system to another one. For example, an R&V system can have for historical reasons two SQL [22] database parts, at which the imaging part is still organized in a folder tree structure.

5.5 Treatment Planning System/DICOM-RT

At this stage, the process of diagnosis and staging as well as the treatment prescription should be concluded and defined. The treatment prescription includes the description of the target volume as well as dose, including fractionation. Additionally, criteria for organs at risk are defined to enable the treatment planning process to commence. Images from the CT scanner, together with defined targets and organs at risk, provide the basis for treatment planning.

5.5.1 CT Scanner

A CT scanner is a device that uses the X-ray computed tomography (X-ray CT) technology to produce tomographic images (virtual "slices") of a patient by computer-processed X-rays. Generally, almost all patients being treated in a radio-oncology department receive a CT scan before treatment. These images are used to exactly localize the tumor region and the surrounding healthy organs (organs at risk). Planning CTs are then used in the treatment planning system (TPS) to calculate the treatment dose of radiation (Fig. 5.3).

Usually, CT scanners use a DICOM modality worklist (MWL). A DICOM modality worklist can be considered as a task manager. This enables the CT scanner

Fig. 5.3 Planning CT with elements for patient positioning on the CT couch

to obtain details of patients (name, date of birth, etc.) and scheduled examinations electronically, avoiding the need to reenter such information and possibly causing mistakes. The DICOM images that the CT scanner creates will use the attributes received from the MWL. These images will be sent via a DICOM node to a PACS, an RO archive, a TPS, or a contouring workstation. This can be done automatically or manually, depending to the workflow procedure used in the department.

5.5.2 Treatment Planning System

The treatment planning system (TPS) has become a key element in the radiotherapy process. Regarding patient safety and success of therapy, its accurate and stable functioning is an issue of highest importance. These systems provide the process in which radiation oncologists, radiation therapist, medical physicists, and medical dosimetrists create radiotherapy treatment plans.

Today, treatment planning is almost entirely computer based using patient-computed tomography (CT) data sets (possibly in combination with magnetic resonance imaging and positron emission tomography). Tools providing multimodality image matching (co-registration or image fusion) are part of modern TPSs. Based on images, a virtual patient is generated to create a simulation of the treatment plan using the anatomical, geometrical, radiological, and dosimetric aspects of therapy. Evaluation of the treatment plan is often done by analyzing dose distribution overlaid on the patients' data set. Dose-volume histograms (DVH) provide clinicians with information of the uniformity of the dose in the target volume as well as distribution of dose in organs at risk.

5.5.3 DICOM-RT

The Digital Imaging and Communications in Medicine (DICOM) standard is now widely implemented in radiology as the standard for diagnostic imaging. It has also been extended for use in various subspecialties. One of the first extensions was implemented in radiation therapy and is known as DICOM-RT. In addition to the protocol used in the DICOM standard, seven DICOM-RT objects, namely – RT Image, RT Structure Set, RT Plan, RT Dose, RT Beams Treatment Record, RT Brachy Treatment Record, and RT Treatment Summary Record – have been created, each with a data model. The data models set the standard for integration of radiation therapy information for an electronic patient record and allow for an exchange between different systems [15]. The radiotherapy objects supplement to the DICOM standard can be downloaded from the NEMA.org website [23]. When compared to DICOM tools, the selection of tools compatible with DICOM-RT data is narrower, yet multiple solutions exist, many of which are open source [24].

For analysis of DICOM-RT data plugins/expansions for widely used statistics, suites such as MATLAB [15] or R [25] are readily available. Dicompyler is an extensible radiation therapy research platform and viewer for DICOM and

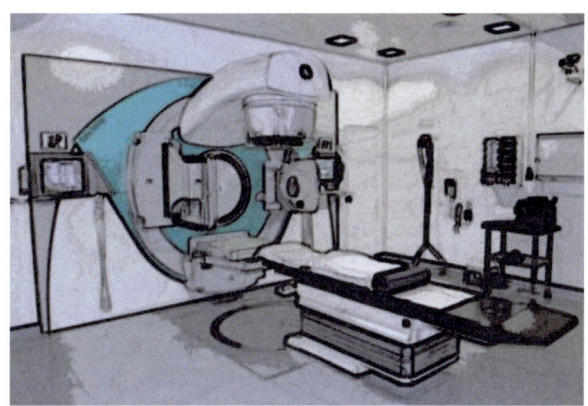

Fig. 5.4 Medical linear accelerator

DICOM-RT [26]. A software toolkit that is available for 3D Slicer [27] enables the import of treatment plans from various sources for visualization, analysis, comparison, and processing [28]. Another problematic issue to be aware of when accessing data on structure sets is that standardized nomenclature is not always used [29, 30].

The Computational Environment for Radiotherapy Research (CERR) [31] is written in MATLAB [32] language. CERR can be used to import and review [33] treatment plans and is compatible with multiple treatment planning systems. CERR is available online [34]. Another tool, RT_Image, which is available online [35], was initially developed for target volume generation based on PET data. In the meantime [15], it has developed making other DICOM-RT structure manipulations possible. DICOMan is a software system that handles DICOM-RT and includes an editor, retriever, and format convertor, among others [36, 37].

5.5.4 Record and Verify System (R&V System), Linear Accelerators

Medical linear accelerators (linacs) generate X-rays and high-energy electrons to treat cancer. Tumor inside the body of the patients. They are mounted on a gantry which allows rotation around the patient and are equipped with a multileaf collimator (MLC) to allow a conformal dose application to the tumor. Additionally, modern digitalized linacs may be equipped with a built-in CT scanner, imaging may be used before or during the irradiation of the patient to ensure the patient is in the correct position. Based on quality assurance requirements, often required by law, the R&V system has the role of guaranteeing the correct transfer of all geometrical und radiological data from the TPS to the linac (Fig. 5.4).

With advancing technology and complex treatment delivery, the need arose for more accurate monitoring and recording of daily treatment delivery [38]. Computerized systems for this purpose have first been described in the late 1970s [39]. As these systems were developed around a specific technical task in radiation oncology, they typically contained only limited patient data, such as basic demographics and scheduling [38]. The formerly limited scope was significantly improved, among others with automated charge billing.

The R&V systems are the link between the TPS and the linac: all parameters such as number of treatment fields, gantry and collimator angle, MLC positions, table position, dose and dose rate, beam quality, and number of treatment sessions are transferred from the TPS to the R&V system. This is done using DICOM-RT (mainly DICOM-RT Plan). R&V systems have a scheduler or a worklist on a daily basis to allow only treatments for a patient that is meant to be irradiated on a certain day. Once a treatment plan has been selected, all details of the treatment field and session will be transferred to the linac. Once irradiated, the completed irradiation is reported back to the R&V system including precise machine parameters (which can be different from the ones calculated in the TPS within a strictly defined tolerance). This is very crucial. If the irradiation is not stored correctly, a certain field or the whole session could be irradiated a second time with potential harm to the patient. This data stored in the database of the R&V system is accessible to analyses (patient, machine, and procedure related).

Reporting tools have traditionally been provided by vendors to enable users to create queries within a simplified layout. Older versions of MOSAIQ (termed Multi-Access) were delivered with Crystal Reports [40]; InfoMaker [41, 42] is still the tool which comes with Aria ver. 11. The migration of both Multi-Access (Pervasive SQL) and Aria (Sybase SQL) to Microsoft SQL (Server 2008 [43]) standard has provided the vendor and the client new possibilities: one can now think about using established functions of business intelligence technologies.

Elekta offers ANALYTIQ and Varian brings InSightive™ Analytics with Aria ver. 13. The underlying computer-based technique of creating multidimensional data cubes can also be implemented by the user within the services of Microsoft SQL Server (Analysis, Reporting) – by using a mirrored nonproductive database. From a technical point of view, it is also reasonable to think about the connection with Microsoft Amalga Unified Intelligence System [44], a unified health platform based on SQL Server 2008. The next generation is now represented by the Caradigm Intelligence Platform [45], based on SQL Server 2012.

Billing systems delivered with the R&V systems may have been developed, e.g., for Aria and the US market; regional adaptation may be required. Depending on the question, these represent databases worth exploring.

5.6 Radiation Oncology Patient Management System (PMS)

Practically, all hospitals have some form of centralized patient management system, the extent of which may vary significantly among institutions. Several vendors provide systems; these include solutions by Siemens, SAP, and many others.

A few vendors have specialized in integrating electronic medical records into a patient management system within a radiation oncology setting. The two most prominent providers are Varian with their product ARIA [18] and Elekta with their product MOSAIQ [19].

5.7 Applications for Quality Assurance (QA)

Besides the three main components of an RO computer system, applications and tools for quality assurance are an important part of a radiation oncology department. QA programs are often regulated by law or specific professional associations and are performed to assure the correctness of the calculation of the radiological patient dose and the mechanical and radiological reliability of the linac and CT scanner.

Concerning the patient dose, two different kinds of applications exist; both retrieve the corresponding data directly from the database of the TPS via DICOM-RT. Some recalculate the patient dose using an alternative dose algorithm (e.g., RadCalc [46], DIAMOND [47], Mobius3D [48]); others use a dose measurement on the linac and compare it to the original dose distribution from the TPS (e.g., Delta4 [49], ArcCHECK [50], OCTAVIUS [51]).

Linac QA is very comprehensive and produces large amounts of data. To perform all required QA tests, many different tools and applications exist which will not be listed and explained here. In order to report at any time the state of the linac QA management systems (e.g., QUALimagiQ [52], AQUILAB [53], PIPSpro [54]) exist, which collect all available QA data in a single database (entered manually or automatically using DICOM or DICOM-RT data).

5.8 Nonspecific Elements

5.8.1 Database/SQL

Not specific to radiation oncology, or even to healthcare, databases typically use standard interfaces. Structured Query Language (SQL) [22] is the most commonly used database language. Depending on the software provider, harvesting information stored in databases may be occluded by limited access granted by the software providers (especially for built-in systems). Furthermore, databases may be accessible, but their table structure is not transparent or evident making reliable analysis difficult.

5.8.2 Hospital Information System (HIS)/HL7

Many radiation oncology departments are integrated within a hospital or other form of collaboration. The minimum information transferred are patient identifiers. The typical data exchange format for such communication is the standard HL7 [55]. It was developed by Health Level Seven, which is a nonprofit organization. It provides a framework for the exchange, integration, sharing, and retrieval of electronic health information. Although versions 3.x exist, the 2.x versions of the standard, which support clinical practice and the management, delivery, and evaluation of health services, are currently the most commonly used [56].

5.8.3 Picture Archiving and Communication System (PACS)/DICOM

PACSs are used to store and recall images; in the typical setting, the main PACS in a hospital would be associated with diagnostic images from the radiology department. Imaging that is relevant to radiation oncology such as magnetic resonance imaging (MRI) or positron emission tomography (PET) would also be stored there. Of note, many PACS systems implemented in hospitals do not support DICOM-RT objects. The standardized format in which this takes place is the Digital Imaging and Communications in Medicine (DICOM) format, which is a standard for handling, storing, printing, and transmitting information in medical imaging. It includes a file format definition and a network communications protocol. The DICOM standard was developed by the members of the National Electrical Manufacturers Association (NEMA) [57].

An introduction and a list of multiple DICOM tools can be found at the website of the Center for Advanced Brain Imaging [58]. Some freeware/open source popular tools available for viewing and manipulating DICOM images include OsiriX [15] and 3D Slicer [27].

5.9 Data Not Specific to Radiation Oncology

5.9.1 Peripheral Sources/Billing Data

Reimbursement for radiation oncology services is based on patient parameters and activities performed. Often, for billing purposes, this information is automatically collected, coded, and transferred to the responsible department. Some departments may have limited interest in collecting data for scientific evaluations, but all departments are interested in getting reimbursed for their work. Billing data may provide information on the number of patients per diagnosis, the type of treatment used, as well as information on the frequency of imaging. Although billing data is derived from regular, core activities of a radiation oncology department, it may be more complete and better structured than other sources. Depending on the organization of a department, information on prescribed medication, data from a radio-oncology ward, and similar sources may also provide valuable data.

5.9.2 High Level/External Sources

Subsets of data produced in a radiation oncology department are often pooled with other sources and collected in national cancer registries [59], such as the Surveillance, Epidemiology and End Results (SEER) program of the National Cancer Institute (NCI) in the United States. Data collected includes information on over 25 cancer entities as well as patient characteristics such as age, sex or ethnicity. SEER data has been used in numerous analysis, although retrospective, for several investigations, this has proven to be very valuable [15]. Virnig et al. used SEER data to analyze radiotherapy use and concluded that SEER data should be combined with Medicare data to obtain the most complete data [60]. The NCI provides several analytical tools on its website, such as the Cansurv software [61].

Vendors are also starting to incorporate crowd wisdom into their products. Based on MOSAIQ (Oncology Electronic Medical Record) and METRIQ (Integrated Oncology Data Management System), Elekta has launched what it calls Data Alliances. These include the NODA (National Oncology Data Alliance), ODA (Oncology Data Alliance), and RODA (Radiation Oncology Data Alliance). They represent "data aggregation and analysis programs." These programs can provide benchmarking reports, based on information from multiple users.

Several centers have taken up the legal, ethical, and administrative challenges [62] of sharing data and initiated projects aimed at making clinical data machine readable and exchangeable. An example of such an initiative is the EUROCAT project including radiation oncology sites in the Netherlands, Belgium, and Germany [63, 64]. By implementing distributed learning models, some problems of data sharing could be reduced while allowing multilevel data exchange across centers.

At John Hopkins University, Oncospace [65] has developed a radiation oncology database, to enable analysis of patient data and exchange. The system was designed to provide data exchange as well as decision support and analysis for multiple issues.

A tool has been developed to allow collaboration in radiotherapy based on the DICOM format [66].

Histogram Analysis in Radiation Therapy (HART) is a MATLAB [32]-based program designed to analyze large quantities of radiotherapy data [67].

5.9.3 Guidelines/Recommendations

On a more abstract level, information on standard operating procedures or departmental guidelines may be available. Although this information is often not complete nor readily machine readable and only implicitly available, with appropriate representation [68], this information may be transformed into formats that allow for its analysis and comparison with other guidelines/treatment algorithms. When high-level evidence from trials is missing and experience does not suffice to answer clinical questions, crowd wisdom (swarm-based medicine) may fill these gaps [69]. Patterns of care investigations and analyses of decision trees may provide valuable information [70].

5.10 Conclusion and Outlook

The most commonly used sources for machine learning in radiation oncology are probably data available in DICOM-RT format and the database of the PMS in use. Mostly, this data is of well-described structure and available digitally. There is room for improvement; standardization in nomenclature within the DICOM-RT format and an improved integration of the radiation oncology PMS with other sources will enrich our analytical capabilities. Besides these typical sources information from dosimetry, billing and cancer registries provide essential information that can enhance core databases.

References

1. Edge SB, Compton CC. The American Joint Committee on Cancer: the 7th edition of the AJCC cancer staging manual and the future of TNM. Ann Surg Oncol. 2010;17(6):1471–4.
2. Wright F, De Vito C, Langer B, Hunter A. Multidisciplinary cancer conferences: a systematic review and development of practice standards. Eur J Cancer. 2007;43(6):1002–10.
3. Cox JD, Stetz J, Pajak TF. Toxicity criteria of the radiation therapy oncology group (RTOG) and the European organization for research and treatment of cancer (EORTC). Int J Radiat Oncol Biol Phys. 1995;31(5):1341–6.
4. Aaronson NK, Ahmedzai S, Bergman B, et al. The European Organization for Research and Treatment of Cancer QLQ-C30: a quality-of-life instrument for use in international clinical trials in oncology. J Natl Cancer Inst. 1993;85(5):365–76.
5. Prather JC, Lobach DF, Goodwin LK, Hales JW, Hage ML, Hammond WE. Medical data mining: knowledge discovery in a clinical data warehouse. In: Proceedings of the AMIA annual fall symposium. American Medical Informatics Association. Bethesda, Maryland, USA. 1997. p. 101.
6. Miller AA. Clinical information systems in oncology – making a difference to patient outcomes. Health Care Inf Rev Online. 2003
7. Röhner F, Schmucker M, Henne K, et al. Integration of the radiotherapy irradiation planning in the digital workflow. Strahlenther Onkol. 2013;189(2):111–6.
8. Jha AK, DesRoches CM, Campbell EG, et al. Use of electronic health records in US hospitals. N Engl J Med. 2009;360(16):1628–38.
9. Fong de los Santos L, Herman MG. Information flow through the radiation oncology process. In: Starkschall G, Siochi RAC, editors. Informatics in radiation oncology. Boca Raton: Taylor & Francis; 2013. p. 63–75.
10. Bleich HL, Slack WV. Designing a hospital information system: a comparison of interfaced and integrated systems. MD Comput. 1991;9(5):293–6.
11. Vorwerk H, Zink K, Wagner DM, Engenhart-Cabillic R. Making the right software choice for clinically used equipment in radiation oncology. Radiat Oncol. 2014;9(1):145.
12. Lenz R, Blaser R, Kuhn KA. Hospital information systems: chances and obstacles on the way to integration. Stud Health Technol Inform. 1999;68:25–30.
13. Brooks KW, Fox TH, Davis LW. A critical look at currently available radiation oncology information management systems. Semin Radiat Oncol. 1997;7(1):49–57. Elsevier.
14. ASTRO. IHE-RO. 2014. https://www.astro.org/Practice-Management/IHE-RO/Index.aspx.
15. Abdel-Wahab M, Rengan R, Curran B, Swerdloff S, Miettinen M, Field C, Ranjitkar S, Palta J, Tripuraneni P. Integrating the healthcare enterprise in radiation oncology plug and play–the future of radiation oncology? Int J Radiat Oncol Biol Phys. 2010;76(2):333–6.
16. Clifton C, Kantarcio M, Doan A, et al. Privacy-preserving data integration and sharing. In: Proceedings of the 9th ACM SIGMOD workshop on research issues in data mining and knowledge discovery. Paris: ACM; 2004. p. 19–26.
17. Ratib O, Swiernik M, McCoy JM. From PACS to integrated EMR. Comput Med Imaging Graph. 2003;27(2):207–15.
18. Varian. ARIA – comprehensive oncology care. http://www.varian.com/euen/oncology/radiation_oncology/aria/.
19. Elekta. Radiation oncology software – MOSAIQ® radiation oncology information system. http://www.elekta.com/healthcare-professionals/products/elekta-software/radiation-oncology.html.

20. Ehrgott M, Holder A. Operations research methods for optimization in radiation oncology. J Radiat Oncol Inf. 2014;6(1):1–41.
21. Han Y, Huh SJ, Ju SG, et al. Impact of an electronic chart on the staff workload in a radiation oncology department. Jpn J Clin Oncol. 2005;35(8):470–4.
22. Bowman JS, Emerson SL, Darnovsky M. The practical SQL handbook: using structured query language. Reading: Addison-Wesley Longman Publishing Co., Inc.; 1996.
23. NEMA.org. Digital Imaging and Communications in Medicine (DICOM) – Supplement 11 – radiotherapy objects. 1997. ftp://medical.nema.org/medical/dicom/final/sup11_ft.pdf.
24. Deasy JO, Apte AP. Open-source informatics tools for radiotherapy research. In: Starkschall G, Siochi RAC, editors. Informatics in radiation oncology. Boca Raton: Taylor & Francis; 2013. p. 147–60.
25. Thompson RF. RadOnc: an R package for analysis of dose-volume histogram and three-dimensional structural data. J Radiat Oncol Inf. 2014;6(1):98–110.
26. Panchal A, Keyes R. SU-GG-T-260: dicompyler: an open source radiation therapy research platform with a plugin architecture. Med Phys. 2010;37(6):3245.
27. Pieper S, Halle M, Kikinis R. 3D slicer. In: IEEE international symposium on Biomedical imaging: nano to macro 2004. IEEE, 2004. p. 632–5.
28. Pinter C, Lasso A, Wang A, Jaffray D, Fichtinger G. SlicerRT: radiation therapy research toolkit for 3D Slicer. Med Phys. 2012;39(10):6332–8.
29. Miller AA. A rational informatics-enabled approach to standardised nomenclature of contours and volumes in radiation oncology planning. J Radiat Oncol Inf. 2014;6(1):53–97.
30. Santanam L, Hurkmans C, Mutic S, et al. Standardizing naming conventions in radiation oncology. Int J f Radiat Oncol Biol Phys. 2012;83(4):1344–9.
31. Deasy JO, Blanco AI, Clark VH. CERR: a computational environment for radiotherapy research. Med Phys. 2003;30(5):979–85.
32. Mathworks. Matlab documentation. http://es.mathworks.com/help/matlab/.
33. Siochi RA, Pennington EC, Waldron TJ, Bayouth JE. Radiation therapy plan checks in a paperless clinic. J Appl Clin Med Phys: Am Coll Med Phys. 2009;10(1):2905.
34. CERR. A computational environment for radiotherapy research. 2014. http://www.cerr.info/about.php.
35. Graves T. RT_Image. 2014. http://rtimage.sourceforge.net/index.html.
36. Yan Y, Dou Y, Weng X, Wallin A. SU-GG-T-256: an enhanced DICOM-RT viewer. Med Phys. 2010;37(6):3244.
37. Yan Y, Weng X, Penagaricano J, Ratanatharathorn V. A universal DICOM wizard to tackle incompatibility problems in the process of IMRT and IGRT. Int J Radiat Oncol Biol Phys. 2008;72(1):S657.
38. Colonias A, Parda DS, Karlovits SM, et al. A radiation oncology based electronic health record in an integrated radiation oncology network. J Radiat Oncol Inf. 2011;3(1):3–11.
39. Fredrickson DH, Karzmark C, Rust DC, Tuschman M. Experience with computer monitoring, verification and record keeping in radiotherapy procedures using a Clinac-4. Int J Radiat Oncol Biol Phys. 1979;5(3):415–8.
40. SAP. SAP Crystal solutions: essential BI for small business. 2014. http://www.sap.com/solution/sme/software/analytics/crystal-bi/index.html.
41. Sybase. Infomaker. 2014. http://infomaker.sharewarejunction.com/.
42. Chong S, Anderson N, Finlay J. SU-E-T-259: implementation of an automated workflow auditing and notification system for radiation oncology. Med Phys. 2011;38(6):3546.
43. MacLennan J, Tang Z, Crivat B. Data mining with Microsoft SQL server 2008. Indianapolis: Wiley; 2011.
44. Plaisant C, Lam S, Shneiderman B, et al. Searching electronic health records for temporal patterns in patient histories: a case study with Microsoft Amalga. In: AMIA annual symposium proceedings; 2008. American Medical Informatics Association; 2008. p. 601.
45. Caradigm. http://www.caradigm.com. Accessed 3.3.2015.
46. Morales J, Cho G. SU-FF-T-213: evaluation of RadCalc V5. 2 as an independent monitor unit checking program for dynamic IMRT plans. Med Phys. Melville, New York, USA. 2009;36(6):2569.

47. Foong P, Looe H, Poppe B. SU-E-T-544: commissioning and clinical evaluation of a secondary check software for 3D conformal and IMRT treatment plans. Med Phys. 2012;39(6):3830–1.
48. Majithia L, DiCostanzo D, Weldon M, Gupta N, Rong Y. SU-E-T-564: validation of photon dose calculation using Mobius3D system compared to AAA and Acuros XB systems. Med Phys. 2013;40(6):335.
49. Bedford JL, Lee YK, Wai P, South CP, Warrington AP. Evaluation of the Delta4 phantom for IMRT and VMAT verification. Phys Med Biol. 2009;54(9):N167.
50. Li G, Zhang Y, Jiang X, et al. Evaluation of the ArcCHECK QA system for IMRT and VMAT verification. Phys Med. 2013;29(3):295–303.
51. Van Esch A, Huyskens DP, Behrens CF, et al. Implementing RapidArc into clinical routine: a comprehensive program from machine QA to TPS validation and patient QA. Med Phys. 2011;38(9):5146–66.
52. Torfeh T, Beaumont S, Guédon J-P, Bonnet D, Denis E, David L. Numerical 3D models used for an evaluation of software tools dedicated to an automatic quality control of EPID images. 2008.
53. Marinello G. Quality assurance for image-guided radiotherapy. 2008. http://www.iaea.org/inis/collection/NCLCollectionStore/_Public/40/003/40003891.pdf.
54. Menon GV, Sloboda RS. Quality assurance measurements of a-Si EPID performance. Med Dosim. 2004;29(1):11–7.
55. Health Level Seven International. http://www.hl7.org/.
56. Shaver D. The HL7 evolution-comparing HL7 versions 2 and 3. Corepoint Health. http://www.corepointhealth.com/sites/default/files/whitepapers/hl7-v2-v3-evolution pdf. Retrieved 2012; p. 16.
57. NEMA.org. Members of the DICOM Standards Committee. http://medical.nema.org/members.pdf.
58. Center for Advanced Brain Imaging. The DICOM standard. http://www.cabiatl.com/mricro/dicom/index.html.
59. Piccirillo JF, Tierney RM, Costas I, Grove L, Spitznagel Jr EL. Prognostic importance of comorbidity in a hospital-based cancer registry. JAMA. 2004;291(20):2441–7.
60. Virnig BA, Warren JL, Cooper GS, Klabunde CN, Schussler N, Freeman J. Studying radiation therapy using SEER-Medicare-linked data. Med Care. 2002;40(8):IV-49–54.
61. Statistical Methodology and Applications Branch DMB, National Cancer Institute. Cansurv. 1.3 ed; 2014. http://surveillance.cancer.gov/cansurv/.
62. Sullivan R, Peppercorn J, Sikora K, et al. Delivering affordable cancer care in high-income countries. Lancet Oncol. 2011;12(10):933–80.
63. Lambin P, Roelofs E, Reymen B, et al. Rapid Learning health care in oncology' – an approach towards decision support systems enabling customised radiotherapy. Radiother Oncol. 2013;109(1):159–64.
64. EuroCat. about EuroCat. 2014. http://www.eurocat.info/information/about.html.
65. McNutt T, Wong J, Purdy J, Valicenti R, DeWeese T. OncoSpace: A new paradigm for clinical research and decision support in radiation oncology. In: 10th international conference on computers in radiotherapy, Amsterdam; 2010.
66. Westberg J, Krogh S, Brink C, Vogelius I. A DICOM based radiotherapy plan database for research collaboration and reporting. J Phys: Conf Ser. 2014;489:012100. IOP Publishing.
67. Pyakuryal A, Myint WK, Gopalakrishnan M, Jang S, Logemann JA, Mittal BB. A computational tool for the efficient analysis of dose-volume histograms for radiation therapy treatment plans. J Appl Clinical Med Phys/Am Coll Med Phys. 2010;11(1):3013.
68. Putora PM, Blattner M, Papachristofilou A, Mariotti F, Paoli B, Plasswilm L. Dodes (diagnostic nodes) for guideline manipulation. J Radiat Oncol Inform. 2010;2(1):1–8.
69. Putora PM, Oldenburg J. Swarm-based medicine. J Med Internet Res. 2013;15(9), e207.
70. Putora PM, Panje CM, Papachristofilou A, dal Pra A, Hundsberger T, Plasswilm L. Objective consensus from decision trees. Radiat Oncol. 2014;9(1):270.

Application of Machine Learning for Multicenter Learning

6

Johan P.A. van Soest, Andre L.A.J. Dekker, Erik Roelofs, and Georgi Nalbantov

Abstract

Advancements in radiation oncology are driving more specific, and thus improved, treatment opportunities. This creates challenges on the assessment of treatment options, as more information is needed to make an informed decision. One of the methods is to use machine-learning techniques to develop predictive models. Although prediction models, embedded in clinical decision support systems (CDSSs), are the foreseen solution, developing/training such prediction models requires large amounts of detailed patient information to reach decisive power. The amount of patients needed to train a reliable prediction model rapidly outgrows the numbers available in a single institution, hence the need for multicenter machinelearning. To be able to learn over multiple centers, several infrastructural prerequisites need to be addressed. First, data needs to be extracted from multiple source systems and represented using standardized terminologies, preferably including the semantics (the actual description) of the represented data. For research and model training purposes, this means that value representations (e.g. "m" or "f" indicating gender) need to be converted into standardized terms (the NCI Thesaurus codes C20197 or C16576, respectively), and that patient-identifiable information (e.g. name, institutional ID, address, etc.) needs to be removed or changed in a non-identifiable way. If datasets from different institutions use the same standardized terminology and data structure, data can be merged. Finally, after merging, prediction models can be learned on the complete dataset, in this chapter known as centralized learning.

J.P.A. van Soest • A.L.A.J. Dekker • E. Roelofs • G. Nalbantov (✉)
MAASTRO Clinic, Dr. Tanslaan 12 6229, PO Box 3035,
6202 ET Maastricht, The Netherlands
e-mail: johan.vansoest@maastro.nl; andre.dekker@maastro.nl;
erik.roelofs@maastro.nl; georgi.nalbantov@maastro.nl

6.1 Introduction

Technical advancements in the fields of physics, radiobiology, and engineering (and indirectly chemistry) are the main drivers for better, and thus more specific, treatment opportunities in radiation oncology. These advancements largely influence treatment methods, especially in regard to treatment planning (IGRT, IMRT, VMAT) and radiation techniques used.

In the current era of *evidence-based medicine* (EBM), all of these advancements need to be validated to be sure whether a specific treatment (plan) is better than the current standard (e.g., in regard to possible patient outcome). However, we also observe that new treatment options do not necessarily improve the outcome for an entire population but might only work for specific groups of patients. The standardized treatment (according to the current guidelines) might be too intense for specific groups of patients (resulting into higher toxicities and/or other radiation-induced complications) or could result in undertreatment of patients. At this point, it becomes interesting to apply machine learning to retrospectively identify prognostic factors (e.g., risk factors) and to develop predictive models to classify patients in distinct groups [17]. These groups can then be used to alter treatment options, e.g., to intensify or temper treatment.

The more subgroups we can identify, the better we can optimize treatment for individual patients, leading towards the next era, called *individualized medicine* (IM). This also imposes challenges on patient subgroup discovery and development of prognostic models as done for many years. Only several large institutions (in terms of patient turnover per year) can perform fine-grained subgroup analysis, as we need a fair number of patients with and without a specific outcome to test hypotheses regarding new treatment options for specific subgroups. Only with these large numbers of patients can we translate results of *individualized medicine* [1] into clinical practice by means of clinical decision support systems (CDSS) [16]. Therefore, we need to collaborate in radiation oncology research and share data to perform machine learning on larger, multicenter datasets.

In this chapter, we will explain the current possibilities of machine learning in a multicenter setting. We will start with the prerequisites and infrastructure fundamentally needed for multicenter machine learning (Sect. 6.2). Afterwards, we will describe the concept of centralized and distributed machine learning, including the benefits and challenges (Sect. 6.3). Finally we will describe several applications/initiatives related to multicenter machine learning (Sect. 6.4) and conclude with a summary of this chapter (Sect. 6.5).

6.2 Prerequisites

When performing multicenter machine learning, several prerequisites are needed to be addressed before actually starting the machine learning process. In this paragraph, we will describe the topics of data extraction (Sect. 6.2.1) and representation (Sect. 6.2.2), network infrastructures (Sect. 6.2.3), and privacy preservation (Sect. 6.2.4).

6.2.1 Data Extraction

Within radiation oncology, data extraction for machine learning is a labor-intensive task, as many data silos exist where data resides. In general, we need to connect to different data sources, extract data from these sources using local querying dialects, and afterwards store the extracted data in a central storage. These steps need to be performed for different information systems used in radiation oncology. We will describe the most common systems in this paragraph. First, we need to include the electronic medical record (EMR), where general patient characteristics are stored (e.g., age, gender, and diagnostic, geographical, and follow-up information such as complication and quality of life scores). Second, medical images (for diagnostic, treatment, and validation purposes) are stored in a picture archiving and communication system (PACS). Although images cannot be used directly in predictive model training, extracted information from these images can be used (Sect. 6.2.1.2). Third, treatment planning-related information (e.g., radiation plan information regarding beams and dose) needs to be incorporated, as the treatment planning system (TPS) stores information in its own database, as well as in the PACS. Fourth, the record and verify system (R&V) holds information regarding the planned treatment (e.g., dose, fractionation, beams) and the actual delivery. This information is also needed during machine learning, e.g. to determine structural differences in the planned and delivered treatment.

Other systems (e.g., sources containing biological data) may apply in specific or future settings; however, we've specified only the general sources of information used for machine learning. In regard to multicenter machine learning, this data extraction leads to the first challenge as different institutions have different systems (in terms of manufacturers and products) in place. All of these systems may store data differently, which requires a customized approach for data extraction for every participating institution.

6.2.1.1 ETL Tooling and Data Warehousing

To (continuously) extract data and store it in a central location, one could consider the use of extraction, transformation, and load (ETL) tooling. This tooling can extract data from different sources (different systems), reconcile data belonging to one patient (transformation), and store the data in a central database: the data warehouse (DWH). This could be useful for large-scale machine learning and research institutions with many smaller-sized trials. As shown by Roelofs et al. [27], implementing a data warehouse can significantly reduce the data collection time, in comparison to manual data extraction and collection. In regard to multicentered settings, this also reduces the number of systems/databases a user/researcher has to include in the data request/retrieval process, thus reducing the time to merge all different datasets. Furthermore, as data are extracted and inserted into the DWH, it should be known what the data represents. The ETL process should therefore be well documented regarding queries, transformations, and the meaning of the stored data in the DWH. In comparison to the DWH, directly querying the source system for research purposes has several disadvantages. These disadvantages are mainly on the topics of query and data validity and query load on production/source systems. When a DWH is in place, query validity should not be an issue (as the data is checked before being incorporated in the DWH). Furthermore, query load issues should be mitigated, as the DWH should run on a different database/server as the production/source systems, and therefore cannot affect clinical operations.

6.2.1.2 Image Biomarker Extraction

As stated in Sect. 6.2.1, the intrinsic information of images (not just the readily available metadata) needs to be extracted from the actual image slices. Extraction of image "features" is not a standard functionality of a PACS; however, features may sporadically be available as TPS systems may store additional information in the metadata of the DICOM images. If features are stored in the metadata, these values are needed to be validated, especially in a multicenter setting where different sites may use different TPS systems, which could implement different algorithms to calculate these features.

When there are no (or only a small number of) features already available, every site in the multicenter setting needs to implement a feature extraction pipeline which calculates variables based on the images available in the local PACS. As the local PACS stores CT and/or PET images, delineated contours (RTSTRUCT), planned (RTPLAN), and delivered (RTDOSE) dose information, the number of features to extract becomes larger. For example, we can extract information regarding the tumor volume, maximum diameter, specific points of the dose-volume histogram (DVH) for target volumes or organs at risk, tumor activity/metabolism, and differences between planned versus delivered dose. Furthermore, *radiomic* analysis on these images produces more than 200 features, based on more advanced image processing algorithms (by calculating intensity distribution metrics based on, e.g., Fourier transformations and wavelets) [15]. Several of these features are potential imaging biomarkers: features which have prognostic and predictive value in terms of to patient outcome or tumor response.

Preferably, this feature extraction pipeline should use common communication protocols, such as DICOM (to receive images) and SQL (to send extracted features to a local database). This increases the possibility to reuse this pipeline in all submitting centers and increases the homogeneity of applications and calculation algorithms used by different centers. Eventually, using equal feature extraction pipelines should result in easier comparison of features/variables between centers. Although we can generalize the applications and algorithms used, including scanning and reconstruction parameters, there is still a large variability at the input of this feature extraction pipeline: differences between delineations of different centers. As shown in literature, differences in delineations may occur between individuals, even within one center [18]. These differences in delineations could result in different outcomes after feature extraction. Especially when two different structures (e.g., rectum and bladder) are close to each other, for example, it might be possible that the delineating individual accidentally delineates the bladder wall as part of the rectum. This results in a higher SUV-mean/max and therefore could compromise the prognostic value of the extracted features.

Based on the examples of delineation differences and calculation applications/algorithms used, it is important to specify the provenance of a specific variable: how did we acquire/extract this information (and which algorithms did we use)? And what are the sources used to extract the information? We will elaborate on these questions in the next paragraph (Sect. 6.2.2).

6.2.2 Data Representation and Semantic Interoperability

To be able to exchange data between participating sites, all sites need to be *syntactically interoperable*. This means that they have to agree which (technical) protocol they use to transfer data; implying that data representation should be equal among participating sites.

Next to standardization of syntactical interoperability, *semantic interoperability* needs to be in place. We will use the definition of Valentini et al. [29] to describe semantic interoperability: "The ability of any communicating entity (not only computers) to share unambiguous meaning. For computers, this is the ability to exchange information and have that information properly interpreted by the receiving system in the same sense as intended by the transmitting system." In general, this means that the receiver cannot interpret information differently, as the sender uses unambiguous terms to describe that information. Therefore, we need to use terminological systems which are known by both sender and receiver. As defined by De Keijzer et al. [7], a terminological system can be a thesaurus, classification, vocabulary, nomenclature, or coding system. A terminological system may pertain to more than one of these systems. For example, ICD-10 [33] is a coding system and vocabulary (as the term is accompanied by a definition); another example is the National Cancer Institute's Thesaurus (NCIT) [28], which (in addition to a vocabulary) also contains a list of synonyms or other relationships. Finally, multiple terminological systems can be embedded in an ontology, where concepts from terminological systems are reused and relations of concepts in a specific domain are described. Furthermore, an ontology can be used as a consensus model to represent data within a specific domain (e.g., radiation oncology) between different participating sites [7].

6.2.2.1 Relational Databases and Ontologies

In regard to multicenter learning, we need to make sure every participating center uses the same database structure to be able to uniformly query (or federate) the data warehouse (DWH) database (Sect. 6.2.1.1). This database structure can be derived by creating a so-called entity-relationship (ER) model, based on the ontology; however, it needs to be adhered by all centers. An example to derive this ER model is the normalized universal approach described by Gali et al. [11]. Next to this database structure, it is important to use the same database system, as different database systems/vendors have different dialects. To mitigate differences in database systems/vendors, it is also possible to use automatic conversion libraries such as Hibernate (http://hibernate.org/), although these systems add another layer of complexity when performing queries and/or data federation.

When adhering to an ontology, values from local systems need to be replaced with standardized values from terminological systems as defined in the ontology. For example, the property biological sex containing the text "male" or "female" needs to be replaced by NCI Thesaurus code C20197 or C16576, respectively. Another participating center may use 0 and 1 or "m" and "f"; however, within the DWH database, all sites should use the NCI Thesaurus codes for semantic interoperability. This conversion of values is typically done in the *transform* step of the ETL process. Therefore, the ETL process needs to be tailored per participating center.

Although data representation is possible within relational databases, it is cumbersome to maintain in a multicenter machine learning setting. As new results give new insights into biological concepts and relationships, the need for extra variables is rapidly growing. Given this fact, it is inevitable that a multicenter network for machine learning will have substantial downtime. For example, when a new concept is added to the ontology, every participating center needs to update their ETL system and DWH database structure, to become up to date with the new ontology version. This may take some time, as administrators of the ETL and DWH system need to validate whether this change is valid, and does not compromise patient de-identification. If one of the queried columns is not available, the Relational Database Management System (RDBMS) will result an error rather than an empty result set. Therefore, it might be that the whole federation/distributed querying system may not work (if proper error handling is not in place). In this example, we used the addition of a column, a relatively easy task which occurs frequently. However, the more complex the changes in the ontology and database structure, the more time and effort it will take to get the network up and running again.

6.2.2.2 Semantic Web, RDF, and Linked Data

One of the solutions to cope with rapidly changing ontologies in a multicenter setting is to move from relational databases to Semantic Web technologies [2, 3]. In this paragraph, we will only discuss the *Resource Description Framework* (RDF), *linked data*, and the *SPARQL protocol and RDF query language* (SPARQL) as a subset of Semantic Web technologies.

Resource Description Framework

RDF is a standard, recommended by the World Wide Web Consortium (W3C) [6], and can be seen as a flexible alternative for the relational database. Where "traditional" relational databases store their data in a structure of tables and columns, the RDF specifies only one table with three columns named *subject*, *predicate*, and *object*. Each row in this single table repository is called a triple, as it only has three cells. Due to this basic difference in structure, the concept of data representation is also different. Because of this fixed table structure, the ontology becomes more important and serves as a data model consensus between centers.

As an example, we have an ontology describing patients and their first name, last name, biological sex, and age. Figure 6.1 shows the visual representation of this ontology. The RDF triples based on this ontology are represented in Table 6.1.

Unique Resource Identifiers and Linked Data

To assure semantic interoperability, we will use the concept of unique resource identifiers (URIs), which is incorporated in the RDF specification. The RDF

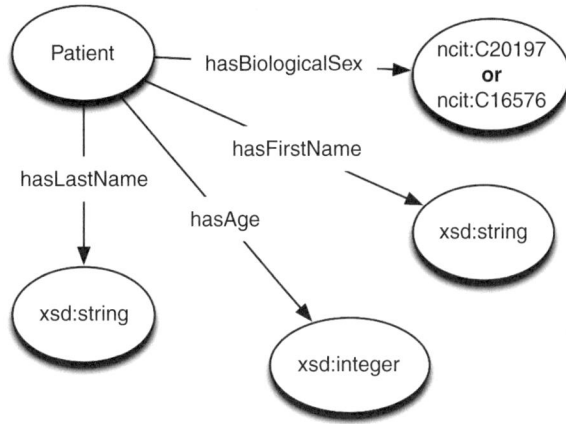

Fig. 6.1 Visual representation of the sample ontology

Table 6.1 RDF representation of a patient based on the ontology of Fig. 6.1

Subject	Predicate	Object
mySet:patient1001	rdf:type	ncit:C16960
mySet:patient1001	myOntology:hasFirstName	"John"^xsd:string
mySet:patient1001	myOntology:hasLastName	"Doe"^xsd:string
mySet:patient1001	myOntology:hasBiologicalSex	ncit:C20197
mySet:patient1001	myOntology:hasAge	67^^xsd:integer

specification states that all resources (concepts and predicates) need to have a URI, which can be a unique resource locator (URL; e.g., http://www.mydomain.org/ontology#hasFirstName) or a unique resource name (URN; e.g., myOntology:hasFirstName). This means that someone needs to own a domain name (e.g., mydomain.org) and is administrator of this domain. If this is the case, he or she can make unique URLs for this domain, for example, to create a unique URI for patient 1001 (e.g., http://www.mydomain.org/rdf#patient1001). If the domain administrator assigns a specific sub-path of the domain to a dataset (called a *namespace*), for example, http://www.mydomain.org/rdf#, then this sub-path can also be substituted by a *prefix*, for example, "mySet". This namespace can then be used to shorten the notation of a unique patient, as shown in Table 6.1. This concept of unique resources also holds for ontologies, where in Table 6.1 the prefix "myOntology" can be used to define the namespace http://www.mydomain.org/ontology# and the prefix "ncit" refers to the unique location of the NCI thesaurus. As everyone should use the same, unique namespaces, the use of URIs enforces semantic interoperability. Therefore, semantic interoperability is enforced within the Resource Description Framework.

Next to the enforcement of semantic interoperability, the use of URIs has a second benefit, namely, the possibility of linked data. As every resource has its unique URI, an RDF store at site A may point to a resource at site B by using the URI of the resource at point B [4]. For example, if a patient underwent a

```
1  PREFIX rdf: <http://www.w3.org/1999/02/22-rdf-syntax-ns#>
2  PREFIX ontology: <http://www.mydomain.org/ontology#>
3  PREFIX ncit: <http://ncicb.nci.nih.gov/xml/owl/EVS/Thesaurus.owl#>
4
5  SELECT ?patient ?firstName ?lastName ?age
6  WHERE {
7     ?patient rdf:type ncit:C16960 .
8     OPTIONAL { ?patient ontology:hasFirstName ?firstName . }
9     OPTIONAL { ?patient ontology:hasLastName ?lastName . }
10    OPTIONAL { ?patient ontology:hasAge ?age . }
11 }
```

Listing 6.1 Basic SPARQL query retrieving patient resources, related first and last names, and age of patient data stored in an RDF store, based on the ontology defined in Fig. 6.1

diagnostic scan at hospital A and was treated in clinic B, then clinic B can specify the treatment and link it to the patient resource with the unique URI used in hospital A.

Querying Using SPARQL

We have described how data can be represented in RDF, and how URIs enforce semantic interoperability and linked data. But how can we retrieve this data from an RDF store? To query these RDF stores, the W3C has adopted the *SPARQL protocol and RDF query language* (SPARQL) [24]. Most RDF stores have integrated a SPARQL endpoint in their RDF store. A SPARQL endpoint is the public interface to receive SPARQL queries and return a result table, all using the HTTP protocol. In contrast to SQL queries, SPARQL queries do not search tables due to the underlying RDF store structure. SPARQL queries perform pattern matching on the triples in the triple store, where variables can be used to retrieve unknown values or to dynamically link values. For example, the query in Listing 6.1 will try to retrieve the first name, last name, and age for all patients. We will shortly describe the lines in this query example.

On line 1–3, the shorthand (prefix) notations for URL locations are defined. Line 5 defines the variables retrieved from the pattern matching; these variables have to start with a question mark. Lines 6–11 define the actual pattern searched for. As shown in Listing 6.1, our basic pattern is to retrieve all patient resources which have a predicate called "rdf:type," which refers to the terminological code of a patient, defined in the NCI Thesaurus (using the prefix "ncit:," which is replaced by the full URL at line 3). Afterwards, we extend our pattern match by including extra properties for every resource linked to the patient resource. If the linked resources of the patient variable have a predicate matching to our specified property (in our ontology), then the variable firstName, lastName, or age will be filled with the found value. If not found, then the query will return the patient resource URI; however, the variables firstName, lastName, or age are not filled in (due to the "OPTIONAL" keyword).

Next to querying one RDF store, a SPARQL query can also be federated to multiple stores. This is an advantage in regard to multicenter learning, as a single query can retrieve data from multiple sources. Due to the structure of RDF stores, data residing in geographically separated RDF stores can easily be merged, as the data

```
 1  PREFIX rdf: <http://www.w3.org/1999/02/22-rdf-syntax-ns#>
 2  PREFIX ontology: <http://www.mydomain.org/ontology#>
 3  PREFIX ncit: <http://ncicb.nci.nih.gov/xml/owl/EVS/Thesaurus.owl#>
 4
 5  SELECT ?patient ?firstName ?lastName ?age
 6  WHERE {
 7          SERVICE <http://endpoint1.mydomain.org/> {
 8                  ?patient rdf:type ncit:C16960 .
 9                  OPTIONAL { ?patient ontology:hasFirstName ?firstName . }
10                  OPTIONAL { ?patient ontology:hasLastName ?lastName . }
11                  OPTIONAL { ?patient ontology:hasAge ?age . }
12          }
13
14          SERVICE <http://endpoint2.mydomain.org/> {
15                  ?patient rdf:type ncit:C16960 .
16                  OPTIONAL { ?patient ontology:hasFirstName ?firstName . }
17                  OPTIONAL { ?patient ontology:hasLastName ?lastName . }
18                  OPTIONAL { ?patient ontology:hasAge ?age . }
19          }
20  }
```

Listing 6.2 An example of horizontal federation in a SPARQL query

```
 1  PREFIX rdf: <http://www.w3.org/1999/02/22-rdf-syntax-ns#>
 2  PREFIX ontology: <http://www.mydomain.org/ontology#>
 3  PREFIX ncit: <http://ncicb.nci.nih.gov/xml/owl/EVS/Thesaurus.owl#>
 4
 5  SELECT ?patient ?firstName ?lastName ?age
 6  WHERE {
 7          SERVICE <http://endpoint1.mydomain.org/> {
 8                  ?patient rdf:type ncit:C16960 .
 9                  OPTIONAL { ?patient ontology:hasFirstName ?firstName . }
10                  OPTIONAL { ?patient ontology:hasLastName ?lastName . }
11                  OPTIONAL { ?patient ontology:hasAge ?age . }
12          }
13
14          SERVICE <http://endpoint2.mydomain.org/> {
15                  ?patient rdf:type ncit:C16960 .
16                  OPTIONAL { ?patient ontology:hasFirstName ?firstName . }
17                  OPTIONAL { ?patient ontology:hasLastName ?lastName . }
18                  OPTIONAL { ?patient ontology:hasAge ?age . }
19          }
20  }
```

Listing 6.3 An example of vertical federation in a SPARQL query

structure is the same for all stores (1 table; 3 columns) and all RDF stores should use URIs. Federation can be done both horizontally (different patients in different RDF stores) or vertically (information of a single patient stored in multiple RDF stores). An application of horizontal federation in SPARQL queries is shown in Listing 6.2; an application of vertical federation is shown in Listing 6.3. In these examples, we will use the "SERVICE" command of SPARQL to identify the execution of a subquery (or pattern match) on a different SPARQL endpoint. In Listing 6.3, we used the exact same pattern query in both services/subqueries (line 7–19). Both subqueries are sent to the respective endpoints, and the subquery results are merged at the federation endpoint. Finally, the requested variables are returned to the requesting application or user. In Listing 6.3, both services have different patterns to match. The first service (line 7–11) searches for all patients and their first/last name on SPARQL endpoint 1. The second service (line 13–15) will reuse the patient resources found in endpoint 1 and tries to find patterns matching the hasAge predicate for these given patient resources. When found, it will use the object linked to the hasAge predicate (in this case a literal of type integer) and store it in the variable "?age". Finally, the query engine will return the output as one table (using the variables of line 5 as columns), including information retrieved from both endpoints.

In this paragraph, we have presented an alternative to the widely known relational databases to represent and retrieve data. The use of Semantic Web technology, and especially RDF, has several advantages over relational databases. Especially the meta-structure of RDF (independent of the modeled domain) and the use of URIs are useful with regard to a flexible storage solution while inherently adopting semantic interoperability and linked data.

On the other hand, using Semantic Web technology has some downsides when used in multicenter machine learning. The main downside is that local institute staff needs to be introduced to Semantic Web technologies, in order to maintain these data repositories and endpoints. Furthermore, development in the field of RDF stores/repositories is an ongoing process and is not yet comparable to relational databases in terms of reliability and performance, especially in daily clinical practice. On the contrary, for research projects (where uptime is less critical), the Semantic Web is more favorable because of its flexibility in storage and data structures.

6.2.3 Network Infrastructure

Up until now, we only described how to extract information from multiple sources (databases, image archives) and to apply standardized terminological systems on the data extracted from these sources. Furthermore, we have described how to represent data using the relational database and semantic web technology. In this paragraph, we will combine the topics of the previous paragraphs (Sects. 6.2.1 and 6.2.2) and explain how we can use them together. First, we will describe the institutional infrastructure, after which we will describe the multicenter infrastructure.

6.2.3.1 Institutional Infrastructure
In this paragraph, we will describe several approaches to represent a single point of access for the outside world (e.g., participating sites in the multicenter machine learning setting). We will discuss five different approaches, namely:

- Traditional ETL and DWH
- Traditional ETL and DWH with an RDF store
- Traditional ETL and DWH with a virtual RDF store

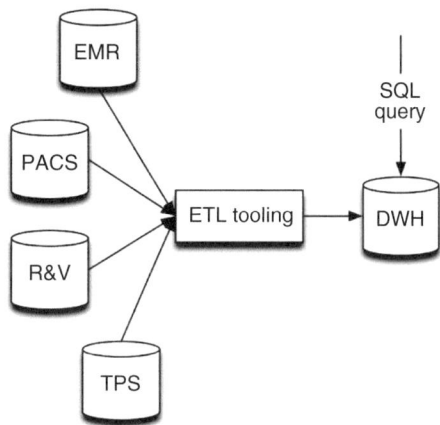

Fig. 6.2 Infrastructure of the traditional ETL and DWH approach

- Virtual RDF store per institute
- Virtual RDF store per source and institute

Traditional ETL and DWH
In the approach using relational databases (Sect. 6.2.2.1), records from different source systems (e.g., EMR, PACS, TPS, and R&V) are merged using an ETL tool (Sect. 6.2.1.1) and converted into the requested data formats following standards used by all collaborating sites (Fig. 6.2). The merged and transformed data are being saved in the DWH database. This database will afterwards be queried when requesting data for machine learning purposes. Therefore, this database needs to be compliant to the ontological structure (among all participating centers). When the ontology is altered, all participating centers need to update the DWH database structure, as well as the transform and/or storage scripts in the ETL tooling.

Traditional ETL and DWH with an RDF Store
This approach uses an RDF store on top of the traditional ETL and DWH approach (Fig. 6.3). It enables the possibility to create an institutional DWH instead of a DWH dedicated for the study. Afterwards, the "Database to RDF" conversion application reads the DWH database and transforms the data it into triples, taking into account a given ontology. This RDF store will afterwards be queried when requesting data for machine learning purposes. Only the "Database to RDF" application needs to follow the rules and data structure defined in the ontology. When the ontology is altered (e.g., adding an extra data element), only this database-to-RDF application needs to be altered (when the information is already available in the DWH). Updating the RDF store is done by clearing and repopulation and is performed at specific time intervals.

Traditional ETL and DWH with a Virtual RDF Store
This approach uses only the database-to-RDF conversion application on top of the traditional ETL and DWH approach (Fig. 6.4). This approach is almost equal to the

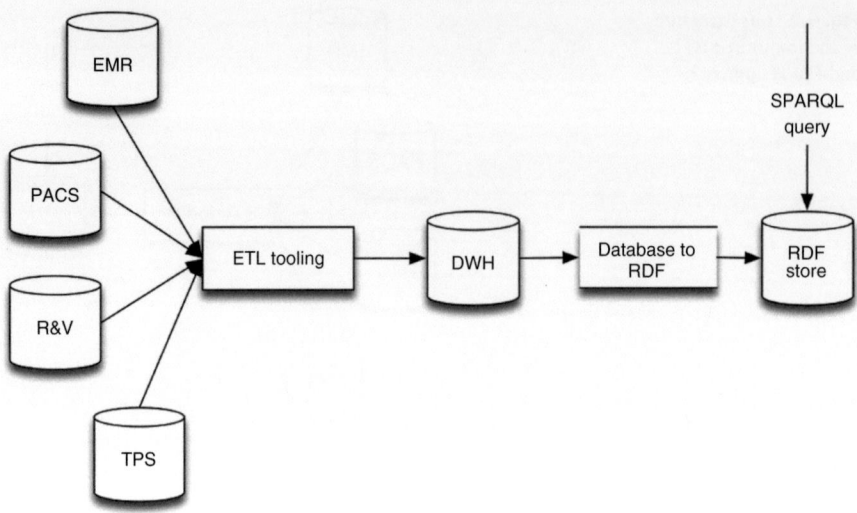

Fig. 6.3 Infrastructure of the approach using a traditional ETL and DWH with an RDF store

Fig. 6.4 Infrastructure of the approach using a traditional ETL and DWH with a virtual RDF store

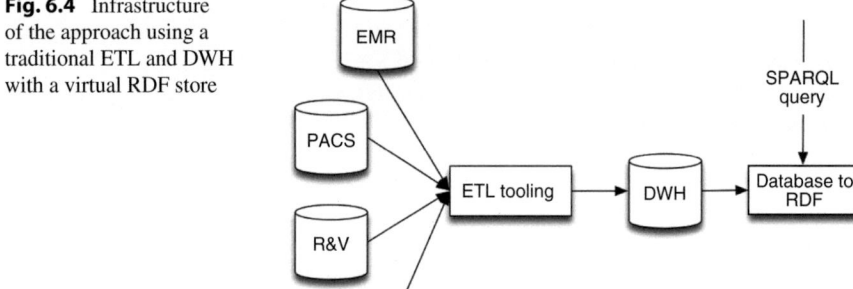

physical RDF store approach (Fig. 6.3); however, it has one difference in converting data from relational databases to RDF.

In this case, the "Database to RDF" application acts as a SPARQL endpoint, accepting SPARQL queries and returning the result of these queries. There is no data stored, as there is no RDF store, only a SPARQL endpoint. When performing a SPARQL query, the database-to-RDF application will transform SPARQL queries into SQL queries and executes these SQL queries on the DWH. In regard to maintenance, this option holds the same requirements as using the physical RDF store. The only difference is the absence of an intermediate RDF store, resulting in real-time results of the data available in the DWH.

Virtual RDF Store per Institute
As the DWH usually is not a real-time representation of the clinically available data, this approach removes the DWH and directly queries the source systems. In this

Fig. 6.5 Infrastructure using only a virtual RDF store

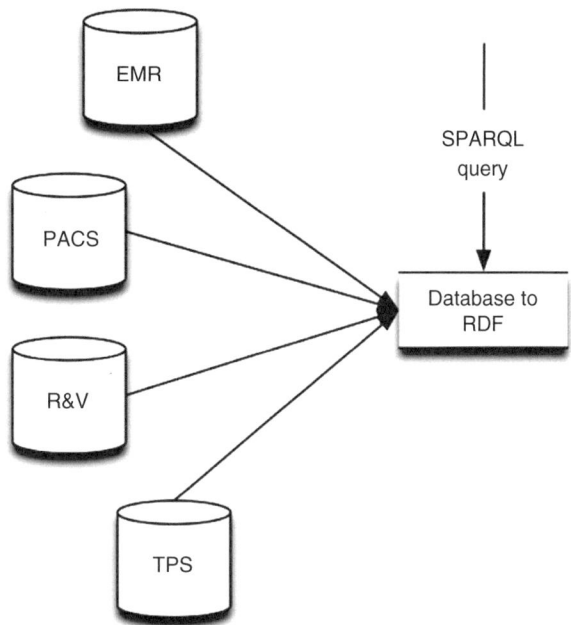

approach, the database-to-RDF application is functioning as a SPARQL endpoint without an RDF store and converts SPARQL queries into SQL queries for the different source systems (Fig. 6.5). It therefore creates challenges for the database-to-RDF application, as it needs to transformation data (to convert local terms to standardized terms), which was previously done by the ETL tooling. If multiple source systems are involved, the database-to-RDF application merges the results from all sources and presents them as a SPARQL query result. The main benefit of this approach is that we can query for real-time data, rather than have to wait before the data is added to the DWH. Furthermore, data redundancy of the intermediate storage (the DWH) is not needed, reducing the need for storage resources. However, the main disadvantage is with regard to performance, as data and queries are transformed on the fly.

Virtual RDF Store per Source and Institute

This approach is almost similar to the "Virtual RDF store per institute" approach, however, with differences in data transformation and federation (Fig. 6.6). First, every local data source will get a SPARQL endpoint, using, for example, the database-to-RDF application. This application will convert the data from the source system into RDF, compliant with the ontology used in the multicenter setting. Afterwards, the central federation endpoint will be used to merge all triples from all database-to-RDF applications/sources (vertical federation). In this setting, one SPARQL query will be sent to the federation endpoint. This federation endpoint will split the SPARQL query into several sub-SPARQL queries and execute these SPARQL queries on the SPARQL endpoints placed on top of the data sources. Afterwards, the federation endpoint will merge the results and return the merged result set to the application/user performing the query. The benefit of this approach is the distribution of computational resources to reduce the query execution time.

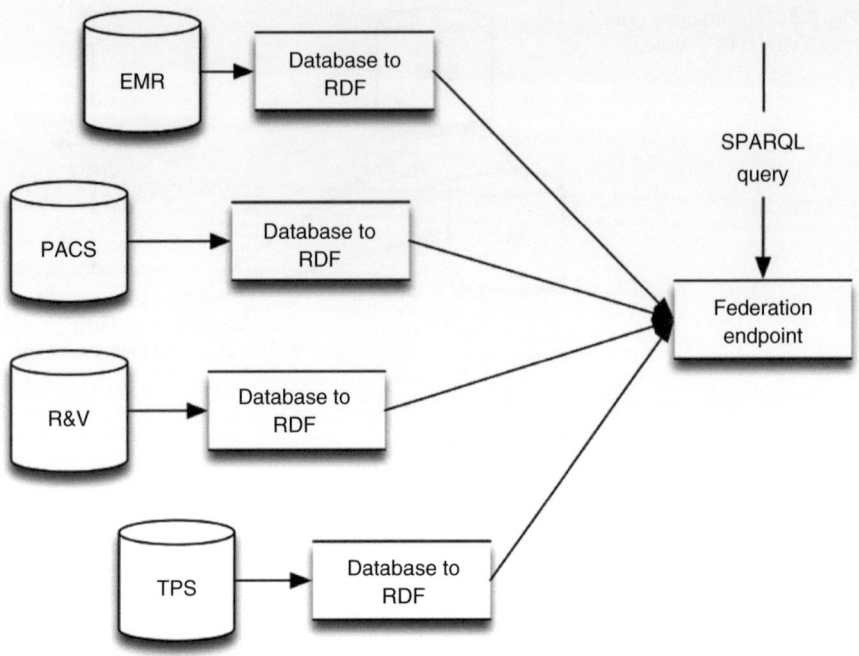

Fig. 6.6 Infrastructure using a virtual RDF store per source and institute

The drawback is that $n+1$ applications (where n is the number of database-to-RDF applications) are need to be maintained and updated when the ontology changes.

6.2.3.2 Multicenter Infrastructure

In the previous paragraph (Sect. 6.2.3.1), we described the institutional infrastructure options to create one façade or data query endpoint for every center. It depends on whether we are using centralized or distributed machine learning (Sect. 6.3) and whether we need an additional computation unit (e.g., a dedicated or virtual server) in each center. Both distributed and centralized approaches can be implemented using relational databases or Semantic Web technology; however, the decision regarding data representation techniques needs to be made upfront and accepted by all participating centers. In this paragraph, we will first describe the centralized machine learning infrastructure and afterwards move towards the distributed infrastructure.

Centralized Multicenter Infrastructure

The general overview for the centralized multicenter infrastructure is shown in Fig. 6.7. The participating sites are displayed as a data store, as we do not need to know what the institutional infrastructure looks like. This approach gives participating centers the opportunity to establish the institutional infrastructure according to local policies. Additional to all institutional entry points, a central machine learning server (performing the computations) and a central federation point need to be set up. The central federation point will perform the horizontal federation between participating centers. To ensure privacy (Sect. 6.2.4), the data stores of the participating

6 Application of Machine Learning for Multicenter Learning

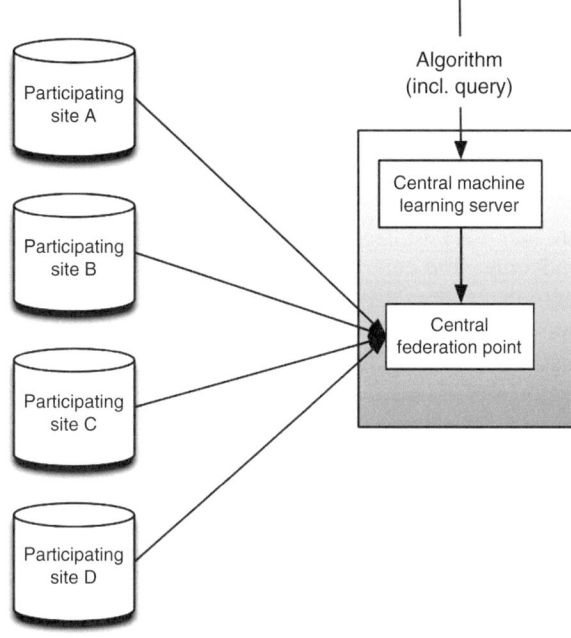

Fig. 6.7 Centralized multicenter infrastructure

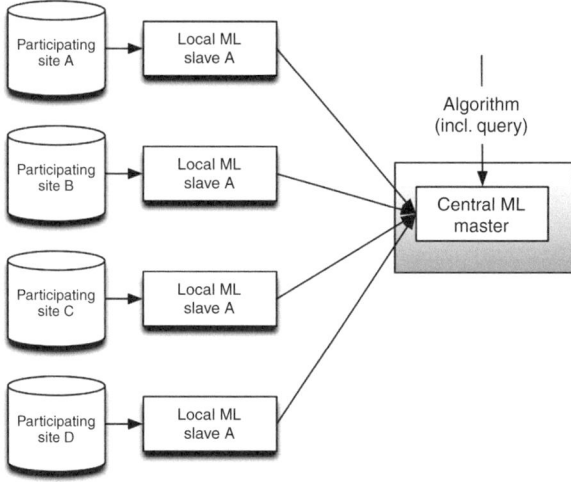

Fig. 6.8 Distributed multicenter infrastructure

centers may limit external access by only allowing access from the central federation point. The central machine learning server will accept and execute algorithms (including queries to execute on the central federation point). After the algorithm has finished, it will return the outcome of the computation to the external source which sent the job (algorithm+query).

Distributed Multicenter Infrastructure

The distributed multicenter infrastructure is different from the centralized version with respect to computational locations. As shown in Fig. 6.8, the central federation point has been removed, and local computation units (machine learning slaves/agents) have been introduced. In this infrastructural setting, the central machine learning server (master server) is a coordinating server. When a job (algorithm + query) is submitted to the central ML master, the algorithm is being split into smaller sub-algorithms. These sub-algorithms and queries are packed into sub-jobs and sent towards the local computation units. They will query the local endpoint and execute the sub-algorithm. After finishing the sub-algorithm, the results are sent back to the central ML master, which gathers the results from all local endpoints. The central master will then determine whether it will perform a new sub-job on all endpoints or aggregate values and sends the final (aggregated) result back to the job-submitter. More information regarding the actual execution of the algorithm in a distributed setting can be found in Sect. 6.3.2.

6.2.4 Privacy Preservation

For both distributed and centralized multicenter infrastructures, privacy preservation is a major topic to take into account. If correctly implemented, the distributed multicenter infrastructure is generally more secure as the results of the algorithm (e.g., a predictive model) are transferred instead of the source data. However, this does not mean that the issues concerning privacy preservation are solved. For example, it is still possible to retrieve metadata about a dataset of one patient. In this section, we will address several options for privacy preservation, ranging from pseudonymization to irreversibly modifying the original datasets. Despite of all the options described below, we have to state that, in our opinion, there is no standard method to ensure privacy preservation. The researcher/designer of the infrastructure will always have to find a balance between the loss of information and the anonymity of participating patients.

Pseudonymization
The first option for privacy preservation is bidirectional pseudonymization of patient identifiers, for example, replacing patient names and hospital's patient identification numbers by study-specific alternatives. This can be achieved by maintaining a two-column table, where one column contains the patients identification number and the second column contains the study identification number for this patient. Variations to this concept may apply, for example, using an extra column to maintain the study where this mapping applies to. Typically, the pseudonymization of hospital to study identification numbers is done during the transform part of the ETL process. Other patient identifying information (e.g., first and last names) can be replaced by the same study ID or may not be incorporated and thus removed during the ETL process.

The second option is to use an unidirectional pseudonymization algorithm, for example, by hashing patient identifiers (e.g., using an SHA-{1–3} algorithm). This hash should be unidirectional, meaning that the pseudonymized patient identifiers cannot be reversed to the original identifiers. Unidirectional pseudonymization might be more appropriate than bidirectional pseudonymization, however might introduce problems when study data are needed to be linked to the actual patients.

For example, when study results show a worse outcome for specific patients and when it is immoral to withhold this information to these patients.

Data Obfuscation
When using strict inclusion criteria with rare variables, it might be that the resulting dataset is very small and patients might become identifiable by combination. For example, if only two patients match some inclusion criteria and the biological sex (which is a requested variable) is different in both patients, we can identify these patients when querying local source systems. This issue holds for both the centralized and distributed multicenter infrastructures. To reduce the chance of compromising the anonymity of patients, Murphy and Chueh [21] introduced a method for data obfuscation where (especially in the case of a small number of events/patients) results are obfuscated by returning a random value within a specific range based on the actual value. This method does not circumvent the problem completely, as someone with bad intentions is able to approximate the original value by sending the same request multiple times. To circumvent these actions, Murphy and Chueh proposed to implement an audit system, where performing the same query multiple times within a specific time span will result in a request denial. In this way, the system returns a value not completely representing the actual value however returns a value within a tolerable margin (when not exceeding the maximum number of requests).

Data Perturbation
The downside on obfuscation is that it does change the distance (e.g., Euclidian distance) between points (e.g., patients or observations) in a k-dimensional space, where every dimension may be a specific variable in the dataset. As the distance changes, it may influence the prediction model training algorithm and train a model that does not represent the actual data and values. This can lead to problems during validation, especially when the validation data is obfuscated, however in another way (due to the randomness in the obfuscation algorithm). Therefore, transformation of data might be a solution, as the whole dataset is transformed while maintaining the distance between points. As shown by Liu et al. [19], this transformation is still not good enough for privacy preservation, as the original data can be derived using independent component analysis (ICA) or overcomplete ICA. To overcome this issue, Liu et al. advise to use their random projection-based multiplicative perturbation (RPBMP) method, which reduces the number of dimensions and transforms the dataset while maintaining statistical information regarding the distance between variables. Using this method, it should not be possible to retrieve the original values and would therefore obstruct the possibility to match variables to individual patients. This RPBMP method is afterwards reused by Yu et al. [35], where they explored differences in dimension reduction options and applied it to a non-small cell lung cancer (NSCLC) dataset. Data perturbation and dimension reduction are potential solutions to preserve privacy in a multicenter setting, although they could lead to issues when performing a risk analysis (identifying variables which influence a specific outcome). The risk analysis then can only determine which *compressed* dimensions are of influence; however, it cannot determine which biological (or source) variables/features are responsible for this influence in patient outcome.

6.3 Centralized and Distributed Machine Learning

When the prerequisites regarding semantic interoperability, data structure, infrastructure, and privacy preservation are in place (Sect. 6.2), we can actually start performing machine learning. In this section, we merely touch upon centralized machine learning (Sect. 6.3.1) in favor of describing distributed machine learning approaches in full (Sect. 6.3.2), which are considered superior for future, large-scope implementations.

6.3.1 Centralized Machine Learning

As described in Sect. 6.2.3.2, the centralized approach only needs one machine learning unit (Fig. 6.7). In this case, the machine learning system will query and retrieve data from the federation data store, irrespectively of knowing where the actual data comes from (except when provenance variables are included in this dataset). As the retrieved dataset is not different in comparison to traditional machine learning approaches, we can use standard machine learning toolboxes such as Weka [12], RapidMiner [13] or others [25]. The disadvantage is that data, with/without privacy preservation in place, is transferred to a central location at time of machine learning algorithm execution. This might contradict the policy of centers with regard to data sharing.

6.3.2 Distributed Machine Learning

The major difference between distributed and centralized machine learning is the transfer of data versus the transfer of training models. In the centralized approach, data is transferred to the machine learning system, whereas in the distributed approach the data stays within the institute. Rather than requesting a dataset, the distributed approach dispatches a sub-process of the machine learning algorithm towards the institutional machine learning unit and returns the result of this sub-process. In this setting, the amount of data per transfer diminishes; however, the data transfer frequency increases. A thorough explanation how distributed machine learning algorithms work is given by Boyd et al. [5] and Wu et al. [34]. From this work of Boyd et al., we reused the MapReduce concept, developed by Dean and Ghemawat [8], to implement the distributed machine learning concept. This is equal to the rationale used by Wu et al.

6.3.2.1 Linear Regression Implementation
The MapReduce concept can be explained by using the linear regression algorithm. Before we explain the MapReduce concept, we will first explain the intuition of the linear regression algorithm, using the nonstandard gradient descent approach. This approach is different from the closed-form solution; however, it enables the possibility for distributed multicenter learning, as we will show afterwards. A trained univariate linear regression classifier can be expressed using the function:

$$f(x) = \alpha + \beta x \tag{6.1}$$

To learn the α and β parameters of this model, the linear regression training algorithm can be described as shown in Algorithm 6.1.

Algorithm 6.1. Linear Regression Training

```
 1: procedure TRAINLINEARREGRESSION (x, y)
 2:     α = random()
 3:     β = random()
 4:     J = cost(α, β, x, y)
 5:     J'_α = costDerivativeAlpha(α, β, x, y)
 6:     J'_β = costDerivativeBeta(α, β, x, y)
 7:     while J' ≤ 0 do
 8:         α = updateVariable(α, J'_α)
 9:         β = updateVariable(β, J'_β)
10:         J = cost(α, β, x, y)
11:         J'_α = costDerivativeAlpha(α, β, x, y)
12:         J'_β = costDerivativeBeta(α, β, x, y)
13:     end while
14:     return α, β
15: end procedure
```

The input parameters for Algorithm 6.1 are x and y, respectively, determining the prediction values and outcomes for which we are training this univariate linear model. On line 2 and 3, initial α and β values are randomly chosen. Afterwards (line 4), the cost (a measure of distance between the calculated outcome, and the actual outcome) for the randomly chosen α and β is calculated using the following function:

$$J(\alpha,\beta) = \frac{1}{2m} * \sum_{i=1}^{n} \left(f\left(x^{(i)}\right) - y^{(i)} \right)^2$$

$$= \frac{1}{2m} * \sum_{i=1}^{n} \left(\left(\alpha + \beta x^{(i)}\right) - y^{(i)} \right)^2 \quad (6.2)$$

In this function, the variable n represents the number of observations used for training. For every observation, both x and y need to be available.

After calculating the cost, we also need the partial derivatives of the cost function. We can write both partial derivatives as

$$J'_\alpha = \frac{\partial}{\partial \alpha} J(\alpha,\beta)$$
$$= \frac{1}{m} * \sum_{i=1}^{n} \left(\left(\alpha + \beta x^{(i)}\right) - y^{(i)} \right) \quad (6.3)$$

$$J'_\beta = \frac{\partial}{\partial \beta} J(\alpha,\beta)$$
$$= \frac{1}{m} * \sum_{i=1}^{n} \left(\left(\alpha + \beta x^{(i)}\right) - y^{(i)} \right) x^{(i)} \quad (6.4)$$

After calculation of these parameters, we can enter the main loop of Algorithm 6.1. In this loop, we'll first update the α and β variables (line 8 and 9) using the following functions:

$$\alpha = \alpha * c * J'_{\alpha} \quad (6.5)$$

$$\beta = \beta * c * J'_{\beta} \quad (6.6)$$

In these functions, the variable c determines the gradient descent rate. After calculating the new α and β, the algorithm continues by calculating the cost function and the partial derivatives of the cost function again (lines 10–12). Afterwards, it will start a new iteration of this loop (line 7), calculating the new α and β, calculating the cost, and calculating the partial derivatives (lines 8–12). This process is repeated until the algorithm reaches one of several termination criteria. Most preferably, the partial derivatives should converge close to 0, as this would indicate that an optimum has been reached. Alternative termination criteria are the number of iterations (as the algorithm does not converge, e.g., due to a large gradient descent rate c) or no significant changes in the calculated cost of the last m iterations. Finally, the algorithm will return both α and β as the outcome of the training algorithm.

6.3.2.2 Cost Function and MapReduce

As shown in Algorithm 6.1, the calculation of the cost and/or the partial derivatives is the only function in the algorithm where the original data is needed. These functions also consume the most of the computational resources and are positively correlated to the number of observations and variables incorporated in the regression model. To reduce the computation time, we can split the original data and refactor the cost and partial derivative functions to multiple machines and/or processing units. When reducing the number of summations (see Eqs. 6.2, 6.3, and 6.4) for every processing unit, we can reduce the overall time to calculate the cost and/or partial derivatives.

This distribution of processing power can be achieved using the MapReduce concept. In the previous example, we can implement the MapReduce concept as shown in Algorithm 6.2.

Algorithm 6.2. MapRedue Implementation of Cost Function

1: **procedure** CALCULATEMAPREDUCECOST (α, β, x, y)
2:    **for** processing unit u in processingUnits **do**
3:       $u.\,startCalculatingSquaredDistance(\alpha, \beta, x, y)$
4:    **end for**
5:
6:    $wait()$
7:
8:    $distanceSq = 0$
9:    $m = 0$
10:   **for** processing unit u in processingUnits **do**

11:     $result \leftarrow u.retrieveSquaredDistances(\alpha,\beta,x,y)$
12:     $distanceSq = distanceSq + result[0]$
13:     $m = m + result[1]$
14: **end for**
15: **return** $calculateCost(distanceSq, m)$
16: **end procedure**

This algorithm first starts with the "Map" part and subsequently performs the "Reduce" part of the MapReduce concept. At first, all registered processing units are invoked to start the following summation calculation (line 2–4):

$$D = \sum_{i=s}^{n}\left(\left(\alpha + \beta x^{(i)}\right) - y^{(i)}\right)^2 \tag{6.7}$$

Note that this summation calculation only calculates the summation over a specific subset (from observation s to n). The result of this squared difference summation is afterwards temporarily stored or directly returned to the initiator.

After the initiator has waited until all processing units have finished their calculation, the "Reduce" part of the algorithm comes in. The initiator (and therefore also the "Reducer") sums all squared distances and the number of individuals processed used for this calculation (line 8–14). Afterwards, the initiator calculates the total cost using the equation:

$$J = \frac{1}{2m} * D \tag{6.8}$$

In this equation, we use the summed squared distances from all processing units (variable D) and the number of observations from all processing units (variable m), resulting in a result equal to Eq. 6.2 (the non-parallelized cost function).

To calculate the partial derivatives of the cost function, we could reuse Algorithm 6.2 and modify Eqs. 6.7 and 6.8 to use the summed squared distances and numbers of observations.

6.3.2.3 MapReduce, Distributed, and Multicenter Machine Learning

As shown above, we are able to distribute the resource-intensive part of the linear regression training algorithm over multiple processing units. As previously stated, processing units can be multiple CPUs or multiple computers. In the latter situation, we need to include the data in the invocation of the summed squared distances calculation or have the data already available at all processing computers (e.g., by mirroring files/databases and/or using network drives). This concept of distributing the computation over multiple machines and mirroring the data is typically done in grid computing. To reduce the data needed to be mirrored, one can consider to split the original data over the involved computers. As the squared difference summation is aggregated during the "Reduce" part of the algorithm, there is no need to have the complete dataset on all computers.

The absence of complete datasets on all involved machines creates the opportunity for distributed multicenter machine learning. When implemented as in Algorithm 6.2, the "Reducer" does not need to know the size of the complete dataset at forehand (variable m). All processing units send back two variables to the Reducer: the summation of the squared distance and the number of observations. This preserves privacy by not sending over the actual patient information, only the aggregated results, while still being able to perform machine learning over large datasets in different institutes. This aligns with the distributed multicenter infrastructure described in Sect. 6.2.3.2. Therefore, we can perform distributed multicenter learning using the MapReduce concept, where the original data does not leave the centers.

6.3.2.4 Training, Testing, and Validation

Now we defined a concept of distributed multicenter machine learning (for training purposes), we can perform testing and validation on this infrastructure. To perform testing and/or validation, the easiest approach is to use one participating institute as the testing dataset and one participating institutes' dataset as the validation dataset. For example, if we have five participating institutes $I \in i_{1-5}$, we can use the first three institutes (i_{1-3}) to execute the distributed training. Afterwards, the master node can send the trained model to i_4 for testing purposes, resulting in performance metrics of the prediction model, for example, determining discriminative and/or accuracy (e.g., c-index, Brier score, Hosmer-Lemeshow test) of the trained model. After model development has finished, an external validation can be performed by sending the model to i_5, which calculates and returns the performance of the trained prediction model.

A second, more elaborate, option is to distribute the testing and validation steps over all participating institutes I. In this case, training would be done on 60 % of all subjects and testing and validation on, respectively, 20 and 20 % of all remaining subjects. If done correctly, assignment into the training, test, or validation set should be determined before starting the model training; however, it should be remembered during the whole process. This adds extra complexity to the computation units within the institutes. These units have to remember for which distributed learning algorithm the computational instruction is and determine which dataset to use. For the testing phase, this approach might be better, as the training and testing datasets are homogeneous. For external validation, it raises the discussion whether an external validation set should be from a completely different center.

Finally, we can perform a k-fold cross-validation, where the number of participating centers can determine the number of folds, where we would use $I_{\text{learn}} \subset I$, with $i_n \notin I_{\text{learn}}$ as the training dataset. In this case, i_n will be the validation set for a specific fold.

6.4 Applications

In the previous paragraphs, we defined the prerequisites and described how to perform distributed machine learning. In this paragraph, we will discuss several initiatives and applications of multicenter learning. It is not mandatory that all applications use the complete set of prerequisites described previously in this chapter.

I2B2

The Informatics for Integrating Biology and the Bedside (I2B2; http://www.i2b2.org) project aims at integrating data from different biomedical disciplines and delivering this data to researchers. The project delivers tools to translate genomic and biologic findings to clinical findings (e.g., diseases or disorders). To be able to achieve this *translational medicine* approach, institutional data sources are federated in the I2B2 DWH using ETL tooling (Sect. 6.2.3.1). The DWH database structure, called the Clinical Research Chart (CRC), is generic for medical purposes, as it does not define specific data fields. The database structure is basically a "star schema" where only patient information and observations are stored [22]. To describe all information in an observation-centered storage, local terminologies, or standardized terminological systems (Sect. 6.2.2), are needed to define different types of observations. Afterwards, researchers can query/request data. When a specific dataset has been queried, this dataset can be stored in a separate database, using the same CRC database structure. In this separate database, researchers can clean/modify the dataset to their needs and execute machine learning algorithms on this dataset.

In regard to multicenter machine learning, I2B2 supports merging multiple research databases using the Shared Health Research Information Network (SHRINE) tool [31], resulting in a federated research database of multiple institute research databases. Therefore, it enables the opportunity for centralized multicenter learning. In this approach, the terminology to define observations can be aligned when merging databases or can be kept separate [23]. In the latter approach, the researcher has to put in more effort in data alignment during the analysis, which is not favorable as it is prone to causing mistakes in the analysis.

EuroCAT

The Euregional Computer-Aided Theragnostics (EuroCAT; http://www.eurocat.info) project aims at reuse of clinical data for research purposes and to improve the speed and quality of clinical research. The project uses a distributed learning approach as described in Sect. 6.3.2, targeted at prediction models for lung cancer. To be able to perform this distributed learning approach, a so-called umbrella protocol was developed by the participating partners. This protocol describes the standardized data collection, including the variables to record (and terminological systems to use), questionnaires, and informed consent document templates. The first version of the EuroCAT system used a DWH and ETL infrastructure at the local institutes, as described in Sect. 6.2.3.1. Afterwards, the DWH was replaced by an RDF store. The EuroCAT system has shown that distributed multicenter machine learning works and produces the same results as centralized learning when implemented correctly [32]. Furthermore, the project has shown that distributed multicenter learning does improve the robustness of prediction models when validating on an external dataset [9].

VATE

The VATE ("VAlidation of High TEchnology based on large database analysis by learning machine") project shares the aim of the EuroCAT project. The major difference is that this project is based on open standards (in regard to IT infrastructure) and uses Semantic Web technologies (e.g., RDF and ontologies) as a basis for data representation. Prior to this project, the involved institutes had developed a data infrastructure for research purposes using open standards [26]. Equal to the EuroCAT project, the VATE project has developed an umbrella protocol for rectal cancer [20]. Different from the EuroCAT project, the variables to record are classified into several levels regarding the completeness of datasets and are maintained in a publically available ontology (http://webprotege.stanford.edu/#Edit:projectId=37ecb757-c801-4309-aa9b-3dbbc7f9f7c3). The rationale behind these rankings and this public umbrella protocol is that everyone who has data regarding rectal cancer patients can join this linked data network when the data is specified according to the ontological rules, irrespective to the number of available variables. Due to the chosen aim of training a Bayesian Network for rectal cancer on the VATE infrastructure, missing data could be imputed or ignored during training, as shown by Jayasurya et al. [14].

PCORnet
The Patient-Centered Outcomes Research Network (PCORnet) is a program aiming at building a national research network linking datasets from clinical production systems from multiple centers, using a standardized data platform [30]. The program comprises 11 clinical data research networks (CDRN) and 18 patient-powered research networks (PPRN). The aim of the CDRNs is comparable to the previously described EuroCAT and VATE projects. The PPRN projects aim at the empowerment of patients. In these PPRNs, patients would supply the data instead of retrieving data from clinical systems. Therefore, the gathered data and research questions addressed by these projects are different from the CDRN projects [10]. The first (short-term) aims for the program are to build and implement the network in all the CDRNs and PPRNs and include one million patients in 18 months after the start of the project. Long-term aims are to perform (distributed) machine learning on the network.

6.5 Summary

In this chapter, we have seen that multicenter machine learning is possible for both a centralized and distributed approach. To be able to set up a multicenter machine learning environment, several biomedical informatics-related issues need to be addressed. The most important issue is semantic interoperability among participating centers. If the participating centers cannot agree on definitions, how do we know whether all data are equally formatted? Second, the infrastructure (both institutional and central) needs to be implemented, together with the chosen data representation. The choice for an infrastructure comes with the choice of a centralized or distributed approach (Sect. 6.2.3.2). Third, privacy preservation needs to be addressed and may influence the choice for a centralized or distributed approach and the preservation measures implemented (e.g., uni- versus bidirectional pseudonymization or data perturbation versus transformation). When all prerequisites are met, the actual machine learning can be performed. In this part, a centralized approach should not be different from traditional machine learning. The distributed machine learning approach

(Sect. 6.3.2) needs some modifications to traditional machine learning algorithms, as local outcomes need to be aggregated and combined at a central location. Therefore, in distributed machine learning, traditional algorithms need to be split into two parts: a central node performing the general algorithm and institutional nodes performing delegated tasks requested by the central node. Finally, we have shown that distributed machine learning is possible in practice. Showing several projects and/or initiatives where data from different locations are used to develop prediction models.

In general, we have shown that distributed machine learning is not only a task for the "traditional" machine learning expert (which is already not the case in healthcare and radiation oncology); however, it also needs other disciplines, such as expertise from the fields of terminology/ontology development, network/infrastructure, and security/privacy.

References

1. Abernethy AP, Etheredge LM, Ganz PA, Wallace P, German RR, Neti C, Bach PB, Murphy SB. Rapid-learning system for cancer care. J Clin Oncol. 2010;28(27):4268–74. doi:10.1200/JCO.2010.28.5478.
2. Allemang D, Hendler JA. Semantic web for the working ontologist effective modeling in RDFS and OWL. 2nd ed. Waltham: Morgan Kaufmann; 2011.
3. Berners-Lee T, Hendler J, Lassila O. The semantic web. Sci Am. 2001;284(5):28–37.
4. Bizer C, Heath T, Berners-Lee T. Linked data-the story so far. Int J Semantic Web Inform Syst. 2009;5(3):1–22.
5. Boyd S. Distributed optimization and statistical learning via the alternating direction method of multipliers. Found Trends Mach Learn. 2010;3(1):1–122. doi:10.1561/2200000016.
6. Brickley D, Guha R. RDF schema 1.1. 2014. URL http://www.w3.org/TR/2014/REC-rdf-schema-20140225/.
7. De Keizer NF, Abu-Hanna A, Zwetsloot-Schonk JHM. Understanding terminological systems. I: terminology and typology. Method Inform Med. 2000;39:16–21.
8. Dean J, Ghemawat S. MapReduce: simplified data processing on large clusters. Commun ACM. 2008;51(1):107–13.
9. Dekker A, Nalbantov G, Oberije C, Wiessler W, Elbe M, Dries W, JanvaryL, Bulens P, Krishnapuram B, Lambin P. Multi-centric learning with a federated IT infrastructure: application to 2-year lung-cancer survival prediction. In: 2nd ESTRO FORUM, Elsevier, Geneva, Switzerland, 2013: p. S35. http://www.estro-events.org/ESTROevents/Documents/FORUM_abstract_bookPRESS_lowres.pdf.
10. Fleurence RL, Curtis LH, Califf RM, Platt R, Selby JV, Brown JS. Launching PCORnet, a national patient-centered clinical research network. J Am Med Inform Assoc. 2014. doi:10.1136/amiajnl-2014-002747.
11. Gali A, Chen C, Claypool K, Uceda-Sosa R. From ontology to relational databases. In: Wang S, Tanaka K, Zhou S, Ling T-W, Guan J, Yang D, et al. (Eds.), Conceptual modeling for advanced application domains. Springer Berlin Heidelberg; 2004. p. 278–89. http://dx.doi.org/10.1007/978-3-540-30466-1_26.
12. Hall M, Frank E, Holmes G, Pfahringer B, Reutemann P, Witten IH. The WEKA data mining software: an update. ACM SIGKDD Explorations Newsletter. 2009;11(1):10–8.
13. Hofmann M, Klinkenberg R. RapidMiner: data mining use cases and business analytics applications. Boca Raton: CRC Press; 2013. ISBN: 978-1482205497
14. Jayasurya K, Fung G, Yu S, Dehing-Oberije C, De Ruysscher D, Hope A, De Neve W, Lievens Y, Lambin P, Dekker ALAJ. Comparison of bayesian network and support vector machine

models for two-year survival prediction in lung cancer patients treated with radiotherapy. Med Phys. 2010;37(4):1401–7. doi:10.1118/1.3352709.
15. Lambin P, Rios-Velazquez E, Leijenaar R, Carvalho S, van Stiphout RG, Granton P, Zegers CM, Gillies R, Boellard R, Dekker A, Aerts HJ. Radiomics: extracting more information from medical images using advanced feature analysis. Eur J Cancer. 2012;48(4):441–6. doi:10.1016/j.ejca.2011.11.036.
16. Lambin P, Roelofs E, Reymen B, Velazquez ER, Buijsen J, Zegers CM, Carvalho S, Leijenaar RT, Nalbantov G, Oberije C, Scott Marshall M, Hoebers F, Troost EG, van Stiphout RG, van Elmpt W, van der Weijden T, Boersma L, Valentini V, Dekker A. 'Rapid learning health care in oncology' – an approach towards decision support systems enabling customised radiotherapy. Radiother Oncol. 2013;109(1):159–64. doi:10.1016/j.radonc.2013.07.007.
17. Lambin P, van Stiphout RGPM, Starmans MHW, Rios-Velazquez E, Nalbantov G, Aerts HJWL, Roelofs E, van Elmpt W, Boutros PC, Granone P, Valentini V, Begg AC, De Ruysscher D, Dekker A. Predicting outcomes in radiation oncology – multifactorial decision support systems. Nat Rev Clin Oncol. 2012;10(1):27–40. doi:10.1038/nrclinonc.2012.196.
18. Leijenaar RTH, Carvalho S, Velazquez ER, van Elmpt WJC, Parmar C, Hoekstra OS, Hoekstra CJ, Boellaard R, Dekker ALAJ, Gillies RJ, Aerts HJWL, Lambin P. Stability of FDG-PET radiomics features: an integrated analysis of test-retest and inter-observer variability. Acta Oncol. 2013;52(7):1391–7. doi:10.3109/0284186X.2013.812798.
19. Liu K, Kargupta H, Ryan J. Random projection-based multiplicative data perturbation for privacy preserving distributed data mining. Knowledge Data Eng IEEE Transact. 2006;18(1):92–106.
20. Meldolesi E, van Soest J, Dinapoli N, Dekker A, Damiani A, Gambacorta MA, Valentini V. An umbrella protocol for standardized data collection (SDC) in rectal cancer: a prospective uniform naming and procedure convention to support personalized medicine. Radiother Oncol. 2014. doi:10.1016/j.radonc.2014.04.008.
21. Murphy SN, Chueh HC. A security architecture for query tools used to access large biomedical databases. In: Proceedings of the AMIA symposium. American Medical Informatics Association. 2002. p. 552.
22. Murphy SN, Mendis M, Hackett K, Kuttan R, Pan W, Phillips LC, Gainer V, Berkowicz D, Glaser JP, Kohane I. Architecture of the open-source clinical research chart from informatics for integrating biology and the bedside. In: AMIA annual symposium proceedings, vol. 2007. American Medical Informatics Association. 2007. p. 552–6.
23. Murphy SN, Weber G, Mendis M, Gainer V, Chueh HC, Churchill S, Kohane I. Serving the enterprise and beyond with informatics for integrating biology and the bedside (i2b2). J Am Med Inform Assoc. 2010;17(2):124–30. doi:10.1136/jamia.2009.000893.
24. Prud'hommeaux E, Seaborne A. SPARQL query language for RDF. 2008. URL http://www.w3.org/TR/rdf-sparql-query/.
25. Ramamohan Y, Vasantharao K, Chakravarti CK, Ratnam ASK. A study of data mining tools in knowledge discovery process. Int J Soft Comput Eng (IJSCE). ISSN. 2012;2(3):2231–307.
26. Roelofs E, Dekker A, Meldolesi E, van Stiphout RG, Valentini V, Lambin P. International datasharing for radiotherapy research: an open-source based infrastructure for multicentric clinical data mining. Radiother Oncol. 2014;110(2):370–4. doi:10.1016/j.radonc.2013.11.001.
27. Roelofs E, Persoon L, Nijsten S, Wiessler W, Dekker A, Lambin P. Benefits of a clinical data warehouse with data mining tools to collect data for a radiotherapy trial. Radiother Oncol. 2013;108(1):174–9. doi:10.1016/j.radonc.2012.09.019.
28. Sioutos N, Coronado SD, Haber MW, Hartel FW, Shaiu WL, Wright LW. NCI thesaurus: a semantic model integrating cancer-related clinical and molecular information. J Biomed Inform. 2007;40(1):30–43. doi:10.1016/j.jbi.2006.02.013.
29. Valentini V, Schmoll HJ, Velde CJH. Multidisciplinary management of rectal cancer questions and answers. Berlin/New York: Springer; 2012.
30. Waitman LR, Aaronson LS, Nadkarni PM, Connolly DW, Campbell JR. The greater plains collaborative: a PCORnet clinical research data network. J Am Med Inform Assoc. 2014. doi:10.1136/amiajnl-2014-002756.

31. Weber GM, Murphy SN, McMurry AJ, MacFadden D, Nigrin DJ, Churchill S, Kohane IS. The shared health research information network (SHRINE): a prototype federated query tool for clinical data repositories. J Am Med Inform Assoc. 2009;16(5):624–30. doi:10.1197/jamia.M3191.
32. Wiessler W, Dekker A, Nalbantov G, Oberije C, Eble M, Dries W, Janvary L, Bulens P, Balaji, K, Lambin P. Privacy-preserving, multi-centric machine learning across institutions and countries: does it work? Elsevier, Geneva. 2013
33. World Health Organization. International statistical classification of diseases and related health problems. Geneva: World Health Organization; 2011.
34. Wu Y, Jiang X, Kim J, Ohno-Machado L. Grid binary LOgistic REgression (GLORE): building shared models without sharing data. J Am Med Inform Assoc. 2012;19(5):758–64. doi:10.1136/amiajnl-2012-000862.
35. Yu S, Fung G, Rosales R, Krishnan S, Rao RB, Dehing-Oberije C, Lambin P. Privacy-preserving cox regression for survival analysis. In: Proceedings of the 14th ACM SIGKDD international conference on Knowledge discovery and data mining. New York: ACM; 2008. p. 1034–42.

Part II

Machine Learning for Computer-Aided Detection

Computerized Detection of Lesions in Diagnostic Images

7

Kenji Suzuki

Abstract

Computer-aided detection (CADe) has been an active research area in medical imaging. As imaging technologies advance, a large number of medical images are produced which physicians/radiologists must read. They may overlook lesions from such a large number of medical images. Consequently, CADe that provides suspicious lesions with radiologists/physicians is developed and becoming indispensable in their decision making to prevent them from overlooking lesions. Machine learning (ML) plays an essential role in CADe, because lesions and organs in medical images may be too complex to be represented accurately by a simple equation; modeling of such complex objects often requires a number of parameters that have to be determined by data. In this chapter, ML techniques used in CADe schemes for lung nodules in chest radiography and thoracic CT and those for the detection of polyps in CT colonography (CTC) are described, which include patch-/pixel-based ML and feature-based (segmented-object-based) ML.

7.1 Introduction

Computer-aided detection and diagnosis (CAD) [26, 27, 39, 41] has been an active research area in medical imaging. CAD is defined as detection/diagnosis made by a physician/radiologist who takes into account the computer output as a "second opinion" [26]. CAD is often categorized into two major groups, computer-aided detection (CADe) and computer-aided diagnosis (CADx). CADe focuses on a

---

K. Suzuki, PhD
Department of Radiology, The University of Chicago, P-104A, 5841
South Maryland Avenue, MC 2026, Chicago, IL 60637, USA
e-mail: suzuki@uchicago.edu
url: http://suzukilab.uchicago.edu/index.htm

© Springer International Publishing Switzerland 2015
I. El Naqa et al. (eds.), *Machine Learning in Radiation Oncology:
Theory and Applications*, DOI 10.1007/978-3-319-18305-3_7

detection task, namely, localization of lesions in medical images. CADx focuses on a diagnosis (characterization) task, for example, distinction between benign and malignant lesions. As imaging technologies advance, a large number of medical images are produced which physicians/radiologists must read. They may overlook lesions from such a large number of medical images. Thus, CAD is becoming indispensable in physicians' decision making. Evidence suggests that CAD can help improve the diagnostic performance of physicians/radiologists [16, 24, 25, 64, 65, 92, 124]. Consequently, many investigators have developed CAD schemes such as those for the detection of lung nodules in chest radiographs [40, 135, 144] and in thoracic CT [3, 7, 121], those for the detection of microcalcifications/masses in mammography [15], breast MRI [42], and breast ultrasound (US) [29] and those for the detection of polyps in CT colonography (CTC) [107, 139, 140].

Machine learning (ML) plays an essential role in CAD, because objects such as lesions and organs in medical images may be too complex to be represented accurately by a simple equation; modeling of such complex objects often requires a number of parameters that have to be determined by data [114, 116, 117]. For example, a lung nodule is generally modeled as a solid sphere, but there are spiculated nodules and ground-glass nodules [66]. Although a polyp in the colon is modeled as a bulbous object, there are polyps that exhibit a flat shape [77, 104]. Thus, diagnostic tasks in medical images essentially require "learning from examples (or data)" to determine a number of parameters in a complex model. Because of its importance and significance, the field of ML in medical imaging became very active. The special issues on ML in medical imaging were published in various journals [102, 112, 113, 138, 161]; a series of international workshops on this topic was held from 2010 [136, 147, 148, 156].

One of the most popular uses of ML in CAD is the classification of lesion candidates into certain classes (e.g., abnormal or normal, lesions or nonlesions, and malignant or benign) based on input features (e.g., area, contrast, and circularity) obtained from segmented candidates (this class of ML is referred to as feature-based ML or segmented-object-based ML). The task of ML is to determine "optimal" boundaries for separating classes in the multidimensional feature space which is formed by the input features [30]. The ML algorithms for classification include linear discriminant analysis [34], quadratic discriminant analysis [34], multilayer perceptron [97, 98], and support vector machines [145, 146]. Such ML algorithms were applied to lung nodule detection in chest radiography [20, 22, 47, 103] and thoracic CT [3, 5, 152, 163], detection of microcalcifications in mammography [31, 35, 157, 169], detection of masses in mammography [158], polyp detection in CT colonography [59, 150, 167], determining subjective similarity measure of mammographic images [82–84], and detection of aneurysms in brain MRI [4].

Recently, as available computational power increased dramatically, patch-/pixel-based ML (PML) [115, 121] emerged in medical image processing/analysis which uses values in image patches (direct pixel values and/or features calculated from the values in the image patches) instead of features calculated from segmented regions as input information; thus, segmentation is not required. Because the PML can avoid errors caused by inaccurate segmentation that often occur for subtle or complex objects, the performance of the PML can potentially be higher for such objects than that of common classifiers (i.e., feature-based MLs).

In this chapter, ML techniques used in CADe schemes of the thorax and colon are described, including CADe schemes for lung nodules in chest radiography and thoracic CT, and those for the detection of polyps in CTC.

7.2 Overview of Architecture of a CADe Scheme

A flowchart for a generic CADe scheme of lesions in diagnostic images is shown in Fig. 7.1. A CADe scheme generally consists of four core steps and two optional steps: (1) segmentation of the organ of interest, (3) detection of lesion candidates in the segmented organ, (4) segmentation and feature analysis of the detected lesion candidates, (5) classification of the lesion candidates by the use of a classifier with features (feature-based ML); optionally (2) enhancement of lesions between steps 1 and 3, and (6) reduction of false-positive (FP) detections after step 5. Segmentation of the organ of interest is the first necessary step that aims to make the rest of the steps focus on that organ. The development of the detection of lesion candidates

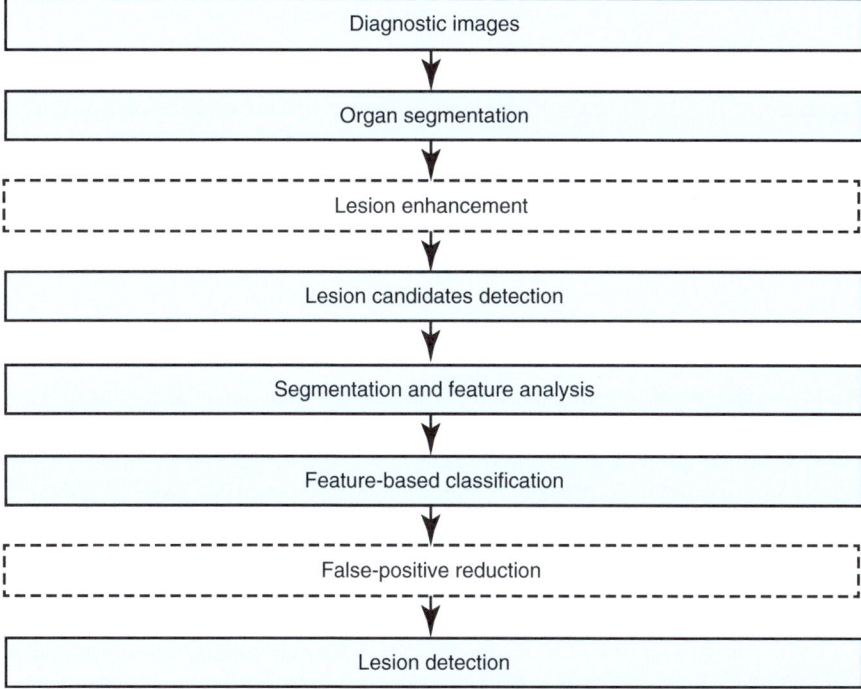

Fig. 7.1 Flowchart for a generic CADe scheme for the detection of lesions in diagnostic images. *Boxes* with *solid lines* indicate four core steps in the CADe scheme, and those with *dashed lines* indicate optional, yet important steps

generally aims to obtain a high sensitivity level, because the sensitivity lost in this step cannot be recovered in the later steps. In the next step, the detected (or localized) lesion candidates are segmented, and connected-component labeling [49, 50, 127] is performed to identify each segmented candidate as an individual isolated object. Pattern features such as gray-level-based features, texture features, and morphologic features are extracted from the segmented candidates. Finally, the detected lesion candidates are classified into lesions or nonlesions by the use of a classifier (or feature-based ML). This final step is very important, because it determines the final performance of a CADe scheme when the additional step of FP reduction is not employed. The development of the classification step aims to remove as many nonlesions (i.e., FPs) as possible while minimizing the removal of lesions (i.e., true-positive detections). The optional steps 3 and 6 are described below.

To improve the performance of CADe schemes, researchers sometimes adopt an additional step that is enhancement of lesions after step 1 of the segmentation of the organ of interest. This additional step aims to improve the sensitivity for the detection of lesion candidates in the subsequent step. It also often helps improve the specificity. Researchers also often adopt an additional step of reduction of FPs at the end of the steps. The FP reduction step aims to improve the specificity of the CADe scheme. Reduction of FPs is very important, because a large number of FPs could adversely affect the clinical application of CADe. A large number of FPs is likely to confound the radiologist's task of image interpretation and thus lower his/her efficiency. In addition, radiologists may lose their confidence in CADe as a useful tool.

After the development of a CADe scheme, the evaluation of the stand-alone performance of the developed scheme is the last step in CADe *development*. CADe *research* does not end by this step: the evaluation of radiologists' performance with the use of the developed CADe scheme is the important last step in CADe *research*.

7.3 Machine Learning (ML) in CADe

7.3.1 Feature-Based (Segmented-Object-Based) ML (Classifiers)

An ML technique is generally used in the step of classification of lesion candidates. The ML technique is trained with sets of input features and correct class labels. This class of ML is referred to as feature-based ML, segmented-object-based ML, or simply as a classifier. Because classifiers (or feature-based ML) are described in detail in many pattern-recognition and computer-vision textbooks, this chapter does not repeat the details of the techniques. Please refer to such textbooks, e.g., [30], [12, 34, 48, 145, 146], for details. The task of ML here is to determine "optimal" boundaries for separating classes in the multidimensional feature space which is formed by input features [30]. A standard classification approach is illustrated in Fig. 7.2. First, lesions (lesion candidates) are segmented by the use of a segmentation method. Next, features are extracted from the segmented lesions. Features may include shape-based (morphologic) features, gray-level-based features (including histogram-based features), and texture features. Some researchers consider texture

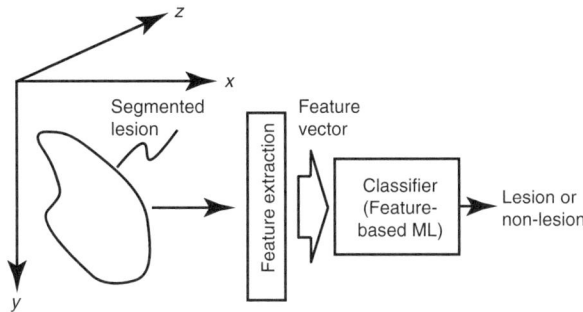

Fig. 7.2 Feature-based (segmented-object-based) ML (classifier) for classification of a detected and segmented lesion candidate

features in the category of the gray-level-based features. Then, extracted features are entered as input to an ML model such as linear discriminant analysis [34], quadratic discriminant analysis [34], a multilayer perceptron (or artificial neural network) [97, 98], and a support-vector machine [145, 146]. When an artificial neural network is used as a classifier, the structure of the artificial neural network may be designed by the use of an automated design method such as sensitivity analysis [110, 125]. The ML model is trained with sets of input features and correct class labels. A class label of 1 is assigned to the corresponding output unit when a training sample belongs to a certain class (e.g., class A), and 0 is assigned to the other output units (e.g., classes B, C, etc.). In the case of two-class classification, one output unit instead of two output units is often used with the output value 0 being class A, and 1 being class B. After training, the class of the unit with the maximum value is determined to be the corresponding class to which an unknown sample belongs.

Feature selection has long been an active research topic in machine learning, because it is one of the main factors that determine the performance of a classifier. In general, multiple or often many features are extracted from segmented lesions as the classifier input. Not all of the features, however, would be useful for a classifier to distinguish between lesions and nonlesions, because some of them might be highly correlated with each other or redundant; some of them may not be strongly associated with the given classification task. For designing a classifier with high performance, it is crucial to select "effective" features. Therefore, feature selection is often used to select "effective" features for a given task. One of the most recent, promising feature selection methods is feature selection under the criterion of the maximal area under the receiver-operating-characteristic curve [160].

7.3.2 Patch-/Pixel-Based Machine Learning (PML)

7.3.2.1 Overview

Recently, as available computational power has increased dramatically, patch-/pixel-based machine learning (PML) [114] emerged in medical image processing/analysis which uses values in image patches (i.e., pixel values and/or features calculated from the image patches), instead of features calculated from segmented regions, as input information; thus, segmentation is not required. PML has been used in the classification of the detected lesion candidates in CADe schemes. Recently in the computer-vision field, deep learning and deep neural networks [11, 54] have been attracting researchers' attentions as a breakthrough technology in computer vision. Deep learning and deep neural networks use PML architecture.

PMLs were first developed for tasks in medical image processing/analysis and computer vision. There are three classes of PMLs: (1) neural filters [126, 129] including neural edge enhancers [128, 130], (2) convolution neural networks (NNs) [62, 68, 69, 71, 73, 88, 100] including shift-invariant NNs [153, 171, 172], and (3) massive-training artificial neural networks (MTANNs) [89, 111, 120, 121, 140] including multiple MTANNs [3, 121, 126, 129, 131, 134], a mixture of expert MTANNs [132, 139], a multiresolution MTANN [120], a Laplacian eigenfunction MTANN (LAP-MTANN) [141], and a massive-training support vector regression (MTSVR) [159]. The class of neural filters was used for image-processing tasks such as edge-preserving noise reduction in fluoroscopy, radiographs and other digital pictures [126, 129], edge enhancement from noisy images [128], and enhancement of subjective edges traced by a physician in cardiac images [130]. The class of convolution NNs was applied to classification tasks such as false-positive (FP) reduction in CAD schemes for the detection of lung nodules in chest radiographs (CXRs) [68, 69, 73], FP reduction in CAD schemes for the detection of microcalcifications [71] and masses [100] in mammography, face recognition [62], and character recognition [88]. The class of MTANNs was used for classification, such as FP reduction in CAD schemes for the detection of lung nodules in CXR [134] and thoracic CT [3, 65, 121], distinction between benign and malignant lung nodules in CT [131], and FP reduction in a CAD scheme for polyp detection in CT colonography [132, 139–141, 159]. The MTANNs were also applied to pattern enhancement and suppression such as separation of bones from soft tissue in CXR [19, 89, 120], and enhancement of lung nodules in CT [111]. There are other PML approaches in the literature. An iterative, pixel-based, supervised, statistical classification method called iterated contextual pixel classification has been proposed for segmenting posterior ribs in CXR [74]. A pixel-based, supervised regression filtering technique called filter learning has been proposed for separation ribs from soft tissue in CXR [75].

7.3.2.2 Massive-Training Artificial Neural Network (MTANN)

An MTANN was developed by extension of neural filters to accommodate various pattern-recognition tasks [121]. A two-dimensional (2D) MTANN was first developed for distinguishing a specific opacity from other opacities in 2D images [121]. The 2D MTANN was applied to reduction of FPs in computerized detection of lung nodules on 2D CT images in a slice-by-slice way [3, 65, 121] and in CXR [134], the separation of ribs from soft tissue in CXR [89, 119, 120], and the distinction between benign and malignant lung nodules on 2D CT slices [131]. For processing of three-dimensional (3D) volume data, a 3D MTANN was developed by extending the

7 Computerized Detection of Lesions in Diagnostic Images

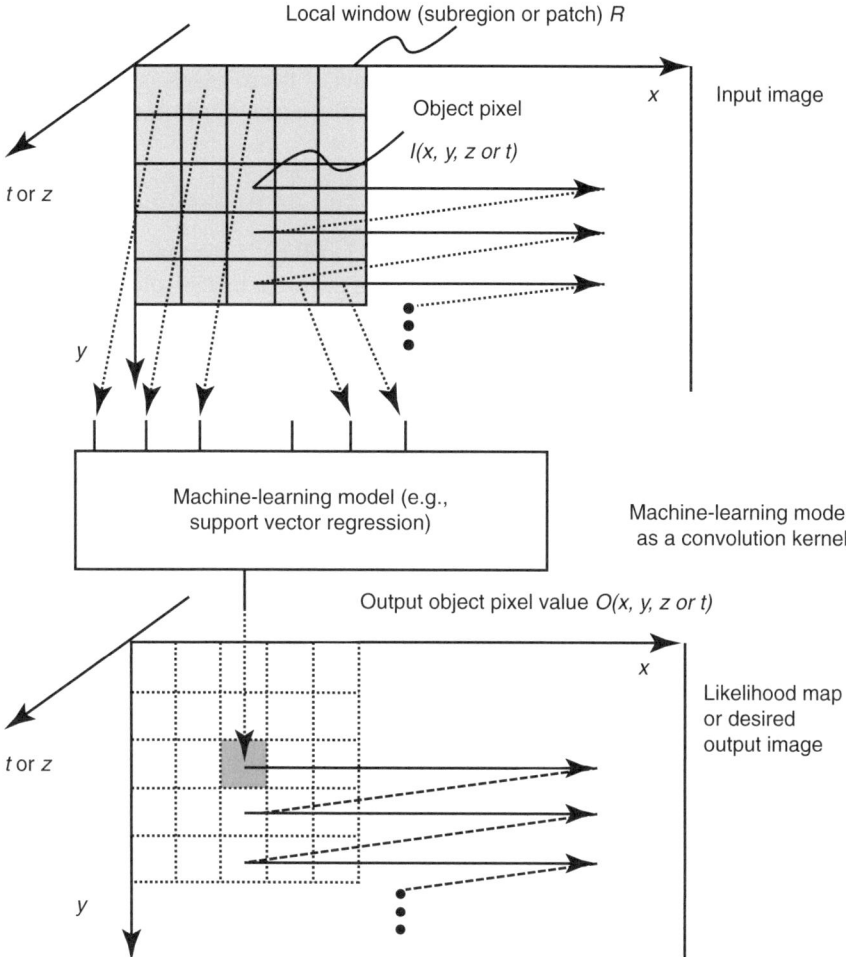

Fig. 7.3 Architecture of an MTANN which is a class of PML

structure of the 2D MTANN, and it was applied to 3D CT colonography data [132, 139–141, 159] in CADe of polyps.

The generalized architecture of an MTANN is shown in Fig. 7.3. An MTANN consists of an ML model (typically a regression model) such as a linear-output ANN regression model [128] and a support vector regression model [159], which is capable of operating on pixel/voxel data directly [128]. The linear-output ANN regression model uses a linear function instead of a sigmoid function as the activation function of the output-layer unit because the characteristics of an ANN were improved significantly with a linear function when applied to the continuous mapping of values in image processing [128]. Note that the activation functions of the hidden layer units are a sigmoid function for nonlinear processing, and those of the input layer units an identity function, as usual. The pixel/voxel values of the input images/volumes may be normalized from 0 to 1. The input to the MTANN consists of pixel/voxel values in a subregion/subvolume (image patch or local window), R, extracted from an input image/volume. The output of the MTANN is a continuous scalar value, which is generally associated with the center voxel in the subregion (image patch), and is represented by

$$O(x,y,z\,\text{or}\,t) = \text{ML}\{I(x-i, y-j, z-k\,\text{or}\,t-k) | (i,j,k) \in R\}, \qquad (7.1)$$

where x, y, and z or t are the coordinate indices, $\text{ML}(\cdot)$ is the output of the ML model, and $I(x, y, z$ or $t)$ is a pixel/voxel value of the input image/volume. A three-layer structure may be selected as the structure of the ANN, because it has been proved that any continuous mapping can be approximated by a three-layer ANN [10, 55]. More layers can be used for efficient solving of a complicated problem. The structure of input units and the number of hidden units in the ANN may be designed by the use of sensitivity-based unit-pruning methods [110, 125]. Other ML models such as support vector regression [145, 146] can be used as a core part of the MTANN. ML regression models rather than ML classification models would be suited for the MTANN framework, because the output of the MTANN are continuous scalar values (as opposed to nominal categories or classes, e.g., 0 or 1). The entire output image/volume is obtained by scanning with the input subvolume (local window) of the MTANN on the entire input image/volume. The input subregion/subvolume and the scanning with the MTANN can be analogous to the kernel of a convolution filter and the convolutional operation of the filter, respectively.

The MTANN is trained with input images/volumes and the corresponding "teaching" (designed) images/volumes for enhancement of a specific pattern and suppression of other patterns in images/volumes. The "teaching" images/volumes are ideal or desired images for the corresponding input images/volumes. For enhancement of lesions and suppression of nonlesions, the teaching volume contains a map for the "likelihood of being lesions," represented by

$$T(x,y,z\,\text{or}\,t) = \begin{cases} \text{a certain distribution} & \text{for a lesion} \\ 0 & \text{otherwise.} \end{cases} \qquad (7.2)$$

To enrich the training samples, a training region, R_T, extracted from the input images is divided pixel by pixel into a large number of overlapping subregions. Single pixels are extracted from the corresponding teaching images as teaching values. The MTANN is massively trained by the use of each of a large number of input subregions (image patches) together with each of the corresponding teaching single pixels, hence the term "massive-training ANN." The error to be minimized by training of the MTANN is represented by

$$E = \frac{1}{P} \sum_c \sum_{(x,y,z\,\text{or}\,t) \in R_T} \{T_c(x,y,z\,\text{or}\,t) - O_c(x,y,z\,\text{or}\,t)\}^2, \qquad (7.3)$$

where c is a training case number, O_c is the output of the MTANN for the cth case, T_c is the teaching value for the MTANN for the cth case, and P is the number of total training voxels in the training region for the MTANN, R_T. The expert 3D MTANN is trained by a linear-output back-propagation (BP) algorithm [128] which was derived for the linear-output ANN model by the use of the generalized delta rule [98]. After training, the MTANN is expected to output the highest value when a lesion is located at the center of the subregion of the MTANN, a lower value as the distance from the subregion center increases, and zero when the input subregion contains a nonlesion.

7.3.3 Difference Between PML and Feature-Based ML (Classifiers)

One of the two major differences between PMLs and ordinary classifiers (i.e., feature-based ML or segmented-object-based ML) is the input information. Ordinary classifiers use features extracted from a segmented object in a given image, whereas PMLs use pixel values in an image patch in a given image as the input information. Although the input information to PMLs can be features (see addition of features to the input information to neural filters in [129], for example), these features are obtained from an image patch pixel by pixel (as opposed to ones from a segmented object or by object). In other words, features for PMLs are features at each pixel in a given image, whereas features for ordinary classifiers are features from a segmented object. In that sense, feature-based classifiers can be referred to as segmented-object-based classifiers. Because PMLs use pixel/voxel values in image patches in images directly instead of features calculated from segmented objects as the input information, segmentation or feature extraction from the segmentation results is not required. Although the development of segmentation techniques has been studied for a long time, segmentation of objects is still challenging, especially for complicated objects, subtle objects, and objects in a complex background. Thus, segmentation errors may occur for such complicated objects. Because with PMLs, errors caused by inaccurate segmentation and inaccurate feature calculation from the segmentation results can be avoided, the performance of PMLs can be higher than that of ordinary classifiers for some cases, such as complicated objects.

The other major difference between PMLs and ordinary classifiers is the output information. The output information from ordinary classifiers, convolution NNs, and the perceptron used for character recognition is nominal class labels such as normal or abnormal (e.g., 0 or 1), whereas that from neural filters, MTANNs, and shift-invariant NNs is pixels or images, namely, continuous values. With the scoring method in MTANNs, output images of the MTANNs are converted to likelihood scores for distinguishing among classes, which allow MTANNs to do classification. In addition to classification, MTANNs can perform pattern enhancement and suppression as well as object detection, whereas the other PMLs cannot.

7.4 CADe in Thoracic Imaging

7.4.1 Thoracic Imaging for Lung Cancer Detection

Lung cancer continues to rank as the leading cause of cancer deaths in the United States and in other countries such as Japan. Because CT is more sensitive than chest radiography in the detection of small nodules and of lung carcinoma at an early stage [52, 60, 81, 105], lung cancer screening programs are being investigated in the United States [53, 142], Japan [60, 105], and other countries with low-dose (LD) CT as the screening modality. Evidence suggests that early detection of lung cancer may allow more timely therapeutic intervention for patients [51, 105]. Helical CT, however, generates a large number of images that must be interpreted by radiologists/physicians. This may lead to "information overload" for the radiologists/physicians. Furthermore, they may miss some cancers during their interpretation of CT images [46, 66]. Therefore, a CADe scheme for the detection of lung nodules in CT images has been investigated as a tool for lung cancer screening.

7.4.2 CADe of Lung Nodules in Thoracic CT

7.4.2.1 Overview

In 1994, Giger et al. [38] developed a CADe scheme for the detection of lung nodules in CT based on comparison of geometric features. They applied their CADe scheme to a database of thick-slice diagnostic CT scans. In 1999, Armato et al. [5, 6] extended the method to include 3D feature analysis, a rule-based scheme, and LDA for classification. They tested their CADe scheme with a database of thick-slice (10 mm) diagnostic CT scans. They achieved a sensitivity of 70 % with 42.2 FPs per case in a leave-one-out cross-validation test. Gurcan et al. [45] employed a similar approach, i.e., a rule-based scheme based on 2D and 3D features, followed by LDA for classification. They achieved a sensitivity of 84 % with 74.4 FPs per case for a database of thick-slice (2.5–5 mm, mostly 5 mm) diagnostic CT scans in a leave-one-out test. Lee et al. [63] employed a simpler approach which is a rule-based scheme based on 13 features for classification. They achieved a sensitivity of 72 % with 30.6 FPs per case for a database of thick-slice (10 mm) diagnostic CT scans.

Suzuki et al. [121] developed a PML technique called an MTANN for reduction of a single source of FPs and a multiple MTANN scheme for reduction of multiple sources of FPs that had not been removed by LDA. They achieved a sensitivity of 80.3 % with 4.8 FPs per case for a database of thick-slice (10 mm) screening LDCT scans of 63 patients with 71 nodules with solid, part-solid, and nonsolid patterns, including 66 cancers in a validation test. This MTANN approach did not require a large number of training cases: the MTANN was able to be trained with ten positive and ten negative cases [17, 99, 123], whereas feature-based classifiers generally require 400–800 training cases [17, 99, 123]. Arimura et al. [3] employed a rule-based scheme followed by LDA or by the MTANN [121] for classification. They tested their scheme with a database of 106 thick-slice (10 mm) screening LDCT scans of 73 patients with 109 cancers, and they achieved a sensitivity of 83 % with 5.8 FPs per case in a validation test (or a leave-one-patient-out test for LDA). Farag et al. [32] developed a template-modeling approach that uses level sets for classification. They achieved a sensitivity of 93.3 % with an FP rate of 3.4 % for a database of thin-slice screening LDCT scans of 16 patients with 119 nodules and 34 normal patients. Ge et al. [36] incorporated 3D-gradient field descriptors and ellipsoid features in LDA for classification. They employed Wilks' lambda stepwise feature selection for selecting features before the LDA classification. They achieved a sensitivity of 80 % with 14.7 FPs per case for a database of 82 thin-slice CT scans of 56 patients with 116 solid nodules in a leave-one-patient-out test. Matsumoto et al. [79] employed LDA with eight features for classification. They achieved a sensitivity of 90 % with 64.1 FPs per case for a database of thick-slice diagnostic CT scans of five patients with 50 nodules in a leave-one-out test.

Yuan et al. [170] tested a commercially available CADe system (ImageChecker CT, LN-1000, by R2 Technology, Sunnyvale, CA; Hologic now). They achieved a sensitivity of 73 % with 3.2 FPs per case for a database of thin-slice (1.25 mm) CT scans of 150 patients with 628 nodules in an independent test. Pu et al. [93] developed a scoring method based on the similarity distance of medial axis-like shapes for classification. They achieved a sensitivity of 81.5 % with 6.5 FPs per case for a database of thin-slice screening CT scans of 52 patients with 184 nodules, including 16 nonsolid nodules. Retico et al. [94] used a voxel-based neural approach (i.e., a class of the MTANN approach) with pixel values in a subvolume as input for classification. They obtained sensitivities of 80–85 % with 10–13 FPs per case for a database of thin-slice screening CT scans of 39 patients with 102 nodules. Ye et al. [163] used a rule-based scheme followed by a weighted SVM for classification. They achieved a sensitivity of 90.2 % with 8.2 FPs per case for a database of thin-slice screening CT scans of 54 patients with 118 nodules including 17 nonsolid nodules in an independent test. Golosio et al. [44] used a fixed-topology ANN for classification, and they evaluated their CADe scheme with a publicly available database from the Lung Image Database Consortium (LIDC) [8]. They achieved a sensitivity of 79 % with four FPs per case for a database of thin-slice CT scans of 83 patients with 148 nodules that one radiologist detected from an LIDC database in an independent test.

Murphy et al. [86] used a k-nearest-neighbor classifier with features selected from 135 features for classification. They achieved a sensitivity of 80 with 4.2 FPs per case for a large database of thin-slice screening CT scans of 813 patients with 1,525 nodules in an independent test. Tan et al. [143] developed a feature-selective classifier based on a genetic algorithm and ANNs for classification. They achieved a sensitivity of 87.5 % with four FPs per case for a database of thin-slice CT scans of 125 patients with 80 nodules that four radiologists agreed from the LIDC database in an independent test. Messay et al. [80] developed a sequential forward selection process for selecting the optimum features for LDA and quadratic discriminant analysis (QDA). They obtained a sensitivity of 83 % with three FPs per case for a database of thin-slice CT scans of 84 patients with 143 nodules from the LIDC

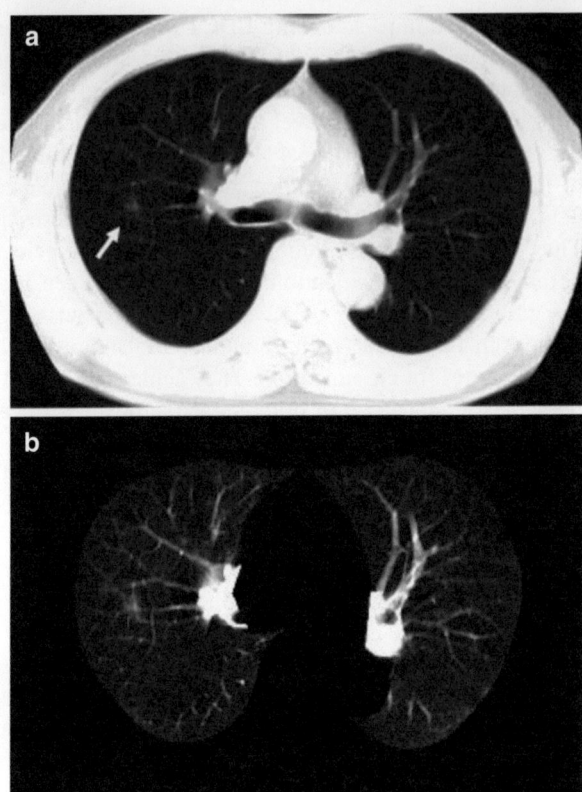

Fig. 7.4 (a) Axial slice of a CT scan of the lungs with a lung cancer (indicated by an *arrow*) and (b) a lung segmentation result

database in a sevenfold cross-validation test. Riccardi et al. [95] used a heuristic approach based on geometric features, followed by an SVM for classification. They achieved a sensitivity of 71 % with 6.5 FPs per case for a database of thin-slice CT scans of 154 patients with 117 nodules that four radiologists agreed on from the LIDC database in a twofold cross-validation test.

Thus, various approaches have been proposed for CADe schemes for lung nodules in CT. Sensitivities for the detection of lung nodules in CT range from 70 to 95 %, with from a few to 70 FPs per case. Major sources of FPs are various-sized lung vessels. Major sources of false negatives are ground-glass nodules, nodules attached to vessels, and nodules attached to the lung wall (i.e., juxtapleural nodules). Ground-glass nodules are difficult to detect, because they are subtle, are of low contrast, and have ill-defined boundaries. The MTANN approach was able to enhance and thus detect ground-glass nodules [121]. The cause of false negatives due to vessel-attached nodules and juxtapleural nodules is mis-segmentation and thus inaccurate feature calculation. Because the MTANN approach does not require segmentation or feature calculation, it was able to detect such nodules [121].

7 Computerized Detection of Lesions in Diagnostic Images

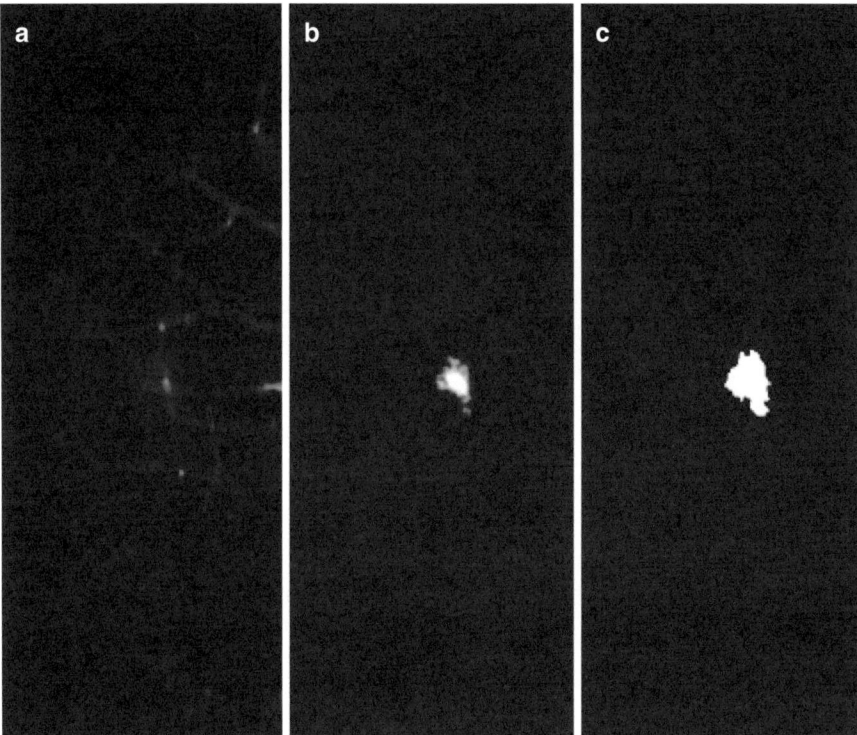

Fig. 7.5 Lesion enhancement by means of a supervised MTANN lesion-enhancement filter. (**a**) Original axial CT slice with a lung nodule. (**b**) Output image of the trained MTANN nodule-enhancement filter. In the output image (**b**), the lung nodule in the original CT image (**a**) is enhanced, whereas normal structures such as lung vessels are suppressed substantially. (**c**) Detection and segmentation of the nodule by using thresholding followed by removal of small regions

7.4.2.2 Illustration of a CADe Scheme

Figure 7.4a illustrates an axial slice of a CT scan of the lungs with a lung cancer. The lung cancer on the CT image is the target that we want to detect with a CADe scheme. As shown in the flowchart in Fig. 7.1, the first step in a CADe scheme is segmentation of the organ of interest, in this case, the lungs. For a high-contrast image with a stable gray scale over different patients like the lung CT image, thresholding often works. To avoid missing nodules attached to the lung walls, mathematical morphology operations are often performed. Figure 7.4b illustrates lung segmentation by simple thresholding followed by mathematical morphology filtering.

To improve the performance of CADe schemes, an optional step of enhancement of lesions is sometimes employed. Suzuki [111] developed a supervised "lesion enhancement" filter based on an MTANN for enhancing lesions and suppressing nonlesions in medical images. Figure 7.5b illustrates the enhancement of a lung nodule in a CT image by means of a trained MTANN lesion-enhancement filter for the original axial CT slice shown in Fig. 7.5a. In the output image, the lung nodule

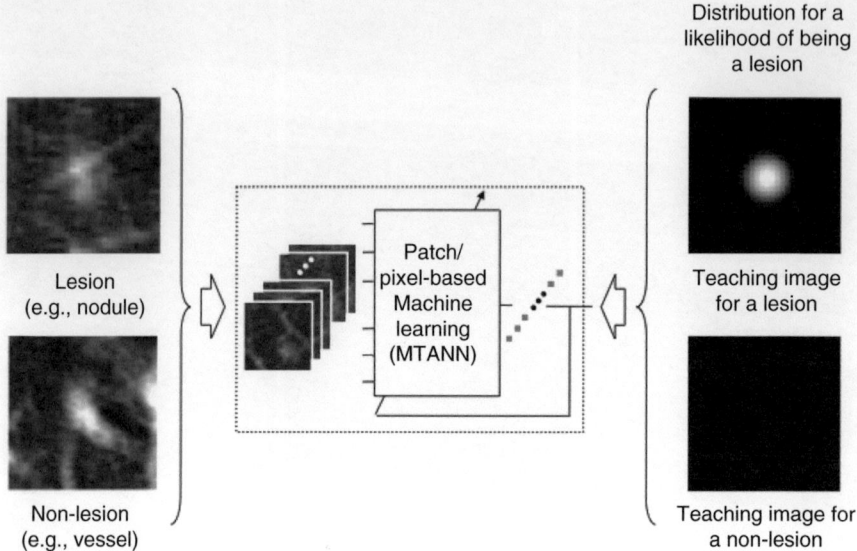

Fig. 7.6 Architecture of an MTANN for FP reduction. The teaching image for a lesion contains a Gaussian distribution; that for a nonlesion contains zero (*completely dark*). After the training, the MTANN expects to enhance lesions and suppress nonlesions

in the original CT image is enhanced, while normal structures such as lung vessels are suppressed substantially. Figure 7.5c shows the detection and segmentation result for the lung nodule by using simple thresholding followed by removal of small regions. After thresholding, connected-component labeling [49, 50, 127] was performed to calculate the area of each isolated region (i.e., connected component). By removing small regions, the lung nodule was detected correctly with no FP detection. By the use of the MTANN lesion-enhancement filter, the performance of the initial nodule candidate detection step was substantially improved from a 96 % sensitivity with 19.3 FPs per section to a 97 % sensitivity with 6.7 FPs per section.

Morphologic and gray-level-based features such as contract, area, and circularity were calculated from the segmented nodule candidates. The extracted features were then inputted to a classifier (feature-based ML) to classify the candidates into nodules or non-nodules. At this stage, there were a lot of FPs (non-nodules) that the classifier had not been able to distinguish from nodules.

To reduce remaining FPs, Suzuki et al. developed an FP reduction technique based on MTANNs [121]. The architecture of the MTANN for FP reduction is shown in Fig. 7.6. For enhancement of nodules (i.e., true positives) and suppression of non-nodules (i.e., FPs) on CT images, the teaching image contains a distribution of values that represent the "likelihood of being a nodule." For example, the teaching volume contains a 3D Gaussian distribution with standard deviation σ_T for a lesion and zero (i.e., completely dark) for nonlesions, as illustrated in Fig. 7.6. This distribution represents the "likelihood of being a lesion":

Fig. 7.7 Scoring method for combining pixel-based output responses from the trained MTANN into a single score for each ROI

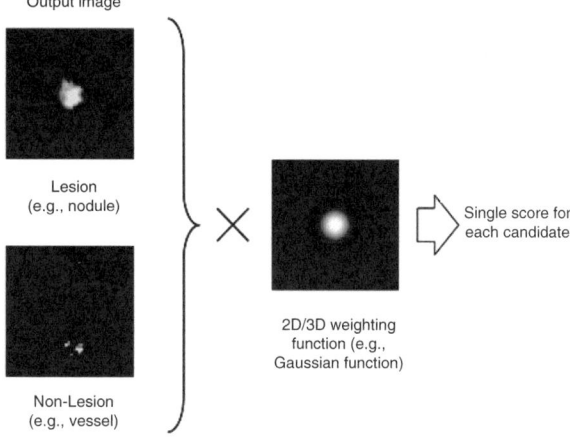

$$T(x,y,z\,or\,t) = \begin{cases} \dfrac{1}{\sqrt{2\pi}\sigma_T} \exp\left\{-\dfrac{\left(x^2 + y^2 + z^2\,or\,t^2\right)}{2\sigma_T^2}\right\} & \text{for a lesion} \\ 0 & \text{otherwise.} \end{cases} \quad (7.4)$$

A scoring method is used for combining of output voxels from the trained MTANNs, as illustrated in Fig. 7.7. A score for a given region-of-interest (ROI) from the MTANN is defined as

$$S = \sum_{(x,y,z\,or\,t) \in R_E} f_W(x,y,z\,or\,t) \times O(x,y,z\,or\,t), \quad (7.5)$$

where

$$f_W(x,y,z\,or\,t) = f_G(x,y,z\,or\,t;\sigma) = \dfrac{1}{\sqrt{2\pi}\sigma} e^{-\dfrac{x^2 + y^2 + z^2\,or\,t^2}{2\sigma^2}} \quad (7.6)$$

is a 3D Gaussian weighting function with standard deviation σ, and with its center corresponding to the center of the volume for evaluation, R_E, and O is the output image of the trained MTANN, where its center corresponds to the center of R_E. The use of the 3D Gaussian weighting function allows us to combine the responses (outputs) of a trained MTANN as a 3D distribution. A 3D Gaussian function is used for scoring, because the output of a trained MTANN is expected to be similar to the 3D Gaussian distribution used in the teaching images. This score represents the weighted sum of the estimates for the likelihood that the ROI (lesion candidate) contains a lesion near the center, i.e., a higher score would indicate a lesion, and a

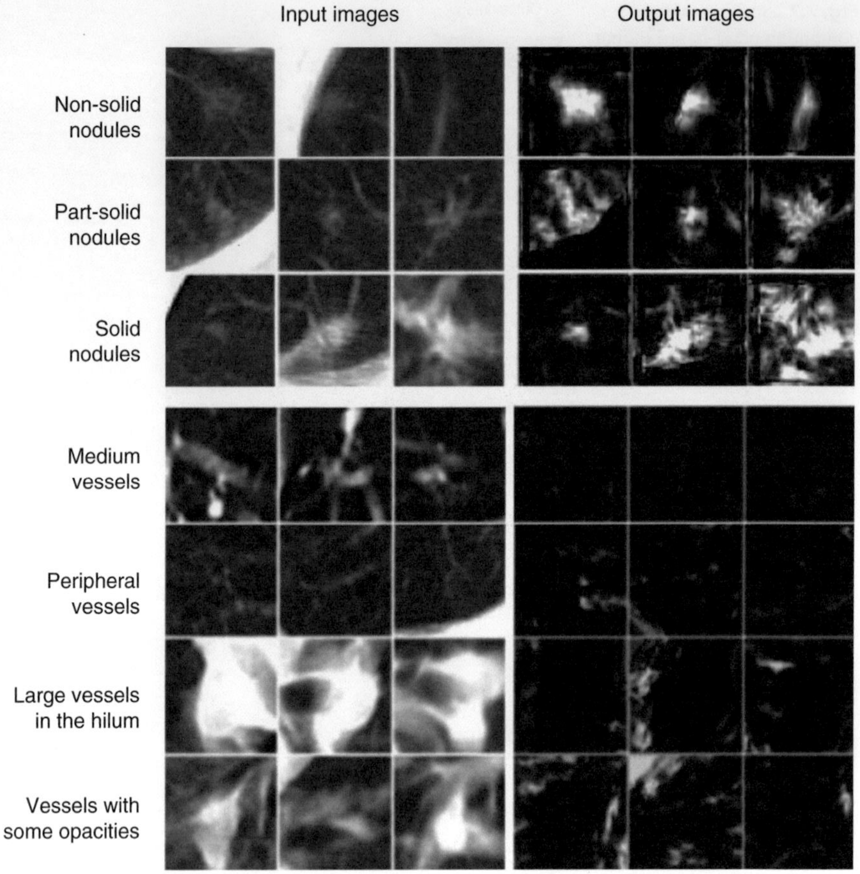

Fig. 7.8 Enhancement of lung nodules and suppression of FPs (i.e., lung vessels) by the use of MTANNs for FP reduction. Once lung nodules are enhanced, and FPs suppressed, FPs can be distinguished from lung nodules by the use of scores obtained from the output images

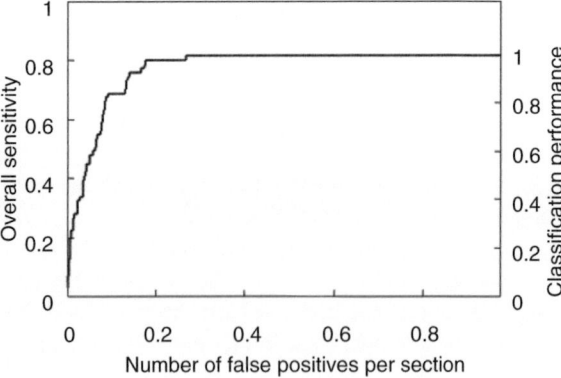

Fig. 7.9 FROC curve indicating the performance of the FP reduction by MTANNs in a CADe scheme for the detection of lung nodules in CT. With the trained MTANNs, FPs were removed without any removal of true positives

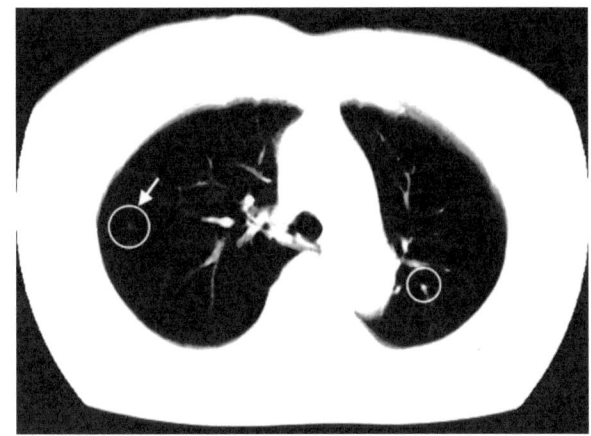

Fig. 7.10 CADe outputs (indicated by *circles*) on an axial CT slice of the lungs. A lung nodule (indicated by an *arrow*) was detected correctly by a CADe scheme with one FP detection (branch of lung vessels) on the right

lower score would indicate a nonlesion. Thresholding is then performed on the scores for distinction between lesions and nonlesions.

The MTANNs were trained to enhance lung nodules and suppress various types of FPs (i.e., non-nodules) such as lung vessels. Figure 7.8 shows the results of the enhancement of various lung nodules such as nonsolid (ground-glass), part-solid (mixed-ground-glass), and solid nodules (a) and those of the suppression of various-sized lung vessels (b). Figure 7.9 shows a free-response receiver operating characteristic (FROC) curve [13] indicating the performance of the trained MTANNs in the CADe scheme. With the MTANNs, the specificity of the CADe scheme was improved from 0.98 to 0.18 FPs per case without sacrificing the original sensitivity of 80.3 %.

Figure 7.10 shows an example of CADe outputs on a CT image of the lungs. A CADe scheme detected a lung nodule correctly with one FP which was a branch of the lung vessels.

7.4.3 CADe of Lung Nodules in CXR

Chest radiographs (CXRs) is the most commonly used imaging examination for chest diseases because they are the most cost-effective, routinely available, and dose-effective diagnostic examination [85, 173]. Because CXRs are widely used, improvements in the detection of lung nodules in CXRs could have a significant impact on early detection of lung cancer. Studies have shown that, however, 30 % of nodules in CXRs were missed by radiologists in which nodules were visible in retrospect. Therefore, CADe schemes [40, 144] for nodules in CXRs have been investigated for assisting radiologists in improving their sensitivity. A wide variety of approaches in CADe schemes for nodule detection in CXRs have been developed. Giger et al. developed a difference-image technique to reduce complex anatomic background structures while enhancing nodule-like structures for initial nodule candidate detection [37, 40]. Lo et al. used a technique similar to the difference-image technique to create nodule-enhanced images, which were then processed by a feature-extraction technique based on edge detection, gray-level thresholding, and sphere profile matching [70, 72]. Then a convolution neural network was employed in the classification step. Penedo et al. then improved the performance of the scheme by incorporating two-level ANNs that employed cross-correlation teaching images and input images in the curvature peak space [91]. Coppini et al. developed a CADe scheme based on biologically inspired ANNs with fuzzy coding [22]. Shiraishi et al. incorporated a localized searching method based on anatomical classification and automated techniques for the parameter setting of three types of ANNs into a CADe scheme [103].

Studies showed that 82–95 % of the missed lung cancers in CXR were partly obscured by overlying bones such as ribs and/or a clavicle [9, 101]. To address this issue, Suzuki et al. [118, 120] developed a multiresolution MTANN for separation of bones such as ribs and clavicles from soft tissue in CXRs. They employed multi-resolution decomposition/composition techniques [2, 106] to decompose an original high-resolution image into different-resolution images. First, one obtains a medium-resolution image $g_M(x, y)$ from an original high-resolution image $g_H(x, y)$ by performing downsampling with averaging, i.e., four pixels in the original image are replaced by a pixel having the mean value for the four pixel values, represented by

$$g_M(x,y) = \frac{1}{4} \sum_{i,j \in R_{22}} g_H(2x-i, 2y-j), \tag{7.7}$$

where R_{22} is a 2-by-2-pixel region. The medium-resolution image is enlarged by upsampling with pixel substitution, i.e., a pixel in the medium-resolution image is replaced by four pixels with the same pixel value, as follows:

$$g_M^U(x,y) = g_M(x/2, y/2). \tag{7.8}$$

Then, a high-resolution difference image $d_H(x, y)$ is obtained by subtraction of the enlarged medium-resolution image from the high-resolution image, represented by

$$d_H(x,y) = g_H(x,y) - g_M^U(x,y). \tag{7.9}$$

These procedures are performed repeatedly, producing further lower-resolution images. Thus, multiresolution images having various frequencies are obtained by the use of the multiresolution decomposition technique.

An important property of this technique is that exactly the same original-resolution image $g_H(x, y)$ can be obtained from the multiresolution images, $d_H(x, y)$ and $g_M(x, y)$, by performing the inverse procedures, called a multiresolution composition technique, as follows:

$$g_H(x,y) = g_M(x/2, y/2) + d_H(x,y). \tag{7.10}$$

Therefore, we can process multiresolution images independently instead of processing original high-resolution images directly; i.e., with these techniques, the

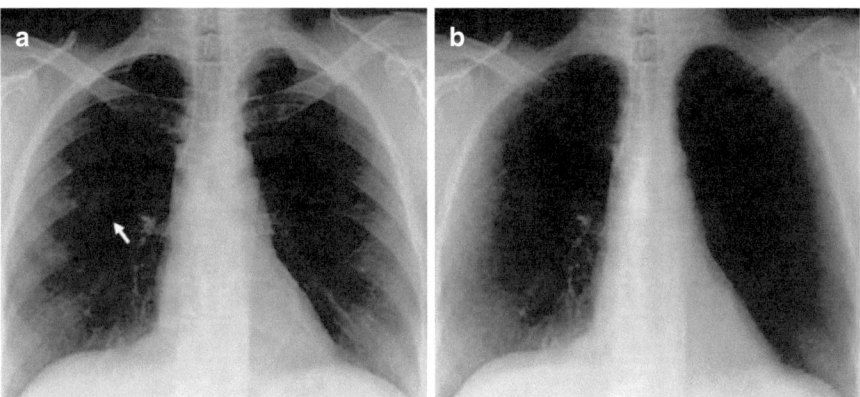

Fig. 7.11 Suppression of bones such as ribs and clavicles from soft tissue in CXR. (**a**) Original CXR with a lung nodule (indicated by an *arrow*). (**b**) Bone suppression imaging (or "virtual" dual-energy radiography) result by means of a multiresolution MTANN

processed original high-resolution image can be obtained by composing of the processed multiresolution images. Each of multiple MTANNs only needs to support a limited spatial frequency rage in each resolution image instead of the entire spatial frequencies in the original image.

First, input CXRs and the corresponding teaching bone images are decomposed into sets of different-resolution images, and then these sets of images are used for training three MTANNs in the multiresolution MTANN. Each MTANN is an expert for a certain resolution, i.e., a low-resolution MTANN is in charge of low-frequency components of ribs, a medium-resolution MTANN is for medium-frequency components, and a high-resolution MTANN for high-frequency components. Each resolution MTANN is trained independently with the corresponding resolution images. After training, the MTANNs produce different-resolution images, and then these images are composed to provide a complete high-resolution image by the use of the multiresolution composition technique. The complete high-resolution image is expected to be similar to the teaching bone image; therefore, the multiresolution MTANN would provide a "bone-image-like" image in which ribs and clavicles are separated from soft tissues. Chen and Suzuki [19] improved the performance of the MTANN "virtual" dual-energy chest radiography by means of anatomically specific multiple MTANNs. Figure 7.11 illustrates suppression of bones from soft tissue in CXR by using the MTANNs [19].

Suzuki et al. developed an FP reduction technique based on MTANNs in a CADe scheme of nodules in CXR. They removed 68 % of the FPs that had not removed by feature-based ML, and the performance of the CADe scheme was substantially improved from 4.5 to 1.4 FPs per image, while maintaining the original sensitivity of 81.3 %.

Chen et al. developed a CADe scheme of lung nodules in CXRs based on feature-based SVM [21]. They improved the performance by using the MTANN virtual dual-energy imaging [18]. They improved the performance substantially from the original sensitivity of 79 % with five FPs per image to a sensitivity of 85 % with the

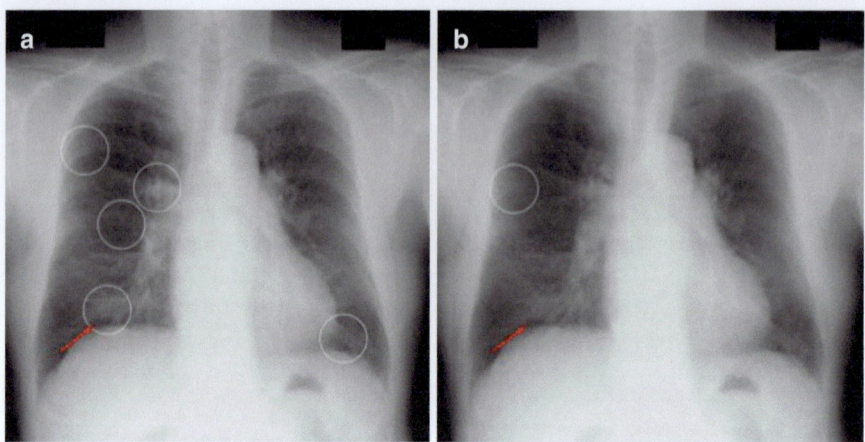

Fig. 7.12 Illustration of the improvement in nodule detection by CADe scheme with our VDE technology. CADe marks are indicated by *circles*. (**a**) False negatives (*arrow*) and false positives of the original CADe scheme. (**b**) True positives (*arrow*) and false positives of the VDE-based CADe scheme with the VDE technology

same FP rate. Figure 7.12 illustrates computer outputs from their CADe scheme without and with the MTANN virtual dual-energy imaging [18].

They compared the performance of their CADe scheme with that of an FDA-approved CADe product with the same database. Their CADe scheme achieved a sensitivity of 81 % with 2.0 FPs per image, whereas the FDA-approved product achieved a substantially inferior performance that was a sensitivity of 67 % at the same FP rate. They also compared the performance with other CADe schemes in literature by using the same publicly available database of the JSRT [154]. Wei et al. reported that their CAD scheme achieved a sensitivity of 80 % with 5.4 FPs per image. Hardie et al. reported that their scheme marked 80 % of nodules with five FPs per image [47]. The performance of Chen Suzuki CADe scheme was substantially higher than that of Hardie's CADe scheme, i.e., it achieved a sensitivity of 78 % at an FP rate of 2.0 per image, whereas Hardie's CADe scheme achieved a sensitivity of 63 % at the FP rate.

7.5 CADe in Colonic Imaging

7.5.1 Colonic Imaging for Colorectal Cancer Detection

Colorectal cancer is the second leading cause of cancer deaths in the United States [56]. Evidence suggests that early detection and removal of polyps (i.e., precursors of colorectal cancer) can reduce the incidence of colorectal cancer [23, 155]. Consequently, the American Cancer Society (ACS) recommends that an individual who is at average risk for developing colorectal cancer, beginning at age 50, should have colorectal cancer screening with examinations including optical colonoscopy and CTC. CTC (or

7 Computerized Detection of Lesions in Diagnostic Images

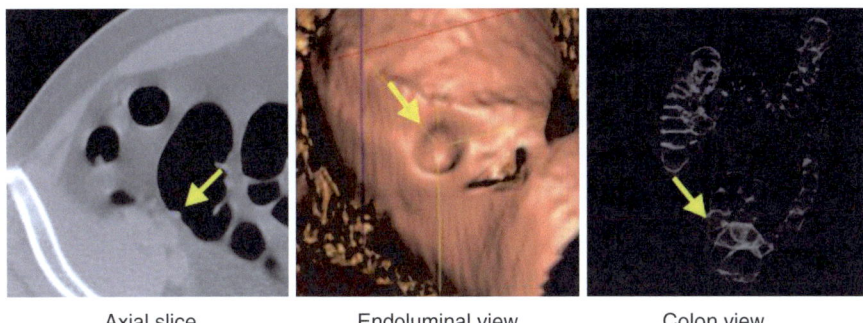

Axial slice   Endoluminal view   Colon view

Fig. 7.13 CADe output (indicated by an *arrow*) for the detection of polyps in an axial slice, an endoluminal view, and a 3D colon view in CTC. A polyp (indicated by an *arrow*) was detected correctly by a CADe scheme

virtual colonoscopy) is a technique for detecting colorectal neoplasms by the use of CT scans of the colon [78]. The diagnostic performance of CTC in detecting polyps, however, varies by experience of radiologists, hospitals, and protocols [33]. Therefore, CADe of polyps has been investigated to address this issue with CTC [122, 164, 165].

7.5.2 Overview of CADe of Polyps in CTC

CADe has the potential to (a) increase radiologists' sensitivity in the detection of polyps, (b) decrease reader variability, and (c) reduce radiologists' reading time when CADe is used during the primary read [164, 165]. A number of researchers have developed CADe schemes for the detection of polyps in CTC [61, 90, 108, 109, 166–168]. Figure 7.13 shows an example of a CADe output for the detection of polyps in CTC. A CADe scheme detected the polyp correctly.

In 2000, Summers et al. [107] developed a CADe scheme for the detection of polyps in CTC based on curvature analysis. In 2001, Yoshida and Nappi [167] developed a CADe scheme based on curvature analysis called a shape index. In 2001, Gokturk et al. [43] employed an SVM with histogram input that is used as a shape signature for classification. Näppi and Yoshida [87] developed a CADe scheme based on LDA or QDA with 54 volumetric features (nine statistics of six features). Acar et al. [1] used edge-displacement fields and QDA for classification. Jerebko et al. [59] used a multilayer perceptron to classify polyp candidates in their CADe scheme and improved the performance by incorporating a committee of multilayer perceptrons [57] and a committee of SVMs [58]. Wang et al. [151] developed a classification method based on LDA with internal features (geometric, morphologic, and textural) of polyps.

Suzuki et al. [140] developed a PML technique called a 3D MTANN by extending the structure of a 2D MTANN [121] to process 3D volume data in CTC. Their CADe scheme was based on a Bayesian ANN with texture and geometric features, followed by 3D MTANNs. They removed FPs due to rectal tubes by using a single

3D MTANN [140] and multiple sources of FPs by developing and using a mixture of expert 3D MTANNs [139].

Li et al. [67] developed a classification method based on an SVM classifier with wavelet-based features. Wang et al. [150] improved the SVM performance by using nonlinear dimensionality reduction (i.e., a diffusion map and locally linear embedding). Yao et al. [162] employed a topographic height map for calculating features for an SVM classifier.

Suzuki et al. [132] tested a CADe scheme based on a Bayesian ANN and MTANNs. They used CTC data of 24 patients, including 23 polyps (6–25 mm) and a mass (35 mm), that had been "missed" by radiologists [28] in a multicenter clinical trial [96]. They achieved a by-polyp (by-patient) sensitivity of 96.4 % (100 %) with 1.1 FPs/patient in a leave-one-lesion-out cross-validation test of the classification part. Suzuki et al. [137, 141] also improved the efficiency of the MTANN approach by incorporating principal-component analysis-based and Laplacian eigenmap-based dimension reduction techniques. Xu and Suzuki [159] showed that other nonlinear regression models such as support vector and nonlinear Gaussian process regression models instead of the ANN regression model could be used as the core model in the MTANN framework.

Zhou et al. [174] developed projection features for an SVM classifier. Wang et al. [149] improved the performance of a CAD scheme by adding statistical curvature features in multiple-kernel learning. They obtained a sensitivity of 83 % with five FPs/patient in a leave-one-out cross-validation test of the classification part.

Thus, various ML approaches have been proposed in CADe schemes for polyps in CTC, which include LDA, QDA, an SVM, ANNs, and a Bayesian ANN.

Existing CADe schemes tend to miss superficially elevated neoplasms (often called flat lesions) [76, 77]. Suzuki et al. developed a CADe scheme for the detection of superficially elevated neoplasms [133]. Detection of superficially elevated neoplasms is very important, because they are histologically aggressive and because they are often missed by radiologists in CTC as well as by gastroenterologists in optical colonoscopy.

7.6 Summary

In this chapter, ML techniques used in CADe schemes for the detection of lung nodules in CXR and thoracic CT and those for the detection of polyps in CTC are described. There are two classes of ML techniques: (1) feature-based (segmented-object-based) ML (classifiers) and (2) patch-/pixel-based ML. Feature-based ML, including LDA, QDA, an ANN, a Bayesian ANN, and an SVM, are mainly used in the 5th step of classification of lesion candidates and the 6th step of FP reduction in a CADe scheme, whereas PML is used mainly in the 6th step of FP reduction, but it can be used in the 1st step of organ segmentation [14] and the

2nd step of lesion enhancement [111]. Thus, ML techniques are indispensable steps in CADe schemes.

Acknowledgments This work would not have been possible without the help and support of countless people. The author is grateful to all members in the Suzuki laboratory, i.e., postdoctoral scholars, computer scientists, visiting scholars/professors, medical students, graduate/undergraduate students, research technicians, research volunteers, and support staff, in the Department of Radiology at the University of Chicago, for their invaluable assistance in the studies, to colleagues and collaborators for their valuable suggestions. CAD technologies, PML technologies, the bone separation technology, and their source code developed at the University of Chicago have been licensed to companies including R2 Technology (Hologic), Riverain Medical (Riverain Technologies), Deus Technology, Median Technologies, AlgoMedica, Mitsubishi Space Software, General Electric, and Toshiba.

References

1. Acar B, Beaulieu CF, Gokturk SB, Tomasi C, Paik DS, Jeffrey Jr RB, Yee J, Napel S. Edge displacement field-based classification for improved detection of polyps in CT colonography. IEEE Trans Med Imaging. 2002;21:1461–7.
2. Akansu AN, Haddad RA. Multiresolution signal decomposition. Boston: Academic Press; 1992.
3. Arimura H, Katsuragawa S, Suzuki K, Li F, Shiraishi J, Sone S, Doi K. Computerized scheme for automated detection of lung nodules in low-dose computed tomography images for lung cancer screening. Acad Radiol. 2004;11:617–29.
4. Arimura H, Li Q, Korogi Y, Hirai T, Katsuragawa S, Yamashita Y, Tsuchiya K, Doi K. Computerized detection of intracranial aneurysms for three-dimensional MR angiography: feature extraction of small protrusions based on a shape-based difference image technique. Med Phys. 2006;33:394–401.
5. Armato 3rd SG, Giger ML, MacMahon H. Automated detection of lung nodules in CT scans: preliminary results. Med Phys. 2001;28:1552–61.
6. Armato 3rd SG, Giger ML, Moran CJ, Blackburn JT, Doi K, MacMahon H. Computerized detection of pulmonary nodules on CT scans. Radiographics. 1999;19:1303–11.
7. Armato 3rd SG, Li F, Giger ML, MacMahon H, Sone S, Doi K. Lung cancer: performance of automated lung nodule detection applied to cancers missed in a CT screening program. Radiology. 2002;225:685–92.
8. Armato 3rd SG, McLennan G, McNitt-Gray MF, Meyer CR, Yankelevitz D, Aberle DR, Henschke CI, Hoffman EA, Kazerooni EA, MacMahon H, Reeves AP, Croft BY, Clarke LP. Lung image database consortium: developing a resource for the medical imaging research community. Radiology. 2004;232:739–48.
9. Austin JH, Romney BM, Goldsmith LS. Missed bronchogenic carcinoma: radiographic findings in 27 patients with a potentially resectable lesion evident in retrospect. Radiology. 1992;182:115–22.
10. Barron AR. Universal approximation bounds for superpositions of a sigmoidal function. IEEE Trans Info Theory. 1993;39:930–45.
11. Bengio Y, Lamblin P, Popovici D, Larochelle H. Greedy layer-wise training of deep networks. Adv Neural Info Process Syst. 2007;19:153.
12. Bishop CM. Neural networks for pattern recognition. New York: Oxford University Press; 1995.

13. Bunch PC, Hamilton JF, Sanderson GK, Simmons AH. A free-response approach to the measurement and characterization of radiographic-observer performance. J Appl Photogr Eng. 1978;4:166–71.
14. Calabrese D, Zhou K, Liu Y, Suzuki K. Improved segmentation of liver in CT with massive-training artificial neural network (MTANN) liver enhancer. In: Proceedings of IEEE engineering in medicine and biology conference (IEEE EMBC), Osaka; 2013.
15. Chan HP, Doi K, Galhotra S, Vyborny CJ, MacMahon H, Jokich PM. Image feature analysis and computer-aided diagnosis in digital radiography. I. Automated detection of microcalcifications in mammography. Med Phys. 1987;14:538–48.
16. Chan HP, Sahiner B, Helvie MA, Petrick N, Roubidoux MA, Wilson TE, Adler DD, Paramagul C, Newman JS, Sanjay-Gopal S. Improvement of radiologists' characterization of mammographic masses by using computer-aided diagnosis: an ROC study. Radiology. 1999;212:817–27.
17. Chan HP, Sahiner B, Wagner RF, Petrick N. Classifier design for computer-aided diagnosis: effects of finite sample size on the mean performance of classical and neural network classifiers. Med Phys. 1999;26:2654–68.
18. Chen S, Suzuki K. Computerized detection of lung nodules by means of "virtual dual-energy" radiography. IEEE Trans Biomed Eng. 2013;60:369–78. doi:10.1109/TBME.2012.2226583.
19. Chen S, Suzuki K. Separation of bones from chest radiographs by means of anatomically specific multiple massive-training ANNs combined with total variation minimization smoothing. IEEE Trans Med Imaging. 2014;33:246–57. doi:10.1109/TMI.2013.2284016.
20. Chen S, Suzuki K, MacMahon H. A computer-aided diagnostic scheme for lung nodule detection in chest radiographs by means of two-stage nodule-enhancement with support vector classification. Med Phys. 2011;38:1844–58.
21. Chen S, Suzuki K, MacMahon H. Development and evaluation of a computer-aided diagnostic scheme for lung nodule detection in chest radiographs by means of two-stage nodule enhancement with support vector classification. Med Phys. 2011;38:1844–58.
22. Coppini G, Diciotti S, Falchini M, Villari N, Valli G. Neural networks for computer-aided diagnosis: detection of lung nodules in chest radiograms. IEEE Trans Inf Technol Biomed. 2003;7:344–57.
23. Dachman AH. Atlas of virtual colonoscopy. New York: Springer; 2003.
24. Dachman AH, Obuchowski NA, Hoffmeister JW, Hinshaw JL, Frew MI, Winter TC, Van Uitert RL, Periaswamy S, Summers RM, Hillman BJ. Effect of computer-aided detection for CT colonography in a multireader, multicase trial. Radiology. 2010;256:827–35. doi:10.1148/radiol.10091890.
25. Dean JC, Ilvento CC. Improved cancer detection using computer-aided detection with diagnostic and screening mammography: prospective study of 104 cancers. AJR Am J Roentgenol. 2006;187:20–8.
26. Doi K. Current status and future potential of computer-aided diagnosis in medical imaging. Br J Radiol. 2005;78 Spec No 1:S3–19.
27. Doi K. Computer-aided diagnosis in medical imaging: historical review, current status and future potential. Comput Med Imaging Graph. 2007;31:198–211.
28. Doshi T, Rusinak D, Halvorsen RA, Rockey DC, Suzuki K, Dachman AH. CT colonography: false-negative interpretations. Radiology. 2007;244:165–73.
29. Drukker K, Giger ML, Metz CE. Robustness of computerized lesion detection and classification scheme across different breast US platforms. Radiology. 2005;237:834–40.
30. Duda RO, Hart PE, Stork DG. Pattern recognition. 2nd ed. Hoboken: Wiley Interscience; 2001.
31. El-Naqa I, Yang Y, Wernick MN, Galatsanos NP, Nishikawa RM. A support vector machine approach for detection of microcalcifications. IEEE Trans Med Imaging. 2002;21:1552–63.
32. Farag AA, El-Baz A, Gimelfarb G, El-Ghar MA, Eldiasty T. Quantitative nodule detection in low dose chest CT scans: new template modeling and evaluation for CAD system design. Med Image Comput Comput Assist Interv. 2005;8:720–8.
33. Fletcher JG, Booya F, Johnson CD, Ahlquist D. CT colonography: unraveling the twists and turns. Curr Opin Gastroenterol. 2005;21:90–8.

34. Fukunaga K. Introduction to statistical pattern recognition. 2nd ed. San Diego: Academic Press; 1990.
35. Ge J, Sahiner B, Hadjiiski LM, Chan HP, Wei J, Helvie MA, Zhou C. Computer aided detection of clusters of microcalcifications on full field digital mammograms. Med Phys. 2006;33:2975–88.
36. Ge Z, Sahiner B, Chan HP, Hadjiiski LM, Cascade PN, Bogot N, Kazerooni EA, Wei J, Zhou C. Computer-aided detection of lung nodules: false positive reduction using a 3D gradient field method and 3D ellipsoid fitting. Med Phys. 2005;32:2443–54.
37. Giger ML, Ahn N, Doi K, MacMahon H, Metz CE. Computerized detection of pulmonary nodules in digital chest images: use of morphological filters in reducing false-positive detections. Med Phys. 1990;17:861–5.
38. Giger ML, Bae KT, MacMahon H. Computerized detection of pulmonary nodules in computed tomography images. Invest Radiol. 1994;29:459–65.
39. Giger ML, Chan HP, Boone J. Anniversary paper: history and status of CAD and quantitative image analysis: the role of Medical Physics and AAPM. Med Phys. 2008;35:5799–820.
40. Giger ML, Doi K, MacMahon H. Image feature analysis and computer-aided diagnosis in digital radiography. 3. Automated detection of nodules in peripheral lung fields. Med Phys. 1988;15:158–66.
41. Giger ML, Suzuki K. Computer-aided diagnosis (CAD). In: Feng DD, editor. Biomedical information technology. Amsterdam/Boston: Academic Press; 2007. p. 359–74.
42. Gilhuijs KG, Giger ML, Bick U. Computerized analysis of breast lesions in three dimensions using dynamic magnetic-resonance imaging. Med Phys. 1998;25:1647–54.
43. Gokturk SB, Tomasi C, Acar B, Beaulieu CF, Paik DS, Jeffrey Jr RB, Yee J, Napel S. A statistical 3-D pattern processing method for computer-aided detection of polyps in CT colonography. IEEE Trans Med Imaging. 2001;20:1251–60.
44. Golosio B, Masala GL, Piccioli A, Oliva P, Carpinelli M, Cataldo R, Cerello P, De Carlo F, Falaschi F, Fantacci ME, Gargano G, Kasae P, Torsello M. A novel multithreshold method for nodule detection in lung CT. Med Phys. 2009;36:3607–18.
45. Gurcan MN, Sahiner B, Petrick N, Chan HP, Kazerooni EA, Cascade PN, Hadjiiski L. Lung nodule detection on thoracic computed tomography images: preliminary evaluation of a computer-aided diagnosis system. Med Phys. 2002;29:2552–8.
46. Gurney JW. Missed lung cancer at CT: imaging findings in nine patients. Radiology. 1996;199:117–22.
47. Hardie RC, Rogers SK, Wilson T, Rogers A. Performance analysis of a new computer aided detection system for identifying lung nodules on chest radiographs. Med Image Anal. 2008;12:240–58. doi:10.1016/j.media.2007.10.004. S1361-8415(07)00103-X [pii].
48. Haykin S. Neural networks. Upper Saddle River: Prentice Hall; 1998.
49. He L, Chao Y, Suzuki K. A run-based two-scan labeling algorithm. IEEE Trans Image Process. 2008;17:749–56. doi:10.1109/TIP.2008.919369.
50. He L, Chao Y, Suzuki K, Wu K. Fast connected-component labeling. Pattern Recognit. 2009;42:1977–87.
51. Heelan RT, Flehinger BJ, Melamed MR, Zaman MB, Perchick WB, Caravelli JF, Martini N. Non-small-cell lung cancer: results of the New York screening program. Radiology. 1984;151:289–93.
52. Henschke CI, McCauley DI, Yankelevitz DF, Naidich DP, McGuinness G, Miettinen OS, Libby DM, Pasmantier MW, Koizumi J, Altorki NK, Smith JP. Early Lung Cancer Action Project: overall design and findings from baseline screening. Lancet. 1999;354:99–105.
53. Henschke CI, Yankelevitz DF, Naidich DP, McCauley DI, McGuinness G, Libby DM, Smith JP, Pasmantier MW, Miettinen OS. CT screening for lung cancer: suspiciousness of nodules according to size on baseline scans. Radiology. 2004;231:164–8.
54. Hinton G, Osindero S, Teh Y-W. A fast learning algorithm for deep belief nets. Neural Comput. 2006;18:1527–54.
55. Hornik K, Stinchcombe M, White H. Multilayer feedforward networks are universal approximators. Neural Netw. 1989;2:359–66.

56. Jemal A, Murray T, Ward E, Samuels A, Tiwari RC, Ghafoor A, Feuer EJ, Thun MJ. Cancer statistics. Cancer J Clin. 2005;55:10–30.
57. Jerebko AK, Malley JD, Franaszek M, Summers RM. Multiple neural network classification scheme for detection of colonic polyps in CT colonography data sets. Acad Radiol. 2003;10:154–60.
58. Jerebko AK, Malley JD, Franaszek M, Summers RM. Support vector machines committee classification method for computer-aided polyp detection in CT colonography. Acad Radiol. 2005;12:479–86.
59. Jerebko AK, Summers RM, Malley JD, Franaszek M, Johnson CD. Computer-assisted detection of colonic polyps with CT colonography using neural networks and binary classification trees. Med Phys. 2003;30:52–60.
60. Kaneko M, Eguchi K, Ohmatsu H, Kakinuma R, Naruke T, Suemasu K, Moriyama N. Peripheral lung cancer: screening and detection with low-dose spiral CT versus radiography. Radiology. 1996;201:798–802.
61. Kiss G, Van Cleynenbreugel J, Thomeer M, Suetens P, Marchal G. Computer-aided diagnosis in virtual colonography via combination of surface normal and sphere fitting methods. Eur Radiol. 2002;12:77–81.
62. Lawrence S, Giles CL, Tsoi AC, Back AD. Face recognition: a convolutional neural-network approach. IEEE Trans Neural Netw. 1997;8:98–113.
63. Lee Y, Hara T, Fujita H, Itoh S, Ishigaki T. Automated detection of pulmonary nodules in helical CT images based on an improved template-matching technique. IEEE Trans Med Imaging. 2001;20:595–604.
64. Li F, Aoyama M, Shiraishi J, Abe H, Li Q, Suzuki K, Engelmann R, Sone S, Macmahon H, Doi K. Radiologists' performance for differentiating benign from malignant lung nodules on high-resolution CT using computer-estimated likelihood of malignancy. Am J Roentgenol. 2004;183:1209–15.
65. Li F, Arimura H, Suzuki K, Shiraishi J, Li Q, Abe H, Engelmann R, Sone S, MacMahon H, Doi K. Computer-aided detection of peripheral lung cancers missed at CT: ROC analyses without and with localization. Radiology. 2005;237:684–90.
66. Li F, Sone S, Abe H, MacMahon H, Armato 3rd SG, Doi K. Lung cancers missed at low-dose helical CT screening in a general population: comparison of clinical, histopathologic, and imaging findings. Radiology. 2002;225:673–83.
67. Li J, Van Uitert R, Yao J, Petrick N, Franaszek M, Huang A, Summers RM. Wavelet method for CT colonography computer-aided polyp detection. Med Phys. 2008;35:3527–38.
68. Lin JS, Lo SB, Hasegawa A, Freedman MT, Mun SK. Reduction of false positives in lung nodule detection using a two-level neural classification. IEEE Trans Med Imaging. 1996;15:206–17. doi:10.1109/42.491422.
69. Lo SB, Lou SA, Lin JS, Freedman MT, Chien MV, Mun SK. Artificial convolution neural network techniques and applications for lung nodule detection. IEEE Trans Med Imaging. 1995;14:711–8. doi:10.1109/42.476112.
70. Lo SC, Freedman MT, Lin JS, Mun SK. Automatic lung nodule detection using profile matching and back-propagation neural network techniques. J Digit Imaging. 1993;6:48–54.
71. Lo SC, Li H, Wang Y, Kinnard L, Freedman MT. A multiple circular path convolution neural network system for detection of mammographic masses. IEEE Trans Med Imaging. 2002;21:150–8. doi:10.1109/42.993133.
72. Lo SC, Lou SL, Lin JS, Freedman MT, Chien MV, Mun SK. Artificial convolution neural network techniques and applications to lung nodule detection. IEEE Trans Med Imaging. 1995;14:711–8.
73. Lo SCB, Chan HP, Lin JS, Li H, Freedman MT, Mun SK. Artificial convolution neural network for medical image pattern recognition. Neural Netw. 1995;8:1201–14.
74. Loog M, van Ginneken B. Segmentation of the posterior ribs in chest radiographs using iterated contextual pixel classification. IEEE Trans Med Imaging. 2006;25:602–11.
75. Loog M, van Ginneken B, Schilham AM. Filter learning: application to suppression of bony structures from chest radiographs. Med Image Anal. 2006;10:826–40.

76. Lostumbo A, Suzuki K, Dachman AH. Flat lesions in CT colonography. Abdom Imaging. 2010;35:578–83. doi:10.1007/s00261-009-9562-3.
77. Lostumbo A, Wanamaker C, Tsai J, Suzuki K, Dachman AH. Comparison of 2D and 3D views for evaluation of flat lesions in CT colonography. Acad Radiol. 2010;17:39–47. doi:10.1016/j.acra.2009.07.004. S1076-6332(09)00400-0 [pii].
78. Macari M, Bini EJ. CT colonography: where have we been and where are we going? Radiology. 2005;237:819–33.
79. Matsumoto S, Kundel HL, Gee JC, Gefter WB, Hatabu H. Pulmonary nodule detection in CT images with quantized convergence index filter. Med Image Anal. 2006;10:343–52. doi:10.1016/j.media.2005.07.001.
80. Messay T, Hardie RC, Rogers SK. A new computationally efficient CAD system for pulmonary nodule detection in CT imagery. Med Image Anal. 2010;14:390–406. doi:10.1016/j.media.2010.02.004.
81. Miettinen OS, Henschke CI. CT screening for lung cancer: coping with nihilistic recommendations. Radiology. 2001;221:592–6.
82. Muramatsu C, Li Q, Schmidt R, Suzuki K, Shiraishi J, Newstead G, Doi K. Experimental determination of subjective similarity for pairs of clustered microcalcifications on mammograms: observer study results. Med Phys. 2006;33:3460–8.
83. Muramatsu C, Li Q, Schmidt RA, Shiraishi J, Suzuki K, Newstead GM, Doi K. Determination of subjective similarity for pairs of masses and pairs of clustered microcalcifications on mammograms: comparison of similarity ranking scores and absolute similarity ratings. Med Phys. 2007;34:2890–5.
84. Muramatsu C, Li Q, Suzuki K, Schmidt RA, Shiraishi J, Newstead GM, Doi K. Investigation of psychophysical measure for evaluation of similar images for mammographic masses: preliminary results. Med Phys. 2005;32:2295–304.
85. Murphy GP, Lawrence W, Lenhard RE, American Cancer Society. American Cancer Society textbook of clinical oncology. 2nd ed. Atlanta: The Society; 1995.
86. Murphy K, van Ginneken B, Schilham AM, de Hoop BJ, Gietema HA, Prokop M. A large-scale evaluation of automatic pulmonary nodule detection in chest CT using local image features and k-nearest-neighbour classification. Med Image Anal. 2009;13:757–70. doi:10.1016/j.media.2009.07.001.
87. Nappi J, Yoshida H. Automated detection of polyps with CT colonography: evaluation of volumetric features for reduction of false-positive findings. Acad Radiol. 2002;9:386–97.
88. Neubauer C. Evaluation of convolutional neural networks for visual recognition. IEEE Trans Neural Netw. 1998;9:685–96.
89. Oda S, Awai K, Suzuki K, Yanaga Y, Funama Y, MacMahon H, Yamashita Y. Performance of radiologists in detection of small pulmonary nodules on chest radiographs: effect of rib suppression with a massive-training artificial neural network. AJR Am J Roentgenol. 2009; 193:W397–402. doi:10.2214/AJR.09.2431. 193/5/W397 [pii].
90. Paik DS, Beaulieu CF, Rubin GD, Acar B, Jeffrey Jr RB, Yee J, Dey J, Napel S. Surface normal overlap: a computer-aided detection algorithm with application to colonic polyps and lung nodules in helical CT. IEEE Trans Med Imaging. 2004;23:661–75.
91. Penedo MG, Carreira MJ, Mosquera A, Cabello D. Computer-aided diagnosis: a neural-network-based approach to lung nodule detection. IEEE Trans Med Imaging. 1998;17: 872–80.
92. Petrick N, Haider M, Summers RM, Yeshwant SC, Brown L, Iuliano EM, Louie A, Choi JR, Pickhardt PJ. CT colonography with computer-aided detection as a second reader: observer performance study. Radiology. 2008;246:148–56.
93. Pu J, Zheng B, Leader JK, Wang XH, Gur D. An automated CT based lung nodule detection scheme using geometric analysis of signed distance field. Med Phys. 2008;35:3453–61.
94. Retico A, Delogu P, Fantacci ME, Gori I, Preite Martinez A. Lung nodule detection in low-dose and thin-slice computed tomography. Comput Biol Med. 2008;38:525–34. doi:10.1016/j.compbiomed.2008.02.001.

95. Riccardi A, Petkov TS, Ferri G, Masotti M, Campanini R. Computer-aided detection of lung nodules via 3D fast radial transform, scale space representation, and Zernike MIP classification. Med Phys. 2011;38:1962–71.
96. Rockey DC, Paulson E, Niedzwiecki D, Davis W, Bosworth HB, Sanders L, Yee J, Henderson J, Hatten P, Burdick S, Sanyal A, Rubin DT, Sterling M, Akerkar G, Bhutani MS, Binmoeller K, Garvie J, Bini EJ, McQuaid K, Foster WL, Thompson WM, Dachman A, Halvorsen R. Analysis of air contrast barium enema, computed tomographic colonography, and colonoscopy: prospective comparison. Lancet. 2005;365:305–11. doi:10.1016/S0140-6736(05)17784-8. S0140673605177848 [pii].
97. Rumelhart DE, Hinton GE, Williams RJ. Learning internal representations by error propagation. Parallel Distrib Process. 1986;1:318–62.
98. Rumelhart DE, Hinton GE, Williams RJ. Learning representations by back-propagating errors. Nature. 1986;323:533–6.
99. Sahiner B, Chan HP, Hadjiiski L. Classifier performance prediction for computer-aided diagnosis using a limited dataset. Med Phys. 2008;35:1559–70.
100. Sahiner B, Chan HP, Petrick N, Wei D, Helvie MA, Adler DD, Goodsitt MM. Classification of mass and normal breast tissue: a convolution neural network classifier with spatial domain and texture images. IEEE Trans Med Imaging. 1996;15:598–610. doi:10.1109/42.538937.
101. Shah PK, Austin JH, White CS, Patel P, Haramati LB, Pearson GD, Shiau MC, Berkmen YM. Missed non-small cell lung cancer: radiographic findings of potentially resectable lesions evident only in retrospect. Radiology. 2003;226:235–41.
102. Shen D, Wu G, Zhang D, Yan P, Suzuki K, Wang F. Machine learning in medical imaging. Comput Med Imaging Graph. 2014;41:1–2.
103. Shiraishi J, Li Q, Suzuki K, Engelmann R, Doi K. Computer-aided diagnostic scheme for the detection of lung nodules on chest radiographs: localized search method based on anatomical classification. Med Phys. 2006;33:2642–53.
104. Soetikno RM, Kaltenbach T, Rouse RV, Park W, Maheshwari A, Sato T, Matsui S, Friedland S. Prevalence of nonpolypoid (flat and depressed) colorectal neoplasms in asymptomatic and symptomatic adults. JAMA. 2008;299:1027–35.
105. Sone S, Takashima S, Li F, Yang Z, Honda T, Maruyama Y, Hasegawa M, Yamanda T, Kubo K, Hanamura K, Asakura K. Mass screening for lung cancer with mobile spiral computed tomography scanner. Lancet. 1998;351:1242–5.
106. Stephane GM. A theory for multiresolution signal decomposition: the wavelet representation. IEEE Trans Pattern Anal Mach Intell. 1989;11:674–93.
107. Summers RM, Beaulieu CF, Pusanik LM, Malley JD, Jeffrey Jr RB, Glazer DI, Napel S. Automated polyp detector for CT colonography: feasibility study. Radiology. 2000;216:284–90.
108. Summers RM, Johnson CD, Pusanik LM, Malley JD, Youssef AM, Reed JE. Automated polyp detection at CT colonography: feasibility assessment in a human population. Radiology. 2001;219:51–9.
109. Summers RM, Yao J, Pickhardt PJ, Franaszek M, Bitter I, Brickman D, Krishna V, Choi JR. Computed tomographic virtual colonoscopy computer-aided polyp detection in a screening population. Gastroenterology. 2005;129:1832–44.
110. Suzuki K. Determining the receptive field of a neural filter. J Neural Eng. 2004;1:228–37. doi:10.1088/1741-2560/1/4/006. S1741-2560(04)85485-5 [pii].
111. Suzuki K. A supervised 'lesion-enhancement' filter by use of a massive-training artificial neural network (MTANN) in computer-aided diagnosis (CAD). Phys Med Biol. 2009;54:S31–45. doi:10.1088/0031-9155/54/18/S03. S0031-9155(09)14266-5 [pii].
112. Suzuki K. Machine learning for medical imaging. Algorithms. 2010. vol. 3. Special issue.
113. Suzuki K. Machine learning for medical imaging. Algorithms. 2012a. vol. 5. Special issue.
114. Suzuki K. Pixel-based machine learning (PML) in medical imaging. Int J Biomed Imaging. 2012b:Article ID 792079, 18 pages.
115. Suzuki K. Pixel-based machine learning in medical imaging. Int J Biomed Imaging. 2012c;2012:792079. doi:10.1155/2012/792079.

116. Suzuki K. A review of computer-aided diagnosis in thoracic and colonic imaging. Quant Imaging Med Surg. 2012d;2:163–76. doi:10.3978/j.issn.2223-4292.2012.09.02.
117. Suzuki K. Machine learning in computer-aided diagnosis of the thorax and colon in CT: a survey. IEICE Trans Info Syst. 2013;E96-D:772–83.
118. Suzuki K, Abe H, Li F, Doi K. Suppression of the contrast of ribs in chest radiographs by means of massive training artificial neural network. Proc SPIE Med Imaging. 2004;5370:1109–19.
119. Suzuki K, Abe H, Li F, Doi K. Suppression of the contrast of ribs in chest radiographs by means of massive training artificial neural network. San Diego: Proceeding- SPIE Medical Imaging (SPIE MI); 2004. p. 1109–19.
120. Suzuki K, Abe H, MacMahon H, Doi K. Image-processing technique for suppressing ribs in chest radiographs by means of massive training artificial neural network (MTANN). IEEE Trans Med Imaging. 2006;25:406–16. doi:10.1109/TMI.2006.871549.
121. Suzuki K, Armato 3rd SG, Li F, Sone S, Doi K. Massive training artificial neural network (MTANN) for reduction of false positives in computerized detection of lung nodules in low-dose computed tomography. Med Phys. 2003;30:1602–17.
122. Suzuki K, Dachman AH. Computer-aided diagnosis in CT colonography. In: Dachman AH, Laghi A, editors. Atlas of virtual colonoscopy. 2nd ed. New York: Springer; 2011. p. 163–82.
123. Suzuki K, Doi K. How can a massive training artificial neural network (MTANN) be trained with a small number of cases in the distinction between nodules and vessels in thoracic CT? Acad Radiol. 2005;12:1333–41.
124. Suzuki K, Hori M, McFarland E, Friedman AC, Rockey DC, Dachman AH. Can CAD help improve the performance of radiologists in detection of difficult polyps in CT colonography? In: Proceedings of RSNA annual meeting, Chicago; 2009. p. 872.
125. Suzuki K, Horiba I, Sugie N. A simple neural network pruning algorithm with application to filter synthesis. Neural Process Lett. 2001;13:43–53.
126. Suzuki K, Horiba I, Sugie N. Efficient approximation of neural filters for removing quantum noise from images. IEEE Trans Signal Process. 2002;50:1787–99.
127. Suzuki K, Horiba I, Sugie N. Linear-time connected-component labeling based on sequential local operations. Comput Vis Image Underst. 2003;89:1–23.
128. Suzuki K, Horiba I, Sugie N. Neural edge enhancer for supervised edge enhancement from noisy images. IEEE Trans Pattern Anal Mach Intell. 2003;25:1582–96.
129. Suzuki K, Horiba I, Sugie N, Nanki M. Neural filter with selection of input features and its application to image quality improvement of medical image sequences. IEICE Trans Info Syst. 2002;E85-D:1710–8.
130. Suzuki K, Horiba I, Sugie N, Nanki M. Extraction of left ventricular contours from left ventriculograms by means of a neural edge detector. IEEE Trans Med Imaging. 2004;23:330–9.
131. Suzuki K, Li F, Sone S, Doi K. Computer-aided diagnostic scheme for distinction between benign and malignant nodules in thoracic low-dose CT by use of massive training artificial neural network. IEEE Trans Med Imaging. 2005;24:1138–50.
132. Suzuki K, Rockey DC, Dachman AH. CT colonography: advanced computer-aided detection scheme utilizing MTANNs for detection of "missed" polyps in a multicenter clinical trial. Med Phys. 2010;37:12–21.
133. Suzuki K, Sheu I, Kawaler E, Ferraro F, Rockey DC, Dachman AH. Computer-aided detection (CADe) of flat lesions in CT colonography (CTC) by means of a spinning-tangent technique. Program of RSNA, Chicago; 2010b. p. 319.
134. Suzuki K, Shiraishi J, Abe H, MacMahon H, Doi K. False-positive reduction in computer-aided diagnostic scheme for detecting nodules in chest radiographs by means of massive training artificial neural network. Acad Radiol. 2005;12:191–201. doi:10.1016/j.acra.2004.11.017. S1076-6332(04)00733-0 [pii].
135. Suzuki K, Shiraishi J, Abe H, MacMahon H, Doi K. False-positive reduction in computer-aided diagnostic scheme for detecting nodules in chest radiographs by means of massive training artificial neural network. Acad Radiol. 2005;12:191–201.

136. Suzuki K, Wang F, Shen D, Yan P. Machine learning in medical imaging (MLMI), Lecture notes in computer science, vol. 7009. Berlin: Springer; 2011. p. 355.
137. Suzuki K, Wu J, Sheu I. Principal-component massive-training machine-learning regression for false-positive reduction in computer-aided detection of polyps in CT colonography, Lecture notes in computer science, machine learning in medical imaging (MLMI), vol. 6357. Beijing: Springer; 2010. p. 182–9.
138. Suzuki K, Yan P, Wang F, Shen D. Machine learning in medical imaging. Int J Biomed Imaging. 2012;2012:123727. doi:10.1155/2012/123727.
139. Suzuki K, Yoshida H, Nappi J, Armato 3rd SG, Dachman AH. Mixture of expert 3D massive-training ANNs for reduction of multiple types of false positives in CAD for detection of polyps in CT colonography. Med Phys. 2008;35:694–703.
140. Suzuki K, Yoshida H, Nappi J, Dachman AH. Massive-training artificial neural network (MTANN) for reduction of false positives in computer-aided detection of polyps: suppression of rectal tubes. Med Phys. 2006;33:3814–24.
141. Suzuki K, Zhang J, Xu J. Massive-training artificial neural network coupled with Laplacian-eigenfunction-based dimensionality reduction for computer-aided detection of polyps in CT colonography. IEEE Trans Med Imaging. 2010;29:1907–17. doi:10.1109/TMI.2010.2053213.
142. Swensen SJ, Jett JR, Hartman TE, Midthun DE, Sloan JA, Sykes AM, Aughenbaugh GL, Clemens MA. Lung cancer screening with CT: Mayo Clinic experience. Radiology. 2003;226:756–61.
143. Tan M, Deklerck R, Jansen B, Bister M, Cornelis J. A novel computer-aided lung nodule detection system for CT images. Med Phys. 2011;38:5630–45. doi:10.1118/1.3633941.
144. van Ginneken B, ter Haar Romeny BM, Viergever MA. Computer-aided diagnosis in chest radiography: a survey. IEEE Trans Med Imaging. 2001;20:1228–41.
145. Vapnik VN. The nature of statistical learning theory. Berlin: Springer; 1995.
146. Vapnik VN. Statistical learning theory. New York: Wiley; 1998.
147. Wang F, Shen D, Yan P, Suzuki K. Machine learning in medical imaging (MLMI), Lecture notes in computer science, vol. 7588. Berlin: Springer; 2012. p. 276.
148. Wang F, Yan P, Suzuki K, Shen D. Machine learning in medical imaging (MLMI), Lecture notes in computer science, vol. 6357. Berlin: Springer; 2010. p. 192.
149. Wang S, Yao J, Petrick N, Summers RM. Combining statistical and geometric features for colonic polyp detection in CTC based on multiple kernel learning. Int J Comput Intell Appl. 2010;9:1–15. doi:10.1142/S1469026810002744.
150. Wang S, Yao J, Summers RM. Improved classifier for computer-aided polyp detection in CT colonography by nonlinear dimensionality reduction. Med Phys. 2008;35:1377–86.
151. Wang Z, Liang Z, Li L, Li X, Li B, Anderson J, Harrington D. Reduction of false positives by internal features for polyp detection in CT-based virtual colonoscopy. Med Phys. 2005;32:3602–16.
152. Way TW, Sahiner B, Chan HP, Hadjiiski L, Cascade PN, Chughtai A, Bogot N, Kazerooni E. Computer-aided diagnosis of pulmonary nodules on CT scans: improvement of classification performance with nodule surface features. Med Phys. 2009;36:3086–98.
153. Wei D, Nishikawa RM, Doi K. Application of texture analysis and shift-invariant artificial neural network to microcalcification cluster detection. Radiology. 1996;201:696–696.
154. Wei J, Hagihara Y, Shimizu A, Kobatake H. Optimal image feature set for detecting lung nodules on chest X-ray images. Tokyo: Computer Assisted Radiology and Surgery; 2002. p. 706–11.
155. Winawer SJ, Fletcher RH, Miller L, Godlee F, Stolar MH, Mulrow CD, Woolf SH, Glick SN, Ganiats TG, Bond JH, Rosen L, Zapka JG, Olsen SJ, Giardiello FM, Sisk JE, Van Antwerp R, Brown-Davis C, Marciniak DA, Mayer RJ. Colorectal cancer screening: clinical guidelines and rationale. Gastroenterology. 1997;112:594–642.
156. Wu G, Zhang D, Shen D, Yan P, Suzuki K, Wang F. Machine learning in medical imaging (MLMI), Lecture notes in computer science, vol. 8184. Berlin: Springer; 2013. p. 262.
157. Wu Y, Doi K, Giger ML, Nishikawa RM. Computerized detection of clustered microcalcifications in digital mammograms: applications of artificial neural networks. Med Phys. 1992;19:555–60.

158. Wu YT, Wei J, Hadjiiski LM, Sahiner B, Zhou C, Ge J, Shi J, Zhang Y, Chan HP. Bilateral analysis based false positive reduction for computer-aided mass detection. Med Phys. 2007;34:3334–44.
159. Xu JW, Suzuki K. Massive-training support vector regression and Gaussian process for false-positive reduction in computer-aided detection of polyps in CT colonography. Med Phys. 2011;38:1888–902.
160. Xu JW, Suzuki K. Max-AUC feature selection in computer-aided detection of polyps in CT colonography. IEEE J Biomed Health Info. 2014;18:585–93. doi:10.1109/JBHI.2013.2278023.
161. Yan P, Suzuki K, Wang F, Shen D. Guest Editors. Special issue on "Machine Learning in Medical Imaging," Machine Vision and Applications, 2012.
162. Yao J, Li J, Summers RM. Employing topographical height map in colonic polyp measurement and false positive reduction. Pattern Recognit. 2009;42:1029–40. doi:10.1016/j.patcog.2008.09.034.
163. Ye X, Lin X, Dehmeshki J, Slabaugh G, Beddoe G. Shape-based computer-aided detection of lung nodules in thoracic CT images. IEEE Trans Biomed Eng. 2009;56:1810–20.
164. Yoshida H, Dachman AH. Computer-aided diagnosis for CT colonography. Semin Ultrasound CT MR. 2004;25:419–31.
165. Yoshida H, Dachman AH. CAD techniques, challenges, and controversies in computed tomographic colonography. Abdom Imaging. 2005;30:26–41.
166. Yoshida H, Masutani Y, MacEneaney P, Rubin DT, Dachman AH. Computerized detection of colonic polyps at CT colonography on the basis of volumetric features: pilot study. Radiology. 2002;222:327–36.
167. Yoshida H, Nappi J. Three-dimensional computer-aided diagnosis scheme for detection of colonic polyps. IEEE Trans Med Imaging. 2001;20:1261–74.
168. Yoshida H, Nappi J, MacEneaney P, Rubin DT, Dachman AH. Computer-aided diagnosis scheme for detection of polyps at CT colonography. Radiographics. 2002;22:963–79.
169. Yu SN, Li KY, Huang YK. Detection of microcalcifications in digital mammograms using wavelet filter and Markov random field model. Comput Med Imaging Graph. 2006;30:163–73.
170. Yuan R, Vos PM, Cooperberg PL. Computer-aided detection in screening CT for pulmonary nodules. Am J Roentgenol. 2006;186:1280–7. doi:10.2214/AJR.04.1969.
171. Zhang W, Doi K, Giger ML, Nishikawa RM, Schmidt RA. An improved shift-invariant artificial neural network for computerized detection of clustered microcalcifications in digital mammograms. Med Phys. 1996;23:595–601.
172. Zhang W, Doi K, Giger ML, Wu Y, Nishikawa RM, Schmidt RA. Computerized detection of clustered microcalcifications in digital mammograms using a shift-invariant artificial neural network. Med Phys. 1994;21:517–24.
173. Zhao H, Lo SC, Freedman M, Wang Y. Enhanced lung cancer detection in temporal subtraction chest radiography using directional edge filtering techniques. In: Proceedings of SPIE medical imaging: image processing, vol 4684, San Diego; 2002.
174. Zhu H, Liang Z, Pickhardt PJ, Barish MA, You J, Fan Y, Lu H, Posniak EJ, Richards RJ, Cohen HL. Increasing computer-aided detection specificity by projection features for CT colonography. Med Phys. 2010;37:1468–81.

Classification of Malignant and Benign Tumors

Juan Wang, Issam El Naqa, and Yongyi Yang

Abstract

Machine learning has a long-standing history of application for computer-aided diagnosis (CADx) purposes and discriminating between different types of benign and malignant lesions. In this chapter, we explain the application of machine learning algorithms for development of classifiers of tumors using features extracted from diagnostic imaging. Examples from our work on mammography using conventional classification approaches and more advanced methods based on content-based image retrieval will be presented and discussed.

8.1 Introduction

In recent years, there have been significant interests and efforts in development of computerized methods for automatically classifying a tumor or lesion being malignant or benign. These methods are collectively known as computer-aided diagnosis (CADx), the purpose of which is to provide a second opinion to assist the radiologists in their diagnosis of detected tumors. Indeed, in the literature, CADx techniques have been studied both for various disease types and for different imaging modalities, ranging from CT in oncology, MRI for brain tumors, mammography for breast cancer, and many others. For instance, the application of CT to early lung

---

J. Wang • Y. Yang (✉)
Department of Electrical and Computer Engineering, Illinois Institute of Technology, 3301 South Dearborn, Chicago, IL 60616, USA
e-mail: wangjuan313@gmail.com; yy@ece.iit.edu

I. El Naqa
Department of Oncology, McGill University, Montreal, QC, Canada

Department of Radiation Oncology, University of Michigan, Ann Arbor, USA
e-mail: issam.elnaqa@mcgill.ca; ielnaqa@med.umich.edu

© Springer International Publishing Switzerland 2015
I. El Naqa et al. (eds.), *Machine Learning in Radiation Oncology: Theory and Applications*, DOI 10.1007/978-3-319-18305-3_8

cancer has been controversial. In a recent randomized clinical trial referred to as the NELSON trial with 15,822 enrolled participants, it was shown that low-dose CT screening can improve the sensitivity and specificity of lung cancer detection [1]. However, this situation has been more challenging in cases of head and neck cancer, where luckily the combination with positron emission tomography (PET) has overcome shortages of CT and revolutionized the management of this cancer [2]. On the other hand, magnetic resonance imaging (MRI), which is more financially expensive but with better soft tissue discrimination and sparing from exposure to ionizing radiation, has risen in recent years in the diagnosis of difficult cases such as prostate [3], brain [4, 5], and breast cancers [6].

In mammography, many CADx techniques have been developed for classification of suspicious breast tumors in mammogram images, including both masses and clustered microcalcifications (MCs). For example, in the early work [7], a three-layer, feed-forward neural network was trained with a back-propagation algorithm for mammographic lesion (including MCs and masses) interpretation. Subsequently, various supervised learning techniques were studied for diagnosis of MC lesions (e.g., [8–13]) and mass lesions (e.g., [14–17]). There also exist several laboratory studies which demonstrate that CADx techniques can either be more accurate than the human readers or help improve their diagnosis accuracy [10, 18–21].

In the rest of this chapter, we will first provide an overview of the major components involved in the development of a CADx framework for tumor classification (Sect. 8.2). Afterward, we will illustrate this framework with some examples of CADx techniques for breast lesions in mammograms (Sect. 8.3). In addition, we will also introduce the use of a visualization tool – based on the technique of multi-dimensional scaling (MDS) – for exploring the similarity among a set of tumors (Sect. 8.4). Such a tool potentially can be useful for one to compare a case under consideration against some similar, known cases in a reference library. We will also discuss some issues and challenges in the development and application of CADx techniques (Sect. 8.5).

8.2 Overview of Classification Framework

When in operation, a CADx framework for tumor classification functions as follows: For a given tumor under consideration, a set of so-called features is first computed from the tumor to quantify its underlying characteristics. These features are typically represented by a vector **x** in an n-dimensional space R^n. Afterward, a mathematical function $f(\mathbf{x})$ is applied to the feature vector **x**, the value of which is used to reflect the likelihood that the tumor is either malignant or benign. The function $f(\mathbf{x})$ is called the decision function or classifier function.

The development of a CADx framework involves the following key components: (1) determine what features **x** to use that are relevant for classification of the tumor, (2) design the classifier function $f(\mathbf{x})$ that is appropriate for the task, and (3) evaluate the accuracy level (i.e., performance) of the classifier output, which is key to the confidence level on the "second opinion."

8.2.1 Perception Modeling

There have been significant improvements over the past decades with respect to developing image quantitative imaging measures, objective image interpretations, feature extraction, and semantic descriptors [22, 23]. However, some major difficulties still remain pertaining to CADx applications. First, it is understood that quantitative measures can vary with the different aspects of perceptual similarity by radiologists of images; the selection of an appropriate similarity measure thus becomes problem dependent. Second, the relation between the low-level visual features and the high-level expert human interpretation of similarity is not well defined when comparing two images; it is thus not exactly clear what features or combination of them are relevant for such judgment [24, 25]. We have been developing perceptual similarity metrics for application in content-based image retrieval (CBIR) of mammogram images [25]. In this approach, the notion of similarity is modeled as a nonlinear function of the image features in a pair of mammogram images containing lesions of interest, e.g., microcalcification clusters (MCCs). If we let vectors **u** and **v** denote the features of two MCCs at issue, the following regression model could be used to determine their similarity coefficient (*SC*):

$$SC(\mathbf{u},\mathbf{v}) = f(\mathbf{u},\mathbf{v}) + \zeta, \qquad (8.1)$$

where $f(\mathbf{u}, \mathbf{v})$ is a function determined using a machine learning approach, which we choose to be support vector machine (SVM) learning [26], and ζ is the modeling error. The similarity function $f(\mathbf{u}, \mathbf{v})$ in Eq. (8.1) is trained using data samples collected in an observer study.

8.2.2 Feature Extraction for Tumor Quantification

The purpose of feature extraction is to describe the content of a tumor under consideration by a set of quantitative descriptors, called features, denoted by vector **x**. Conceptually, these features should be relevant to the disease condition of the tumor. For example, they may be used to quantify the size of the tumor, the geometric shape of the tumor, the density of the tissue, etc., depending on the tumor type and specific application.

In the literature, there have been many types of features studied for classification of benign and malignant tumors. For example, in [27], effective thickness and effective volume were defined on the physical properties of MCs in mammogram images and were demonstrated to be useful for diagnosis. In [28], image intensity and texture features were extracted from post-contrast T1-weighted MR images and were shown to be helpful for brain tumor classification. In [29], wavelet features were compared with Haralick features [30] for MC classification.

While the reported features are many, they can be divided into two broad categories: (1) boundary-based features and (2) region-based features. Boundary-based features are used to describe the properties of the geometric boundary of a tumor. They include, for example, the perimeter, Fourier descriptors, and boundary moments [31]. In contrast, region-based features are derived from within a tumor region, which include the shape, texture, or the frequency domain information of the tumor. Some examples of region-based features are the tumor size, image moment features [31], wavelet-based features [29], and texture features [28].

To ensure good classification performance, the features extracted from a tumor are desired to have certain properties pertinent to the application. For example, a common requirement is that the features should be invariant to any translation or rotation in a tumor image. Other considerations in extracting or designing quantitative features include the effects of the image resolution and gray-level quantization used for the image. The image resolution can affect those features related to the size of a tumor, such as its area and perimeter. The quantization level in an image can affect those features related to the image intensity, such as image moments and features derived from the gray-level co-occurrence matrix (GLCM) [28]. Therefore, prior to feature extraction, the tumor images need to be preprocessed properly in order to avoid any discrepancy in resolution and quantization.

With a great number of features available, as described above, an important task in a CADx framework is how to determine a set of discriminative features in a tumor classification problem. These features are desired to have good differentiating power between benign and malignant tumors. One approach is to exploit the working knowledge of the clinicians and select those features that are closely associated with what the clinicians use in their diagnosis of the lesions [10]. For example, for MC lesions, the size and shape of the MCs and their spatial distribution are all known to be important, because the MCs tend to be more irregular and have a bigger cluster in a malignant lesion [13]. Alternatively, to determine the most salient features for use in the classification, one may employ a systematic feature selection procedure during the training stage of the classifier. The commonly used feature selection procedures in the literature include the filter algorithm [32], wrapper algorithm [33], and embedded algorithm [34].

8.2.3 Design of Decision Function Using Machine Learning

The problem of classifying benign or malignant tumors is a classical two-class classification problem, with benign tumors being one class and malignant ones being the other. For a given tumor characterized by its feature vector $\mathbf{x}$, a decision function $f(\mathbf{x})$ is designed to determine which class, malignant or benign, $\mathbf{x}$ belongs to. Naturally, a fundamental problem is how to design the decision function for a given tumor type. A common approach to this problem is to apply supervised learning, in which a pattern classifier is first trained on a set of known cases, denoted as $\{(\mathbf{x}_i, \mathbf{y}_i), i = 1, \ldots, N\}$, where a training sample is described by its feature vector $\mathbf{x}_i$, and y_i is its known class-label (1 for malignant tumor and −1 for benign tumor). Once trained, the classifier is applied subsequently to classify other cases (unseen during training).

Broadly speaking, depending on its mathematical form, the decision function $f(\mathbf{x})$ is categorized into linear and nonlinear classifiers. A linear classifier is represented as

$$f(\mathbf{x}) = \mathbf{w}^T \mathbf{x} + b \qquad (8.2)$$

where $\mathbf{w}$ is the discriminant vector and b is the bias, which are parameters determined from the training samples. In contrast, a nonlinear classifier $f(\mathbf{x})$ has a more complex mathematical form and is no longer a linear function in terms of the feature variables $\mathbf{x}$. One such example is the feed-forward neural network, in which (nonlinear) sigmoid activation functions are used at the individual nodes within the network.

Because of their simpler form, linear classifiers are easier to train and less prone to over-fitting compared to their nonlinear counterpart. Moreover, it is often easier to examine and interpret the relationship between the classifier output and the individual feature variables in a linear classifier than that in a nonlinear one. Thus, linear classifiers can be favored for certain applications. On the other hand, because of their more complex form, nonlinear classifiers can be more versatile and achieve better performance than linear ones when the underlying decision surface between the two classes is inherently nonlinear in a given problem.

Regardless of their specific form, the classifier functions typically involve a number of parameters, which need to be determined before they can be applied to classifying an unknown case. There have been many different algorithms designed for determining these parameters from a set of training samples, which are collectively known as supervised machine learning algorithms.

Consider, for example, the case of linear classifiers in Eq. (8.2). The parameters $\mathbf{w}$ and b can be determined according to the following different optimum principles: (1) logistic regression [35], in which the log-likelihood function of the training data samples is maximized under a logistic probability model; (2) linear discriminant analysis (LDA) [36], in which the optimal decision boundary is determined under the assumption of multivariate Gaussian distributions for the data samples from the two classes; and (3) support vector machine (SVM) [37], in which the parameters are designed to achieve the maximum separation margin between the two classes (among the training samples).

Similarly, there also exist many methods for designing nonlinear classifiers. One popular type of nonlinear classifiers is the kernel-based methods [38]. In a kernel-based method, the so-called kernel trick is used to first map the input vector $\mathbf{x}$ into a higher-dimensional space via a nonlinear mapping; afterward, a linear classifier is applied in this mapped space, which in the end is a nonlinear classifier in the original feature space. One such example is the popular nonlinear SVM classifier. Other kernel-based methods include kernel Fisher discriminant (KFD), kernel principle component analysis (KPCA), and relevance vector machine (RVM) [39].

Another type of commonly used nonlinear CADx classifiers is the committee-based methods. These methods are based on the idea of systematically aggregating the output of a series of individual weak classifiers to form a (more powerful) decision function. Adaboost [40] and random forests [41] are well-known examples of such committee-based methods. For example, in Adaboost, the training set is modified successively to obtain a sequence of weak classifiers; the output of each weak classifier is adjusted by a weight factor according to its classification error on the training set to form an aggregated decision function [40].

8.2.4 CADx Classifier Training and Performance Evaluation

In concept, a CADx classifier should be trained and evaluated by using the following three sets of data samples: a training set, a validation set, and a testing set. The training set is used to obtain the model parameters of a classifier (such as **w** and b in the linear classifier in Eq. (8.2)). The validation set is usually independent from the training set and is used to determine the tuning parameters of a classifier if it has any. For example, in kernel SVM, one may need to decide the type of the kernel function to use. Finally, the testing set is used to evaluate the performance of the resulting classifier. It must be independent from both the training and validation sets in order to avoid any potential bias.

Ideally, when the number of available data samples is large enough, the training, validation, and testing sets in the above should be kept to be mutually exclusive. However, in practice, the data samples are often scarce, making it impossible to obtain independent training, validation, and testing sets, which is often true when clinical cases are used. To deal with this difficulty, a k-fold cross-validation procedure is often used instead. The procedure works as following: first, the available n data samples are divided randomly into k roughly equal-sized subsets; subsequently, each of the k subsets is held out in turn for testing while the rest $(k-1)$ subsets are used together for training. In the end, the performance is averaged over the k held-out testing subsets to obtain the overall performance. A special case of the k-fold cross-validation procedure is when $k = n$, which is also called a leave-one-out procedure (LOO). It is known that a smaller k yields a lower variance but also a larger bias in the estimated performance. In practice, $k = 5$ or 10 is often used as a good compromise in cross validation [42, 43].

When there are parameters needed to be tuned in a classifier model, a double loop cross-validation procedure [44] can be applied to avoid any potential bias. A double loop cross-validation procedure has a nested structure of two loops (the inner and outer loops). The outer loop is the same as the standard k-fold cross validation above, which is used to evaluate the performance of the classifier. The inner loop is to further perform a standard k'-fold cross validation using only the training set of samples in each iteration of the outer loop, which is used to select the tuning parameters.

For evaluating the performance of a CADx classifier, a receiver-operating characteristic (ROC) analysis is now routinely used. An ROC curve is a plot of the classification sensitivity (i.e., true-positive fraction) as the ordinate versus the specificity (i.e., false-positive fraction) as the abscissa. For a given classifier, an ROC curve is obtained by continuously varying the threshold associated with its decision function over its operating range. As a summary measure of overall diagnostic performance, the area under an ROC curve (denoted by AUC) is often used. A larger AUC value means better classification performance.

8.3 Application Examples in Mammography

8.3.1 Mammography

Mammography is an imaging procedure in which low-energy X-ray images of the breast are taken. Typically, they are in the order of 0.7 mSv. A mammogram can detect a cancerous or precancerous tumor in the breast even before the tumor is large enough to feel. Despite advances in imaging technology, mammography remains the most cost-effective strategy for early detection of breast cancer in clinical practice. The sensitivity of mammography could be up to approximately 90 % for patients without symptoms [45]. However, this sensitivity is highly dependent on the patient's age, the size and conspicuity of the lesion, the hormone status of the tumor, the density of a woman's breasts, the overall image quality, and the interpretative skills of the radiologist [46]. Therefore, the overall sensitivity of mammography could vary from 90 to 70 % only [47]. Moreover, it is very difficult to distinguish mammographically benign lesions from malignant ones. It has been estimated that one third of regularly screened women experience at least one false-positive (benign lesions being biopsied) screening mammogram over a period of 10 years [48]. A population-based study included about 27,394 screening mammograms that were interpreted by 1,067 radiologists showed that the radiologists had substantial variations in the false-positive rates ranging from 1.5 to 24.1 % [49]. Unnecessary biopsy is often cited as one of the "risks" of screening mammography. Surgical, needle-core, and fine-needle aspiration biopsies are expensive, invasive, and traumatic for the patient.

8.3.2 Computer-Aided Diagnosis (CADx) of Microcalcification Lesions in Mammograms

Clustered microcalcifications (MCs) can be an important early sign of breast cancer in women. They are found in 30–50 % of mammographically diagnosed cases. MCs are calcium deposits of very small dimension and appear as a group of granular bright spots in a mammogram (e.g., Fig. 8.1). Because of their subtlety in appearance in mammogram images, accurate diagnosis of MC lesions as benign or malignant is a very challenging problem for radiologists. Studies show that a false-positive diagnostic imaging study leads to unnecessary biopsy of benign lesions, yielding a positive predictive value of only 20–40 % [50].

Because of their importance in cancer diagnosis, there has been intensive research in the development of CADx techniques for clustered MCs, of which the purpose is to provide a second opinion to radiologists in their diagnosis to

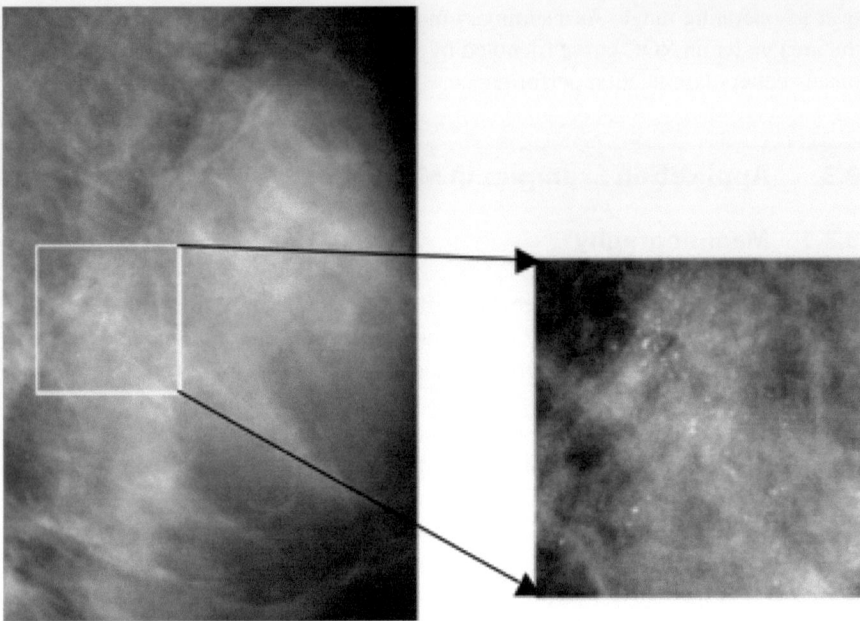

Fig. 8.1 A mammogram image (*left*) and its magnified view (*right*), where MCs are visible as granular bright spots

improve the performance and efficiency [18]. In the literature, various machine learning methods such as LDA, logistic regression, ANN, and SVM have been used in the development of CADx classifiers for clustered MCs. For example, in [51], an LDA classifier was used for classification of benign and malignant MCs based on their visibility and shape features. This approach was subsequently extended to morphology and texture features in [52]. In [53], it was demonstrated an ANN-based approach could improve the diagnosis performance of radiologists for MCs. In [13], FKD, ANN, SVM, RVM, and committee machines were explored in a comparison study, wherein the SVM was shown to yield improved performance over the others. Collectively, the reported research results demonstrate that CADx has the potential to improve the radiologists' performance in breast cancer diagnosis [54].

In the development of CADx techniques in the literature, various types of features have been investigated for characterizing MC lesions [9, 16, 55–58]. These features are defined to characterize the gray-level properties (e.g., the brightness, contrast, and gradient of individual MCs and the texture in the lesion region) or geometric properties of the MC lesions (e.g., the size and shape of the individual MCs, the number of MCs, the area, shape, and spatial distribution of a cluster). They are extracted either from the individual MCs or the entire lesion region. The features from individual MCs are often summarized using statistics to characterize an MC cluster.

CADx Example: Machine Learning Methods for MC Classification In this section, we demonstrate the use of two CADx classifiers for clustered MCs, one is a linear classifier based on logistic regression, and the other is a nonlinear SVM classifier with a RBF kernel [13]. In logistic regression, the parameters **w** and b in Eq. (8.2) are determined through maximization of the following log-likelihood function:

$$L(\mathbf{w},b) = \sum_{i=1}^{N} \log p(y_i, \mathbf{x}_i; \mathbf{w}, b) \quad (8.3)$$

where the probability term is given by

$$p(y_i = 1, \mathbf{x}_i; \mathbf{w}, b) = \left[1 + \exp(-\mathbf{w}^T \mathbf{x}_i - b)\right]^{-1}. \quad (8.4)$$

For the nonlinear SVM classifier, it can be represented as

$$f(\mathbf{x}) = \mathbf{w}^T \Phi(\mathbf{x}) + b \quad (8.5)$$

where **w** is the discriminant vector, b is the bias, and $\Phi(\mathbf{x})$ is a nonlinear mapping function which is implicitly defined by a kernel function (RBF in our case).

Based on the maximum marginal criterion, the parameters **w** and b in (8.5) are determined as following:

$$\begin{aligned} \min J(\mathbf{w}, \xi) &= \frac{1}{2} \|\mathbf{w}\|^2 + C \sum_{i=1}^{N} \xi_i \\ \text{s.t. } y_i f(\mathbf{x}_i) &\geq 1 - \xi_i, \xi_i \geq 0, i = 1, 2, \cdots, N \end{aligned} \quad (8.6)$$

For testing these classifiers, we used a dataset of 104 cases (46 malignant, 58 benign), all containing clustered MCs. This dataset was collected at the University of Chicago. It contains some cases that are difficult to classify; the average classification performance by a group of five attending radiologists on this dataset yielded a value of only 0.62 in the area under the ROC curve [10]. The MCs in these mammograms were marked by a group of expert readers.

For this dataset, a set of eight features were extracted to characterize MC clusters [10]: (1) the number of MCs in the cluster, (2) the mean effective volume (area times effective thickness) of individual MCs, (3) the area of the cluster, (4) the circularity of the cluster, (5) the relative standard deviation of the effective thickness, (6) the relative standard deviation of the effective volume, (7) the mean area of MCs, and (8) the second highest shape-irregularity measure. These features were selected such that they have meanings that are closely associated with features used by radiologists in clinical diagnosis of MC lesions.

To evaluate the classifiers, a leave-one-out (LOO) procedure was applied to the 104 cases, and the ROCKIT software was used to calculate the performance AUC. The logistic regression classifier achieved AUC=0.7174. In contrast, the SVM achieved AUC=0.7373. These results indicate that the classification performance of the classifiers is far from being perfect, which illustrates the difficulty in diagnosis of MC lesions in mammograms.

8.3.3 Adaptive CADx Boosted with Content-Based Image Retrieval (CBIR)

In recent years, CBIR has been studied as a diagnostic aid in tumor classification [59, 60], of which the goal is to provide radiologists with examples of lesions with known pathology that are similar to the lesion being evaluated. A CBIR system can be viewed as a CADx tool to provide evidence for case-based reasoning. With CBIR, the system first retrieves a set of cases similar to a query, which can be used to assist a decision for the query [61]. For example, in [62, 63], the ratio of malignant cases among all retrieved cases was used as a prediction for the query. In [64], the similarity levels between the query and retrieval cases were used as weighting factors for prediction.

We have been investigating an approach of using retrieved images to boost the classification of a CADx classifier [65–67]. In conventional CADx, a pattern classifier was first trained on a set of training cases and then applied to subsequent testing cases. Deviating from approach, for a given case to be classified (i.e., query), we first obtain a set of known cases with similar features to that of the query case from a reference database and use these retrieved cases to adapt the CADx classifier so as to improve its classification accuracy on the query case. Below, we illustrate this approach using a linear classifier with logistic regression [65].

Assume that a baseline classifier $f(\mathbf{x})$ in the form of Eq. (8.2) has been trained with logistic regression as in Eq. (8.2) on a set of training samples: $\{(\mathbf{x}_i, y_i), i=1,\ldots,N\}$. Now, consider a query lesion $\mathbf{x}$ to be classified. Let $\{(\mathbf{x}_i^{(r)}, y_i^{(r)}), i=1,\ldots,N_r\}$ be a set of N_r retrieved cases which are similar to $\mathbf{x}$. In our case-adaptive approach, we use the retrieved samples $\{(\mathbf{x}_i^{(r)}, y_i^{(r)}), i=1,\ldots,N_r\}$ to adapt the classifier $f(\mathbf{x})$. Specifically, the objective function in (8.3) is modified as

$$L(\mathbf{w},b) = \sum_{i=1}^{N} \log p(y_i, \mathbf{x}_i; \mathbf{w}, b) + \sum_{i=1}^{N_r} \beta_i \log p(y_i^{(r)}, \mathbf{x}_i^{(r)}; \mathbf{w}, b) \quad (8.7)$$

In (8.7), the weighting factors β_i are adjusted according to the similarity of $\mathbf{x}_i^{(r)}$ to the query $\mathbf{x}$. The idea is to put more emphasis on those retrieved samples that are more similar to the query, with the goal of refining the decision boundary of the classifier in the neighborhood of the query. Indeed, the first term in (8.7) simply corresponds to the log-likelihood function in (8.3), while the second term can be viewed as a weighted likelihood of those retrieved similar samples. Intuitively, the retrieved samples are used to steer the pretrained classifier from (8.3) to achieve more emphasis in the neighborhood of the query $\mathbf{x}$. Note that the objective function in (8.7) has the same mathematical form as that in the original optimization problem in (8.3), which can be solved efficiently by the method of iteratively reweighted least square (IRLS) [35].

In our study, we implemented the following strategy for adjusting β_i according to the similarity level of a retrieved sample $\mathbf{x}_i^{(r)}$ to the query $\mathbf{x}$:

$$\beta_i = 1 + k \frac{\alpha_i}{\max_{j=1,\ldots,N_r} \{\alpha_j\}}, \quad i=1,\ldots,N_r \quad (8.8)$$

where α_i denotes the similarity measure between $\mathbf{x}_i^{(r)}$ and $\mathbf{x}$, and $k > 0$ is a parameter used to control the degree of emphasis on the retrieved samples relative to other training samples. The choice of the form in Eq. (8.8) is such that the weighting factor increases linearly with the similarity level of a retrieved case to $\mathbf{x}$, with the most similar case among the retrieved receiving maximum weight $1+k$, which corresponds to k times more influence than the existing training samples in the objective function in Eq. (8.8).

As a similarity measure for retrieved cases, we used the Gaussian RBF kernel function

$$\alpha_i = \exp\left(-\left\|\mathbf{x}_i^{(r)} - \mathbf{x}\right\|^2 / \gamma^2\right), \quad i = 1, \ldots, N_r \qquad (8.9)$$

where γ is a scaling factor controlling the sensitivity of α_i with respect to the distance between the query and a retrieved case. In our experiments, the parameter γ was set to the 10th percentile of the distance between every possible image pairs in the training set. Such a choice is out of the consideration that most of the cases in a database are typically not similar to each other. Those cases with a large distance away from query $\mathbf{x}$ will receive a low similarity measure consequently.

To demonstrate this approach, a set of 589 cases (331 benign, 258 malignant), all containing MC lesions, were extracted from the benign and cancer volumes in the DDSM database maintained at the University of South Florida [68]. The extracted mammogram images were adjusted to correspond to the same optical density and to have a uniform resolution of 0.05 mm/pixel. To quantify the MC lesions in these mammogram images, we first applied an MC detection algorithm using an SVM classifier [25] to automatically locate the MCs in each lesion region provided by the dataset. To help suppress the false positives in the detection, the images were first processed with the isotropic normalization technique prior to the detection [69]. The detected MCs were grouped into clusters.

Afterward, a set of descriptive features was computed for the clustered MCs in the dataset; the following nine features were used [65]: (1) area of the cluster, (2) compactness of the cluster, (3) density of the cluster represented by the number of MCs in a unit area, (4) standard deviation of the inter-distance between neighboring MCs, (5) number of MCs in the cluster, (6) sum of the size of all MC objects in the cluster, (7) mean of the average brightness in each MC object, (8) mean of the intensity standard deviation in each MC object, and (9) the compactness of the 2nd most irregular MC object in the cluster. These features were used to form a vector $\mathbf{x}$ for each lesion in the dataset.

To evaluate the classification performance, a subset of 120 cases (70 benign, 50 malignant) was randomly selected from the dataset for training the baseline classifier, and the remaining 469 cases were used for testing the adaptive classifier. An LOO procedure was applied for each testing case, for which all the remaining cases were used for retrieval. In Fig. 8.2, we show the performance results achieved by the case-adaptive classifier and the baseline classifier; for the adaptive classifier, the AUC value is shown with different numbers of retrieved cases N_r. From Fig. 8.2, it can be seen that the best performance (AUC=0.7755) was obtained by the adaptive

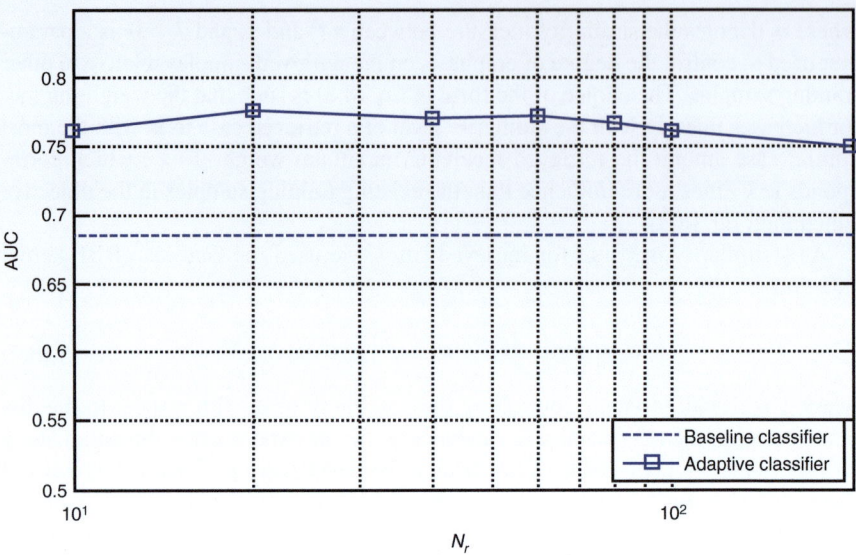

Fig. 8.2 Classification performance (AUC) achieved by the case-adaptive linear classifier. The number of retrieved cases Nr was varied from 10 to 200. For comparison, results are also shown for the baseline classifier

classifier when $N_r = 20$, compared to AUC=0.6848 for the baseline classifier (p-value<0.0001). The performance is also noted to deteriorate somewhat with increased N_r. This is because the number of similar cases for a given query is typically small due to the limited number of cases in the reference library. With large N_r, some of the retrieved cases will become less similar to the query and will not help the classification on the query.

8.4 MDS as a Visualization Tool of Example Lesions

As an alternative approach to CADx, retrieving a set of known lesion similar to the one being evaluated might be of value in assisting radiologists in their diagnosis. In recent years, such an approach has been studied by researchers and applied for different lesion types and imaging modalities [25, 64, 66, 70–73]. For this purpose, we have been studying the use of multidimensional scaling (MDS) for representation and analysis of similar lesions in a large dataset. In a retrieval framework, MDS can be used to study how a query tumor might be related to a set of similar images retrieved from a reference library [66]. When used as a visualization tool, MDS allows one to browse and explore intuitively the distribution of benign and malignant MC lesions in a dataset and to examine how this distribution might be related to the features of the tumors [12, 13, 71, 74].

8.4.1 Multidimensional Scaling (MDS) Technique

MDS is a data embedding technique for representation and analysis of a set of objects based on their mutual similarity (or dissimilarity) measurements [75]. The basic idea of MDS is to represent the objects of interest as points in a low-dimensional (typically 2D or 3D) space such that the geometric distances between the points in this space are in accordance with the similarity measurements between the corresponding objects. The resulting representation in this lower-dimensional space enables one to visualize the relationship among the objects in a rather intuitive manner.

Specifically, consider a set of N objects. The MDS seeks to embed these objects in a lower-dimensional space (R^2 or R^3) as a set of data points $\mathbf{x}_i$ and $i = 1,\cdots,N$, such that the Euclidean distance $d(\mathbf{x}_i, \mathbf{x}_j)$ between a pair of points $\mathbf{x}_i$ and $\mathbf{x}_j$ is proportional to their pairwise proximity measure δ_{ij}. This is accomplished by minimizing the following objective function:

$$\sigma^2 = \frac{\sum w_{ij} \left[d\left(\mathbf{x}_i, \mathbf{x}_j\right) - \delta_{ij} \right]^2}{\sum w_{ij} \delta_{ij}} \qquad (8.10)$$

where w_{ij} are weight factors (specified by users). The quantity σ is known as Stress-1, which measures the goodness of fit of the MDS model.

In our application, we use MDS to represent tumors from mammogram images as points in a 2D plane, wherein the similarity between a pair of tumor images is defined according to their perceptual similarity. Thus, tumors that are in close vicinity of each other in the MDS plot correspond to those that are perceptually similar.

8.4.2 Exploring Similar MC Lesions with MDS

In order to explore how perceptually similar cases with clustered MCs may relate to one another in terms of their underlying characteristics (from disease condition to image features), we conducted an observer study to collect similarity scores from a group of readers on a set of 2,000 image pairs, which were selected from 222 cases based on their image features. Afterward, we applied MDS to embed all the cases in a 2D plot, in which the potential relationship among the different cases is exhibited according to their similarity ratings. Such a plot allows one to study how neighboring cases (i.e., cases similar to each other) may relate to one another. In particular, we will examine the relationships among the cases in several aspects, including (1) case pathology, (2) spatial distribution patterns of their clustered MCs, and (3) image pairs of clustered MCs that are highly similar.

Dataset The dataset used in this study was collected by the Department of Radiology at the University of Chicago. It consists of 365 mammogram images from 222 cases (110 malignant, 112 malignant), of which all have been proven by biopsy containing lesions with MCs. These images are of dimension 1,024 × 1,024 or 512 × 512 pixels, digitized with a spatial resolution of 0.1 mm/pixel. Among the 222 cases, 143 have images in both craniocaudal (CC) and mediolateral-oblique (MLO) views. The MCs in each mammogram were manually identified by a group of experienced radiologists. These MCs were used as ground truth in our study.

Since we are mainly interested in the pairs of images that are similar, we first apply a selection procedure based on the image features of the MCs in these cases to identify those potentially similar image pairs for reader scoring. For this purpose, a set of nine image features [10, 76] is used for quantifying the MCs; these features are commonly used for classification of MC lesions in computer-aided diagnosis (CADx). Specifically, they are (1) image features describing individual MCs, including the standard deviation of the image contrast values of MCs and the maximum and the standard deviation of the sizes of MCs; (2) spatial clustering features of MCs, including the number of MCs in a cluster, the area of the cluster, and the compactness of the cluster; and (3) texture-based features, including the energy, contrast, and correlation derived from the gray-level co-occurrence matrices. The cases in the dataset are then selected for pairing based on the feature values (Euclidean distance) of their MCs. In the end, a total of 2,000 image pairs were selected.

Subsequently, based on the similarity scores collected on the 2,000 image pairs (described below), we further select a subset of 1,000 image pairs from them, the purpose being to refine the set of potentially similar pairs for further reader scoring. These pairs are selected based on both the similarity scores from the readers and the Euclidean distances of all nine features.

Reader Study The reader study was carried out by a group of five radiologists for the 1,000 image pairs, based on their perceptual similarity, using a discrete scale from 0 (most dissimilar) to 10 (most similar). These five radiologists are MQSA-qualified breast imagers with between 2 and 20 years of experience. To reduce the effects of reader fatigue, the set of image pairs is randomly divided into four separate scoring sessions. Similarly, a separate reader study was carried out by a group of five non-radiologists for the 2,000 image pairs (which were used for further pair selection as described above). These readers were researchers in breast imaging with a minimum of 5 years of experience. A total of ten separate sessions were used in scoring.

Because of the subjective nature in interpretation of clustered MCs in mammogram images, readers can vary in their similarity scores. To suppress such apparent differences, we first transformed the similarity scores from individual readers into z-scores. Afterward, the scores were averaged among the readers for the set of 1,000 image pairs (denoted by S_1), which were scored by both radiologists and non-radiologists, and similarly for the other set of 1,000 image pairs (denoted by S_2), which were scored by only non-radiologists. The average scores are further transformed into z-scores.

MDS Plot To explore how perceptually similar cases with clustered MCs relate to each other, we apply the MDS technique to embed the different cases in the dataset in a 2D plot based on their similarity scores.

Consider a pair of cases i and j with similarity score SC_{ij}. In the MDS placement, they will be separated by proximity

$$\delta_{ij} = \frac{1}{3.75 + SC_{ij}} \tag{8.11}$$

where a constant offset 3.75 (over three standard deviations) is added to ensure that d_{ij} is positive.

Due to the fact that similarity scores are available for only those image pairs scored by the readers, the weighted MDS technique is used, in which those image pairs not scored are assigned a weight of 0; for the scored image pairs, the weight is adjusted according to the level of similarity and the readers for scoring as follows: for image pair p consisting of cases i and j,

$$w_p = \begin{cases} 1 & \text{if } p \in S_2 \text{ and } SC_{ij} \leq 0 \\ 1.5 & \text{if } p \in S_1 \text{ and } SC_{ij} \leq 0 \\ 2 & \text{if } p \in S_2 \text{ and } SC_{ij} > 0 \\ 3 & \text{if } p \in S_1 \text{ and } SC_{ij} > 0 \end{cases} \tag{8.12}$$

The rationale for such a choice is to assign a higher weight value to pairs that are more similar and scored by more readers.

In Fig. 8.3, we show the MDS embedding all the 222 cases in the dataset according to their similar scores. While at the first sight there is no apparent separation between cancer and benign cases, it is evident that there are more cancer cases (and fewer benign cases) in the right half of the plot than in the left half. More importantly, cases of same disease tend to be clustered together locally. For example, while cancer cases are scattered in different regions throughout the plot, they are also distributed in small clusters in which a cancer case is closely surrounded by other cancer cases; the same is true for benign cases.

Furthermore, to explore how the readers' notion of similarity may relate to the image features of the clustered MCs, we also show in Fig. 8.3 the spatial distribution patterns of the clustered MCs for some sample cases, where the spatial locations of the individual MCs are indicated by "+" signs. It can be seen that the neighboring cases tend to have MC clusters similar in size and shape and that the MC clusters in the right half of the plot tend to be larger and irregular.

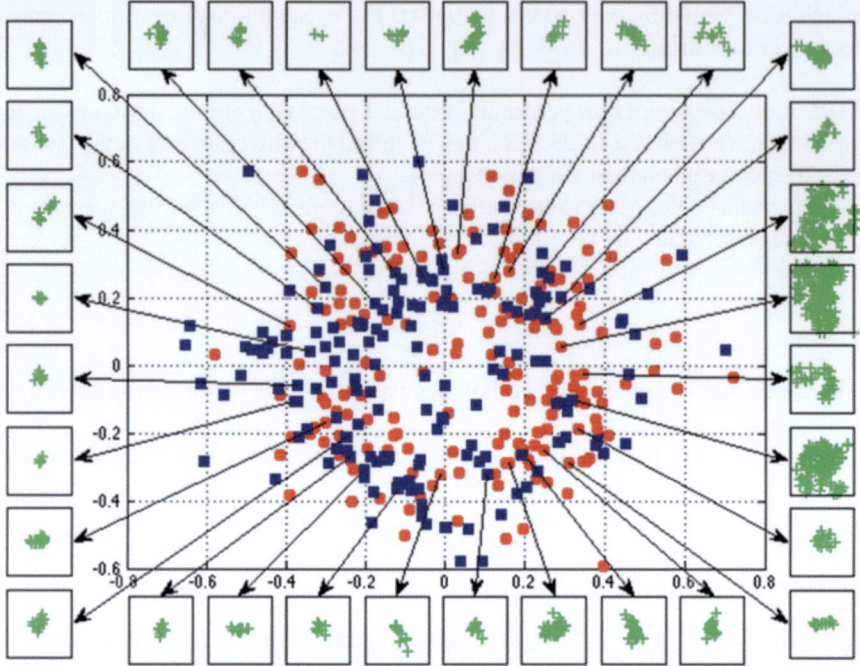

Fig. 8.3 MDS embedding of perceptually similar cases in the dataset, wherein cancer cases are denoted by "red dots" and benign cases are represented by "blue squares." The spatial distribution patterns of clustered MCs are shown for some sample cases, where the spatial MC locations are indicated by the "green plus" signs

8.5 Issues and Recommendations

Despite that there have been many great computerized methods developed for use in CADx schemes as a diagnostic aid to improving radiologists' diagnostic accuracy, some significant, challenging issues still remain to be addressed. Below, we discuss a few of them, which are by no means meant to be complete.

Thanks to intense research and development efforts, multiple laboratory observer studies have shown that CADx schemes can help improve the diagnostic accuracy in differentiating between benign and malignant tumors. For example, in mammography, radiologists with CADx can improve their biopsy recommendation by sending more cancer cases and fewer benign cases to biopsy [17–19, 54, 77]. However, so far, CADx schemes have not yet been introduced clinically.

In CADx, the computer predicts the likelihood that a lesion is malignant, which is presented to the radiologist as a second opinion. One difficulty in implementing CADx clinically is that a CADx classifier is often criticized for being a "black box" approach in its decision. When presented with a numerical value, such as the likelihood of malignancy, but without additional supporting evidence, it may be difficult for a radiologist to incorporate optimally this number into his or her decision. As an alternative aid, image retrieval has been studied as a CADx tool in recent years. We conjecture that by integrating a retrieval system with the CADx classifier, the retrieved images could serve as supporting evidence to the CADx classifier, which may facilitate the interpretation of the likelihood of malignancy by the radiologists.

In the literature, the CADx schemes are often, if not always, developed with different datasets which are limited by the number of cases available. The heterogeneity among the different datasets will inevitably lead to variability when evaluating the performance of a CADx scheme. Thus, it is desirable to establish common benchmark databases which are large enough to be representative of a disease population. In practice, this can be an expensive process. It will ensure that a CADx scheme can be optimized and tested without any bias so that it can generalize well when applied to cases outside the database.

Finally, while research and development has led to improvement in CADx performance, as a diagnostic aid, the accuracy level achieved by CADx classifiers is rather moderate for certain tumor types due to the inherent difficulty of the problem (e.g., MC lesions). There is still need for development of more salient features and CADx algorithms in order to improve the classification accuracy. This may include the use of additional features acquired from multimodality imaging.

Conclusions

In this chapter, we presented the application of machine learning algorithms in CADx systems. Particularly, we presented examples of its application in mammography to differentiate benign and malignant cases. A main critique of traditional CADx approaches when implemented clinically is that a CADx classifier could be perceived as a "black box" approach in its decision. As an alternative aid, CBIR has been studied as a CADx tool in recent years. We conjecture that with the integration of a retrieval system and a CADx classifier, the retrieved images could serve as supporting evidence to the CADx classifier, which may facilitate the interpretation of the likelihood of malignancy by the radiologists in clinical practice. In this chapter, we presented the process of developing and validating such system exploiting both supervised and unsupervised machine learning algorithms.

Acknowledgement This work was supported in part by NIH grant EB009905.

References

1. Horeweg N, Scholten ET, de Jong PA, van der Aalst CM, Weenink C, Lammers J-WJ, et al. Detection of lung cancer through low-dose CT screening (NELSON): a prespecified analysis of screening test performance and interval cancers. Lancet Oncol. 2014;15:1342–50. doi:10.1016/S1470-2045(14)70387-0.
2. Agarwal V, Branstetter BF 4th, Johnson JT. Indications for PET/CT in the head and neck. Otolaryngol Clin North Am. 2008;41:23–49. doi:http://dx.doi.org/10.1016/j.otc.2007.10.005.

3. Thompson J, Lawrentschuk N, Frydenberg M, Thompson L, Stricker P, USANZ. The role of magnetic resonance imaging in the diagnosis and management of prostate cancer. BJU Int. 2013;112 Suppl 2:6–20. doi:10.1111/bju.12381.
4. Leung D, Han X, Mikkelsen T, Nabors LB. Role of MRI in primary brain tumor evaluation. J Natl Compr Canc Netw. 2014;12:1561–8.
5. Young RJ, Knopp EA. Brain MRI: tumor evaluation. J Magn Reson Imaging. 2006;24:709–24. doi:10.1002/jmri.20704.
6. Pilewskie M, King TA. Magnetic resonance imaging in patients with newly diagnosed breast cancer: a review of the literature. Cancer. 2014;120:2080–9. doi:10.1002/cncr.28700.
7. Wu Y, Giger M, Doi K, Vyborny C, Schmidt R, Metz C. Artificial neural networks in mammography: application to decision making in the diagnosis of breast cancer. Radiology. 1993;187(1):81–7.
8. Andreadis II, Spyrou GM, Nikita KS. A comparative study of image features for classification of breast microcalcifications. Meas Sci Tech. 2011;22(11):114005.
9. Cheng HD, Cai X, Chen X, Hu L, Lou X. Computer-aided detection and classification of microcalcifications in mammograms: a survey. Pattern Recognit. 2003;36:2967–91.
10. Jiang Y, Nishikawa RM, Wolverton EE, Metz CE, Giger ML, Schmidt RA, Vyborny CJ. Malignant and benign clustered microcalcifications: automated feature analysis and classification. Radiology. 1996;198:671–8.
11. Sakka E, Prentza A, Koutsouris D. Classification algorithms for microcalcifications in mammograms (review). Oncol Rep. 2006;15(4):1049–55.
12. Wei L, Yang Y, Nishikawa RM, Wenick MN, Edwards A. Relevance vector machine for automatic detection of clustered microcalcifications. IEEE Trans Med Imaging. 2005;24(10):1278–85.
13. Wei L, Yang Y, Nishikawa RM, Jiang Y. A study on several machine-learning methods for classification of malignant and benign clustered microcalcifications. IEEE Trans Med Imaging. 2005;24(3):371–80.
14. Bozek J, Mustra M, Delac K, Grgic M. A survey of image processing algorithms in digital mammography. Recent Adv Multimedia Signal Process Commun. 2009;231:631–57.
15. Cheng HD, Shi XJ, Min R, Hu LM, Cai XP, Du HN. Approaches for automated detection and classification of masses in mammograms. Pattern Recognit. 2006;39(4):646–68.
16. Elter M, Horsch A. CADx of mammographic masses and clustered microcalcifications: a review. Med Phys. 2009;36:2052–68.
17. Huo Z, Giger ML, Vyborny CJ, Wolverton DE, Metz CE. Computerized classification of benign and malignant masses on digitized mammograms: a study of robustness. Acad Radiol. 2000;7(12):1077–84.
18. Chan H, Sahiner B, Helvie MA, Petrick N, Roubidoux MA, Wilson TE, Adler DD, Paramagul C, Newman JS, Sanjay-Gopal S. Improvement of radiologists' characterization of mammographic masses by using computer-aided diagnosis: an ROC study. Radiology. 1999;212:817–27.
19. Horsch K, Giger ML, Vyborny CJ, Lan L, Mendelson EB, Hendrick RE. Classification of breast lesions with multimodality computer-aided diagnosis: observer study results on an independent clinical data set. Radiology. 2006;240:357–68.
20. Huo Z, Giger ML, Vyborny CJ, Wolverton DE, Schimidt RA, Doi K. Automated computerized classification of malignant and benign masses on digitized mammograms. Acad Radiol. 1998;5(3):155–68.
21. Jiang Y, Nishikawa RM, Schmidt RA, Toledano AY, Doi K. Potential of computer-aided diagnosis to reduce variability in radiologists' interpretations of mammograms depicting microcalcifications. Radiology. 2001;220:787–94.
22. Müller H, Michoux N, Bandon D, Geissbuhler A. A review of content-based image retrieval systems in medical applications–clinical benefits and future directions. Int J Med Inform. 2004;73:1–23.
23. Bustos B, Keim D, Saupe D, Schreck T. Content-based 3D object retrieval. IEEE Comput Graph Appl. 2007;27:22–7.

24. Bhanu B, Peng J, Qing S. Learning feature relevance and similarity metrics in image database. In: IEEE workshop proceedings on content-based access of image and video libraries. Washington, DC; 1998. p. 14–8.
25. El-Naqa I, Yang Y, Galasanos NP, Nishikawa RM, Wernick MN. A similarity learning approach to content based image retrieval: application to digital mammography. IEEE Trans Med Imaging. 2004;23:1233–44. doi:10.1109/TMI.2004.834601.
26. Vapnik V. Statistical learning theory. New York: Wiley; 1998.
27. Jiang Y, Nishikawa RM, Giger ML, Doi K, Schmidt R, Vyborny C. Method of extracting signal area and signal thickness of microcalcifications from digital mammograms. Proc SPIE. 1992;1778:28–36.
28. Sachdeva J, Kumar V, Gupta I, Khandelwal N, Ahuja CK. Segmentation, feature extraction, and multiclass brain tumor classification. J Digit Imaging. 2013;26(6):1141–50.
29. Soltanian-Zadeh H, Rafiee-Rad F, Pourabdollah-Nejad S. Comparison of multiwavelet, wavelet, Haralick, and shape features for microcalcification classification in mammograms. Pattern Recognit. 2004;37:1973–86.
30. Haralick R, Shanmugam K, Dinstein I. Textural features for image classification. IEEE Trans Syst Man Cybern. 1973;3:610–21.
31. Gonzalez RC, Woods RE. Digital image processing. Upper Saddle River: Prentice Hall; 2002.
32. Yu L, Liu H. Feature selection for high-dimensional data: a fast correlation-based filter solution. ICML. 2003;3:856–63.
33. Kohavi R, John GH. Wrappers for feature subset selection. Artif Int. 1997;97(1):273–324.
34. Perkins S, Lacker K, Theiler J. Grafting: fast, incremental feature selection by gradient descent in function space. J Mach Learn Res. 2003;3:1333–56.
35. Bishop CM. Pattern recognition and machine learning. New York: Springer; 2006.
36. Friedman J, Hastie T, Tibshirani R. The elements of statistical learning. New York: Springer; 2009.
37. Cortes C, Vapnik V. Support-vector networks. Mach Learn. 1995;20(3):273–97.
38. Cristianini N, Shawe-Taylor J. An introduction to support vector machines and other kernel-based learning methods. Cambridge/New York: Cambridge University Press; 2000.
39. Tipping ME. Sparse Bayesian learning and the relevance vector machine. J Mach Learn Res. 2001;1:211–44.
40. Freund Y, Schapire RE. A decision-theoretic generalization of on-line learning and an application to boosting. J Comput Syst Sci. 1997;55(1):119–39.
41. Dietterich TG. An experimental comparison of three methods for constructing ensembles of decision trees: bagging, boosting, and randomization. Mach Learn. 2000;40(2):139–57.
42. Breiman L, Spector P. Submodel selection and evaluation in regression. The x-random case. Int Stat Rev. 1992;60:291–319.
43. Kohavi R. A study of cross-validation and bootstrap for accuracy estimation and model selection. IJCAI. 1995;14(2):1137–45.
44. Mertens BJ, de Noo ME, Tollenaar RAEM, Dcclder AM. Mass spectrometry proteomic diagnosis: enacting the double cross-validatory paradigm. J Comput Biol. 2006;13(9):1591–605.
45. Mushlin AI, Kouides RW, Shapiro DE. Estimating the accuracy of screening mammography: a meta-analysis. Am J Prev Med. 1998;14:143–53.
46. Urbain JL. Breast cancer screening, diagnostic accuracy and health care policies. CMAJ. 2005;172:210–1.
47. Kolb TM, Lichy J, Newhouse JH. Comparison of the performance of screening mammography, physical examination, and breast US and evaluation of factors that influence them: an analysis of 27,825 patient evaluations. Radiology. 2002;225:165–75.
48. Elmore JG, Barton MB, Moceri VM, Polk S, Arena PJ, Fletcher SW. Ten-year risk of false positive screening mammograms and clinical breast examinations. N Engl J Med. 1998; 338(16):1089–96.
49. Tan A, Freeman Jr DH, Goodwin JS, Freeman JL. Variation in false-positive rates of mammography reading among 1067 radiologists: a population-based assessment. Breast Cancer Res Treat. 2006;100:309–18.

50. Sickles EA, Miglioretti DL, Ballard-Barbash R, Geller BM, Leung JWT, Rosenberg RD, Smith-Bindman R, Yankaskas BC. Performance benchmarks for diagnostic mammography. Radiology. 2005;235:775–90.
51. Chan H, Wei D, Lam K, Lo S, Sahiner B, Helvie M, Adler D. Computerized detection and classification of microcalcifications on mammograms. SPIE. 1995;2434:612–20.
52. Chan H, Sahiner B, Lam KL, Petrick N, Helvie MA, Goodsitt MM, Adler DD. Computerized analysis of mammographic microcalcifications in morphological and texture feature space. Med Phys. 1998;25:2007–19.
53. Markopoulos C, Kouskos E, Koufopoulos K, Kyriakou V, Gogas J. Use of artificial neural networks (computer analysis) in the diagnosis of microcalcifications on mammography. Eur J Radiol. 2001;39(1):60–5.
54. Jiang Y, Nishikawa RM, Schmidt RA, Metz CE, Giger ML, Doi K. Improving breast cancer diagnosis with computer-aided diagnosis. Acad Radiol. 1999;6:22–33.
55. Nishikawa RM. Current status and future directions of computer-aided diagnosis in mammography. Comput Med Imaging Graph. 2007;31:224–35.
56. Rangayyan RM, Fabio JA, Desautels JL. A review of computer-aided diagnosis of breast cancer: toward the detection of subtle signs. J Franklin Inst. 2007;344:312–48.
57. Sampat MP, Markey MK, Bovik AC. Computer-aided detection and diagnosis in mammography. Chap. 10.4. In: Handbook of image & video processing. 2nd ed. Amsterdam/Boston: Elsevier Academic Press; 2005.
58. Wang J, Yang Y. Spatial density modeling for discriminating between benign and malignant microcalcification lesions. In: IEEE international conference on image processing. San Francisco. 2013;133–6.
59. Muller H, Michoux N, Bandon D, Geissbuhler A. A review of content-based image retrieval system in medical applications-clinical benefits and future directions. Int J Med Info. 2004;73:1–23.
60. Rahman M, Want T, Desai B. Medical image retrieval and registration: towards computer assisted diagnostic approach. In: Proceedings of IDEAS workshop on medical information systems: the Digital Hospital. Canada; 2004. p. 78–89.
61. Holt A, Bichindaritz I, Schmidt R, Perner P. Medical applications in case-based reasoning. Knowl Eng Rev. 2005;20:289–92.
62. Bilska-Wolak A, Floyd E. Development and evaluation of a case-based reasoning classifier for prediction of breast biopsy outcome with BI-RADS™ lexicon. Med Phys. 2002;29:2090.
63. Floyd CE, Lo J, Tourassi GD. Case-based reasoning computer algorithm that uses mammographic findings for breast biopsy decisions. Am J Roentgenol. 2000;175(5):1347–52.
64. Zheng B, Lu A, Hardesty LA, Sumkin JH, Hakim CM, Ganott MA, Gur D. A method to improve visual similarity of breast masses for an interactive computer-aided diagnosis environment. Med Phys. 2006;33:111–7.
65. Jing H, Yang Y. Case-adaptive classification based on image retrieval for computer-aided diagnosis. In: IEEE international conference on image processing. Hong Kong; 2010. p. 4333–6.
66. Wei L, Yang Y, Nishikawa RM, Jiang Y. Learning of perceptual similarity from expert readers for mammogram retrieval. In: IEEE international symposium on biomedical imaging. Arlington; 2006. p. 1356–9.
67. Wei L, Yang Y, Nishikawa RM. Microcalcification classification assisted by content-based image retrieval for breast cancer diagnosis. Pattern Recognit. 2009;42:1126–32.
68. Heath M, Bowyer K, Kopans D, Moore R, Kegelmeyer WP. The digital database for screening mammography. The fifth international workshop on digital mammography. Toronto; 2001. p. 212–8.
69. Mcloughlin KJ, Bones PJ, Karssemeijer N. Noise equalization for detection of microcalcification clusters in direct digital mammogram images. IEEE Trans Med Imaging. 2004;23(3):313–20.
70. Aisen A, Broderick L, Winer-Muram H, Brodley C, Kak A, Pavlopoulou C, Dy J, Shyu C, Marchiori A. Automated storage and retrieval of thin-section CT images to assist diagnosis: system description and preliminary assessment. Radiology. 2003;228:265–70.

71. Muramatsu C, Nishimura K, Endo T, Oiwa M, Shiraiwa M, Doi K, Fujita H. Representation of lesion similarity by use of multidimensional scaling for breast masses on mammograms. J Digit Imaging. 2013;26(4):740–7.
72. Tourassi GD, Harrawood B, Singh S, Lo JY, Floyd CE. Evaluation of information-theoretic similarity measures for content-based retrieval and detection of masses in mammograms. Med Phys. 2007;34:140–50.
73. Yang L, Jin R, Mummert L, Sukthankar R, Goode A, Zheng B, Hoi SCH, Satyanarayanan M. A boosting framework for visuality-preserving distance metric learning and its application to medical image retrieval. IEEE Trans Pattern Aana Mach Intell. 2010;32(1):30–44.
74. Wang J, Jing H, Wernick MN, Nishikawa RM, Yang Y. Analysis of perceived similarity between pairs of microcalcification clusters in mammograms. Med Phys. 2014;41(5):051904.
75. Borg I, Groenen PJF. Modern multidimensional scaling: theory and application. New York: Springer; 2005.
76. Karahahiou AN, Boniatis IS, Skiadopoulos SG, Sakellaropoulos FN, Arikidis NS, Likaki EA, Panayiotakis GS, Costaridou LI. Breast cancer diagnosis: analyzing texture of tissue surrounding microcalcifications. IEEE Trans Inf Technol Biomed. 2008;12:731–8.
77. Hadjiiski L, Chan HP, Sahiner B, Helvie MA, Roubidoux MA, Blane C, Paramagul C, Petrick N, Bailey J, Klein K, Foster M, Patterson S, Alder D, Nees A, Shen J. Improvement in radiologists' characterization of malignant and benign breast masses on serial mammograms with computer-aided diagnosis: an ROC study. Radiology. 2004;233:255–65.

Part III

Machine Learning for Treatment Planning

Image-Guided Radiotherapy with Machine Learning

9

Yaozong Gao, Yanrong Guo, Yinghuan Shi, Shu Liao, Jun Lian, and Dinggang Shen

Abstract

In the past decades, many machine learning techniques have been successfully developed and applied to the field of image-guided radiotherapy (IGRT). In this chapter, we will present some latest developments in the application of machine learning techniques to this field. In particular, we focus on the recently developed machine learning methods for delineating male pelvic structures for the treatment of prostate cancer. In the first few sections, we will present and discuss

---

Y. Gao • Y. Guo
UNC IDEA Group, Department of Radiology, Biomedical Research Imaging Center (BRIC),
University of North Carolina, Chapel Hill, NC 27599, USA
e-mail: yzgao@cs.unc.edu; yrguo@mail.unc.edu

Y. Shi
State Key Laboratory for Novel Software Technology, Department of Computer Science
and Technology, Nanjing University, Nanjing, Jiangsu, China, 210023
e-mail: syh@nju.edu.cn

S. Liao
Syngo, Siemens Medical Solutions, Malvern, PA 19355, USA
e-mail: liaoshu.cse@gmail.com

J. Lian
Department of Radiation Oncology, University of North Carolina,
Chapel Hill, NC 27599, USA
e-mail: jun_lian@med.unc.edu

D. Shen (✉)
UNC IDEA Group, Department of Radiology, Biomedical Research Imaging Center (BRIC),
University of North Carolina, Chapel Hill, NC 27599, USA

Department of Radiology, University of North Carolina,
124 Mason Farm Road, Chapel Hill, NC 27599, USA
e-mail: dgshen@med.unc.edu

automatic and semiautomatic methods for CT prostate segmentation in the IGRT workflow. In the last section, we will present our extension of some recently developed machine learning approaches to segment the prostate in MR images.

9.1 Background

IGRT aims to deliver therapeutic doses to cancerous tissues based on imaging of patient. It involves multiple complicated procedures including image acquisition, tumor/normal structure delineation, dose optimization, quality assurance, and treatment delivery. In such a long and complicated workflow, potential mistakes at one or more of these steps can compromise the expected treatment outcome and even harm the patients. In particular, the uncertainty of normal and tumor tissue segmentation is most likely to alter radiation therapy and its outcome [1]. The quality of manual contouring results highly depend on the expertise of the clinician. However, this task is prone to errors and often associated with large inter-operator variation [2]. Furthermore, manual segmentation is time consuming and can also be a hard task for a clinician to distinguish between different complex anatomies with poor imaging contrast, using the naked eyes. Large variation in tumor or normal structure anatomy during the course of fractional radiation therapy necessitates the re-optimization of the original treatment plans accordingly. Treatment methods developed for IGRT target accurate radiation of the tumor tissue while sparing the neighboring normal tissues. However, the adaption stage demands the repetition of multiple steps in the pipeline of radiation therapy, including manual segmentation. When adaptive planning is based on the in-room setup CT of the patient on the treatment day, the segmentation task becomes even more challenging and time consuming since the in-room imaging device, such as cone-beam CT (CBCT), generates inferior image quality to regular planning CT, which is often used for the initial scanning of the patient treatment design. This may be an obstacle for IGRT being widely employed in clinical settings. However, devising sophisticated segmentation methods may overcome this grand challenge and, thereby, improve tumor treatment.

In the last two decades, many improvements from the sophisticated dose optimization algorithm to the advanced linear accelerator targeted a more accurate, efficient, and safer radiation treatment. Computer-aided segmentation is believed to provide a solution to ease the load and the difficulty of tumor and normal tissue segmentation. However, this seems to be a difficult mathematical and image processing problem because of limited image quality, scanning protocol-dependent structure appearance, intersubject variation, and deformation of anatomical boundaries of organs. CT is routinely used for the treatment planning of prostate radiotherapy; however, low soft tissue contrast makes the differentiation between the prostate and the rectum/bladder very difficult. Meanwhile, MR imaging is becoming more widely used in radiotherapy because it is radiation free and provides a better image contrast than CT. However, the appearance of structure is subject to variability across different acquisition techniques. The image segmentation algorithm that works well for an MR image of a particular acquisition protocol may perform *poorly* on another image acquired with a different protocol.

The methods used in auto-segmentation of radiation therapy can be categorized into three main categories according to whether prior knowledge is used [3]. The first category of methods does not utilize information from the previously acquired images and simply uses the voxel intensity and image gradient for structure delineation. The second category of methods, powered by prior knowledge, morphology, or the appearance of organs in previous images, is proved to be more robust and accurate than the first category of methods. The third category of methods (also called hybrid approaches) combines different segmentation algorithms to achieve better performance.

It is a challenging problem on how to retrieve and use prior knowledge effectively. In the literature, three types of knowledge-based segmentation algorithms have been reported, including atlas-based, statistical model-based, and machine learning-based methods [3]. In the atlas-based segmentation, an average image and also the delineated structures are used as a reference or an atlas. The contours on the new image were obtained by transforming the reference image to the new image [4]. Model-based algorithms need to build a statistical shape model or a statistical appearance model that provides an anatomically plausible surface. The characteristics of the model are trained from a set of images with a correct delineation of structures [5]. Machine learning techniques have been employed for classifying different structures and learning image content and tissue appearance from the prior images. In general, the model, trained by features of voxel neighborhood, is more flexible and universal than the other two kinds of segmentation methods [6–8]. In the next section, we will introduce some previous CT prostate segmentation methods and their limitations.

9.2 Previous Methods

Despite of the importance of organ segmentation in IGRT, it remains quite challenging. For example, in the CT-guided prostate radiation therapy, the CT prostate segmentation is still a difficult problem due to the following three reasons: First, unlike the planning CT image, the treatment CT images are of lower quality because they are typically acquired with non-diagnostic CT scanner. As a result, the image contrast of a treatment CT is relatively lower, compared to a regular CT. Figure 9.1 shows several typical treatment CTs and their prostate contours (red). Second, due to existence of bowel gas and filling (as indicated by red arrows in Fig. 9.1), the image appearance of treatment CTs can change drastically. Third, the unpredicted daily prostate motion [9] further complicates the precise prostate segmentation.

Many methods have been proposed to address this paramount yet compelling segmentation problem. For example, Freedman et al. [10] proposed to segment the prostate in CT images by matching the probability distributions of photometric variables (e.g., voxel intensity). Costa et al. [11] proposed the coupled 3D deformable models by considering the nonoverlapping constraint from the bladder. Foskey et al. [12] proposed a deflation method to explicitly eliminate bowel gas before 3D deformable registration. Chen et al. [13] incorporated the anatomical constraints from the rectum to assist the deformable segmentation of the prostate. Haas et al. [14] used 2D flood fill with the shape guidance to localize the prostate in CT images. Ghosh et al. [15] proposed a genetic algorithm with prior knowledge in the form of texture and shape. Although these methods have shown the effectiveness in CT prostate segmentation, their segmentation accuracy is very limited (usually with the overlap ratio (Dice similarity coefficient (DSC)) around 0.8), which can be explained by two factors. First, most of these methods rely only on the image intensity/gradient information to localize the prostate boundary. As shown in Fig. 9.1, due to the indistinct prostate boundary in CT images, simple intensity features are neither reliable nor sufficient to accurately localize the prostate. While several deformable segmentation methods [11, 13] utilized the spatial relationship of the prostate to its nearby organs (e.g., the rectum and bladder) to prevent the over-segmentation of the prostate, these strategies improve only the robustness, not the accuracy of the segmentation. Second, the majority of these methods overlook the information that is inherent in the IGRT workflow. In fact, at each treatment day, several CT scans of the same patient have already been acquired and segmented in the planning day and the previous treatment days. These valuable patient-specific images can be exploited to largely improve patient-specific prostate segmentation.

In the following sections, we will first introduce four recently developed machine learning methods to address the aforementioned challenges by automatically

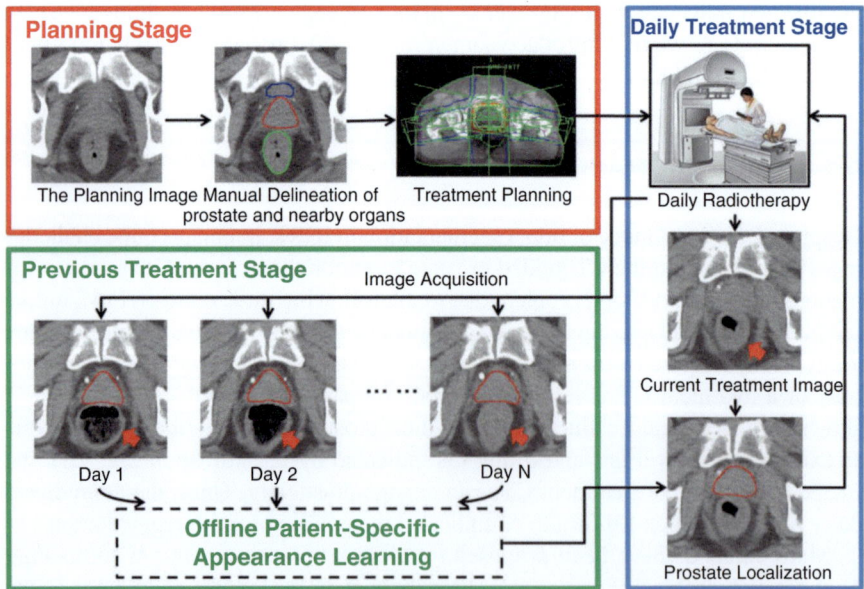

Fig. 9.1 Illustration of image-guided radiotherapy. *Red* contours indicate the prostates

learning the effective features from the previously acquired data of the same patient for accurate prostate localization in CT images. Then, an extension to segmentation of MR prostate images is further presented.

9.3 Learning-Based Prostate Segmentation in CT and MR Images

9.3.1 Learning-Based Landmark Detection for Fast Prostate Localization in Daily Treatment CTs

In this section, we will introduce an application of boosting techniques in automatic landmark detection for fast prostate localization. Moreover, we will show how the accuracy of landmark detection and prostate localization can be further improved within the IGRT setting by adopting a novel method – namely, incremental learning with selective memory (ILSM) – to gradually incorporate the patient-specific information into the general population-based landmark detectors during the treatment course.

In the machine learning field, boosting refers to a technique which sequentially trains a list of weak classifiers to form a strong classifier [16]. One of the most successful applications of boosting is the robust real-time face detection method developed by Paul Viola and Michael J. Jones [16]. In their method, they combined Haar features with the boosting algorithm to learn a cascade of classifiers for efficient face detection. This idea can also be used to detect the anatomical landmark in medical images, which we call learning-based landmark detection. In such methods, the landmark detection is formulated as a classification problem. Specifically, for each image, voxels close to the specific landmark are positive and all others are negatives. In the training stage, the cascade learning framework is applied to learn a sequence of classifiers for gradually separating negatives from positives (Fig. 9.2). Compared to learning a single classifier, cascade learning has shown better classification accuracy and runtime efficiency [17, 18]. Mathematically, cascade learning can be formulated as:

Input: Positive voxel set X_P, negative voxel set X_N, and label set $L = \{+1, -1\}$.
Classifier: $C(x): \mathbb{F}(x) \to L$, where $\mathbb{F}(x)$ denotes the appearance features of a voxel x.
Initial set: $X_0 = X_P \cup X_N$.
Objective: Optimize C_k, $k = 1, 2, \cdots, K$, such that

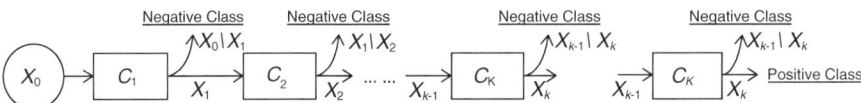

Fig. 9.2 Illustration of cascade learning

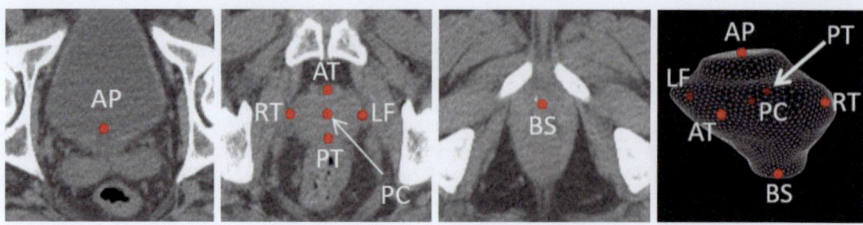

Fig. 9.3 Seven prostate landmarks: prostate center (*PC*), right lateral point (*RT*), left lateral point (*LF*), posterior point (*PT*), anterior point (*AT*), base center (*BS*), and apex center (*AP*)

$$X_0 \supseteq X_1 \supseteq \cdots \supseteq X_k \supseteq \cdots \supseteq X_K, \ X_K \supseteq X_P, \text{ and } \|X_K \cap X_N\| \leq \tau \|X_P\|$$

where $X_k = \{x \mid x \in X_{k-1} \text{ and } C_k(x) = +1\}$ and τ controls the tolerance ratio of false positives.

Here, the cascade classifiers C_k, $k = 1, 2, \cdots, K$, are optimized sequentially. As shown in Eq. (9.1), C_k is optimized to minimize the false positives left over by the previous $k-1$ classifiers:

$$C_k = \underset{C}{\operatorname{argmin}} \left\| \{x \mid x \in X_{k-1} \cap X_N \text{ and } C(x) = +1\} \right\|, \quad s.t. \ \forall x \in X_P, C(x) = +1 \quad (9.1)$$

where $\|\cdot\|$ denotes the cardinality of a set. It is worth noting that the constraint in Eq. (9.1) can be simply satisfied by adjusting the threshold of classifier C_k to make sure that all positive training samples are correctly classified. This cascade learning framework is general to any image feature and classifier. In the conventional cases, extended Haar wavelets [18–20] and AdaBoost classifier [16] are typically employed.

Once the cascade classifiers $\{C_k(x)\}$ are learned, they have captured the appearance characteristics of the specific anatomical landmark. Given a testing image, the learned cascade is applied to each voxel. The voxel with the highest classification score after going through the entire cascade is selected as the detected landmark. To increase the efficiency and robustness of the detection procedure, a multiscale scheme is further adopted. Specifically, the detected landmark in the coarse resolution serves as the initialization for landmark detection in a following finer resolution, in which the landmark is only searched within a local neighborhood centered by the initialization. In this way, in the fine resolution, the search space is limited to a small neighborhood around the coarse-level detection, instead of the entire image domain, thus, making the detection procedure more robust to local minima.

We can adopt the learning-based landmark detection method to detect seven key landmarks of the prostate (Fig. 9.3), which are the prostate centroid, the apex center, the base center, and four extreme points in the middle slice, respectively.

After seven landmarks are automatically detected in the new treatment image, we can use them to align the patient-specific prostate shapes delineated in the previous planning and treatment days onto the current treatment image for fast localization. Specifically, a rigid transform is estimated between the detected landmarks in the current treatment image and the corresponding ones in the previous planning or treatment image for shape alignment. By aligning all previous patient-specific prostate shapes onto the current treatment image, a simple label fusion technique like majority voting can be used for fusing different labeling results to derive a final segmentation of the prostate. To take into account those wrongly detected landmarks, the RANSAC point-set matching algorithm can also be combined with multi-atlas segmentation to improve the robustness (for details, please refer to [21]). This landmark-based prostate localization strategy is very fast and takes only a few seconds to localize the whole prostate in a new treatment image.

Using cascade learning, one can learn anatomy detectors from the training images of different patients (*population-based learning*). However, since intra-patient anatomy variations are much less noticeable than inter-patient variations, patient-specific appearance information available in the IGRT workflow should be exploited in order to improve the detection accuracy for an individual patient. Unfortunately, the number of patient-specific images is often very limited, especially at the beginning of IGRT. To overcome this problem, one may apply random spatial/intensity transformations to produce more "synthetic" training samples with larger variability. However, these artificially created transformations may not capture the real intra-patient variations, e.g., the uncertainty of bowel gas and filling (Fig. 9.4). As a result, cascade learning, using only patient-specific data (*pure patient-specific learning*), often suffers from overfitting. One can also mix population and patient-specific images for training (*mixture learning*). However, since patient-specific images are the "minority" in the training samples, detectors trained by mixed samples might not capture patient-specific characteristics very well.

9.3.1.1 Incremental Learning with Selective Memory (ILSM)

To address the above problem, a novel learning scheme, namely, incremental learning with selective memory (ILSM), is proposed to combine the general information in the population images with the individual information in the patient-specific images. Specifically, population-based landmark detectors serve as an initial appearance model and are subsequently "personalized" by the limited patient-specific data. ILSM consists of *backward pruning* to discard obsolete population appearance information and *forward learning* to incorporate the online-learned patient-specific appearance characteristics.

Notation Denote $D^{\text{pop}} = \{C_k^{\text{pop}}, k = 1, 2, \cdots, K^{\text{pop}}\}$ as the population-based landmark detector learned by using the cascade learning framework. X_P^{pat} and X_N^{pat} are positives and negatives from the patient-specific training images (i.e., previous planning or treatment images), respectively. $D(x)$ denotes the class label (landmark vs non-landmark) of voxel x predicted by landmark detector D.

Backward Pruning The general appearance model learned from the population is not necessarily applicable to the specific patient. More specifically, the anatomical landmarks in the patient-specific images (positives) may be classified as negatives by the population-based anatomy detectors, i.e., $\exists k \in \{1, 2, \cdots, K^{\text{pop}}\}, \exists x \in X_P^{\text{pat}}, C_k^{\text{pop}}(x) = -1$. In order to discard these parts of the population appearance model that do not fit the patient-specific characteristics, we propose *backward pruning* to tailor the population-

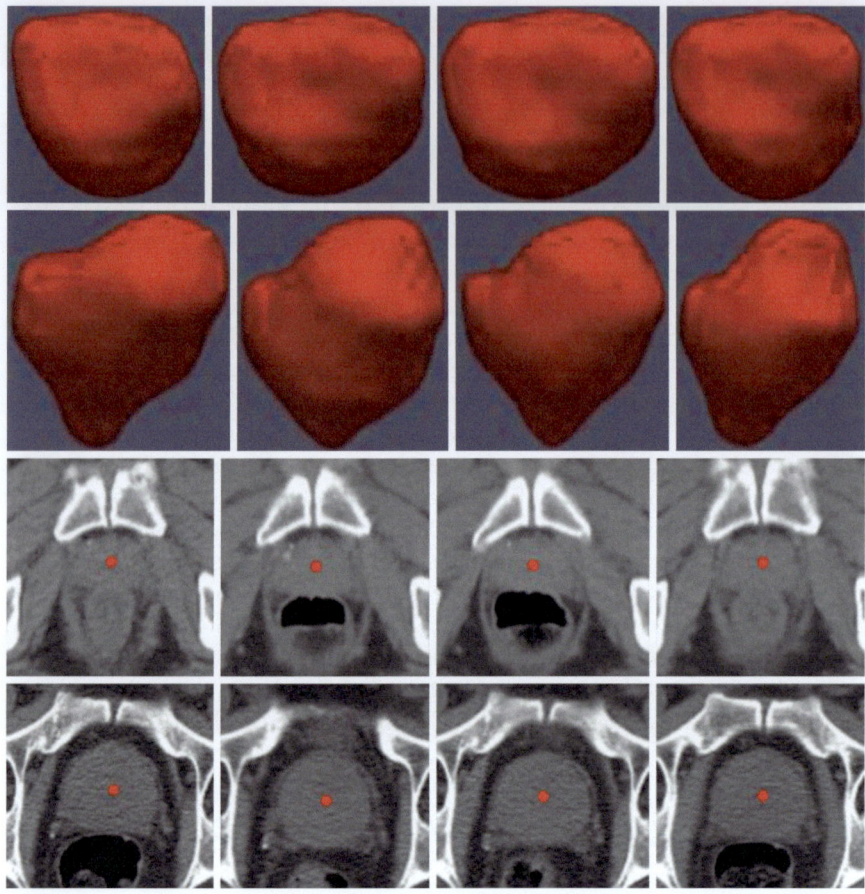

Fig. 9.4 Inter- and intra-patient prostate shape and appearance variations. *Red points* denote the prostate center. Each *row* represents prostate shapes and images for the same patient

based detector. As shown in Algorithm 9.1, in backward pruning, the cascade is pruned from the last level until all patient-specific positives successfully pass through the cascade. This is equivalent to searching for the maximum number of cascade levels that could be preserved from the population-based anatomy detector (Eq. 9.2):

$$K^{bk} = \max\left\{k \mid C_i^{pop}(x) = +1, \forall i \leq k, \forall x \in X_P^{pat}\right\} \quad (9.2)$$

Algorithm 9.1. Backward Pruning Algorithm
Input: $D^{pop} = \left\{C_k^{pop}, k = 1, 2, \cdots, K^{pop}\right\}$
          – the population-based detector
    X_P^{pat} – patient-specific positive samples
Output: D^{bk} – the tailored population-based detector
Init: $k = K^{pop}$, $D^{bk} = D^{pop}$.
while $\exists x \in X_P^{pat} : D^{bk}(x) = -1$ **do**

$$D^{\text{bk}} = D^{\text{bk}} \setminus C_k^{\text{pop}}; k = k-1$$
end while
$K^{\text{bk}} = k$
return $D^{\text{bk}} = \left\{ C_k^{\text{pop}}, k = 1, 2, \cdots, K^{\text{bk}} \right\}$

Forward Learning Once the population cascade has been tailored, the remaining cascade of classifiers encodes the population appearance information that is consistent with the patient-specific characteristics. Yet, until now no real patient-specific information has been incorporated into the cascade. More specifically, false positives might exist in the patient-specific samples, i.e., $\exists x \in X_N^{\text{pat}}$, $\forall k \leq K^{\text{bk}}$, $C_k^{\text{pop}}(x) = +1$. In the forward learning stage, we use the remaining cascade from the backward pruning algorithm as an initialization, and re-apply the cascade learning to eliminate the patient-specific false positives left over by the previously inherited population classifiers. As shown in Algorithm 9.2, a greedy strategy is adopted to sequentially optimize a set of additional patient-specific classifiers $\left\{ C_k^{\text{pat}}, k = 1, 2, \cdots, K^{\text{pat}} \right\}$.

Algorithm 9.2. Forward Learning Algorithm
Input: $D^{\text{bk}} = \left\{ C_k^{\text{pop}}, k = 1, 2, \cdots, K^{\text{bk}} \right\}$
   – the tailored population-based detector
X_P^{pat} – patient-specific positive samples
X_N^{pat} – patient-specific negative samples
Output: D^{pat} – patient-specific detector
Init: $k = 1$, $D^{\text{pat}} = D^{\text{bk}}$
$$X_0 = \left\{ x \mid x \in X_N^{\text{pat}} \bigcup X_P^{\text{pat}}, D^{\text{bk}}(x) = +1 \right\}$$
while $\left\| X_{k-1} \cap X_N^{\text{pat}} \right\| > \tau \left\| X_P^{\text{pat}} \right\|$ **do**
   Train the classifier by minimizing Eq. 9.3 below
$$C_k^{\text{pat}} = \arg\min_c \left\| \left\{ x \mid x \in X_{k-1} \cap X_N^{\text{pat}}, C(x) = +1 \right\} \right\| \quad \text{s.t.} \forall x \in X_P^{\text{pat}}, C(x) = +1 \quad (9.3)$$
$X_k = \left\{ x \mid x \in X_{k-1}, C_k^{\text{pat}}(x) = +1 \right\}$
$D^{\text{pat}} = D^{\text{pat}} \bigcup C_k^{\text{pat}}$; $k = k+1$
end while
$K^{\text{pat}} = k - 1$
return $D^{\text{pat}} = \left\{ C_k^{\text{pop}}, k = 1, 2, \cdots, K^{\text{bk}} \right\} \bigcup \left\{ C_k^{\text{pat}}, k = 1, 2, \cdots, K^{\text{pat}} \right\}$

$\|\cdot\|$ *denotes the cardinality of a set. τ is the parameter controlling the tolerance of false positives.*

After backward pruning and forward learning, the personalized anatomy detector includes two groups of classifiers (Fig. 9.5). While $\left\{ C_k^{\text{pat}}, k = 1, 2, \cdots, K^{\text{pat}} \right\}$ encodes patient-specific characteristics, $\left\{ C_k^{\text{pop}}, k = 1, 2, \cdots, K^{\text{bk}} \right\}$ contains population information that is individualized to this specific patient. This population information effectively remedies the limited variability from the small number of patient-specific training images.

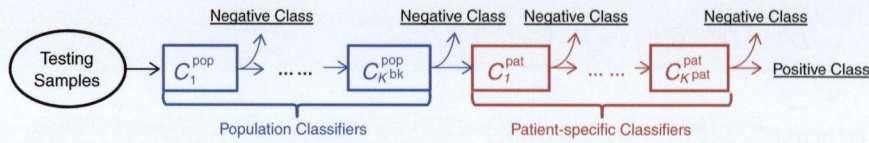

Fig. 9.5 Incrementally learned landmark detector

Table 9.1 Difference between ILSM and four learning-based methods

		POP	PPAT	MIX	IL	ILSM
Training images	Population	√		√	√	√
	Patient specific		√	√	√	√
Learning strategies	Cascade learning	√	√	√	√	√
	Backward pruning					√
	Forward learning				√	√

POP population-based learning, *PPAT* pure patient-specific learning, *MIX* population and patient-specific mixture learning, *IL* incremental learning without backward pruning, *ILSM* proposed incremental learning with selective memory

9.3.2 Experimental Results

Our experimental dataset was acquired from the University of North Carolina at Chapel Hill. In total, we have 25 patients with 349 CT images. Each patient has one planning scan and multiple treatment scans. The prostates and seven landmarks in all CT images have been manually delineated by an expert to serve as the ground truth. We use fivefold cross-validation to evaluate our method and compare it with other methods. To emulate the real clinical setting, for prostate localization in the treatment day $N+1$, we use N previous treatment images and also the planning image as patient-specific training data (Fig. 9.1). From our observations, we found that, when N reaches 4, there was negligible accuracy gained from performing additional ILSMs. Therefore, after treatment day 4, we do not perform ILSM to further refine the patient-specific landmark detectors; instead, we directly adopt the existing detectors for prostate localization. The following reported performances of ILSM are computed using *up to* 5 patient-specific training images (4 treatment images + 1 planning image). Details about the algorithm parameters can be found in [21].

9.3.3 Comparison with Traditional Learning-Based Approaches

To illustrate the effectiveness of our learning framework, we compared ILSM with four different learning-based approaches. All of these methods localize the prostate through learning-based anatomy detection with the same features, classifiers, and cascade framework. Their differences lie in the training images and learning strategies, which are shown in Table 9.1. Note that for all patient-specific training images, artificial transformations are applied to increase the intersubject variability.

9 Image-Guided Radiotherapy with Machine Learning

Table 9.2 Quantitative comparison of landmark detection error (mm) between ILSM and four learning-based methods

	POP	PPAT	MIX	IL	ILSM
PC	6.69 ± 3.65	4.89 ± 5.64	6.03 ± 3.03	5.87 ± 4.01	**4.73 ± 2.69**
RT	7.85 ± 8.44	6.09 ± 9.00	5.72 ± 4.04	6.33 ± 4.82	**3.76 ± 2.80**
LF	6.89 ± 4.63	5.39 ± 7.61	5.61 ± 3.63	5.90 ± 4.54	**3.69 ± 2.69**
PT	7.04 ± 5.04	8.66 ± 13.75	6.18 ± 4.76	6.74 ± 5.05	**4.78 ± 4.90**
AT	6.60 ± 4.97	4.54 ± 5.06	5.38 ± 4.55	5.68 ± 4.97	**3.54 ± 2.19**
BS	6.12 ± 2.97	5.63 ± 7.44	6.63 ± 3.98	5.61 ± 2.94	**4.68 ± 2.71**
AP	10.42 ± 6.03	8.94 ± 16.07	8.77 ± 5.00	9.50 ± 7.17	**6.28 ± 4.60**
Average	7.37 ± 5.52	6.31 ± 10.13	6.33 ± 4.32	6.52 ± 5.09	**4.49 ± 3.49**
p-value	$< 10^{-5}$	$< 10^{-5}$	$< 10^{-5}$	$< 10^{-5}$	n/a

The last row shows the *p*-values of two-sample *t*-test when comparing landmark errors of the four learning-based methods with those of ILSM. The best performance of each measurement is shown in bold lettering

Table 9.2 compares the four learning-based approaches with ILSM on landmark detection errors. We can see that ILSM outperforms other four learning-based approaches on all seven anatomical landmarks. Table 9.3 compares the four learning-based approaches with ILSM on overlap ratios (DSC). To exclude the influence of multi-atlas RANSAC, only a single shape atlas (i.e., the planning prostate shape) is used for localization. Here, "acceptance rate" denotes the percentage of images where an algorithm performs with higher accuracy than inter-operator variability (DSC = 0.81) [22]. According to our experienced clinician, these results can be accepted with minimal manual editing (<20 %). We can see that ILSM achieves the best localization accuracy among all methods. Not surprisingly, by utilizing patient-specific information, all three methods (i.e., PPAT, MIX, and IL) outperform POP. However, their performances are still inferior to ILSM, which shows the effectiveness of ILSM in combining both population- and patient-specific characteristics.

9.3.4 Sparse Representation-Based Classification for Treatment Image Segmentation

Sparse representation as an emerging technique has become the focus of much recent research in machine learning [23, 24], signal processing [25], and computer vision [26, 27]. It has been successfully applied in many fields, such as compressive sensing [28] and face recognition [29], and has achieved considerable improvements over previous methods in those fields. In this section, we present a sparse representation-based classification method to segment the prostate from treatment images.

Sparse representation models data with linear combinations of a few elements from a learned dictionary. Like the traditional data representation methods

Table 9.3 Quantitative comparisons on prostate localization between ILSM and four learning-based methods

	POP (S)	PPAT (S)	MIX (S)	IL (S)	ILSM (S)	ILSM (M)
Mean DSC	0.81±0.10	0.84±0.15	0.83±0.09	0.83±0.09	0.87±0.06	**0.88±0.06**
Acceptance rate (%)	66	85	74	77	90	**91**

S indicates the localization results obtained by using a single shape atlas from the planning image. *M* indicates the localization results obtained by using all shape atlases from the previous planning and treatment images. The best performance of each measurement is shown in bold lettering

(e.g., wavelet and Fourier transform), the sparse representation method has a set of basis elements, which column-wisely form a dictionary. These basis elements do not need to be orthogonal or predefined, which largely differentiates them from the traditional data representation methods. Therefore, the dictionary for sparse representation is usually learned through a process called *dictionary learning* so that the learned dictionary can be well tailored with respect to a specific task (e.g., reconstruction and classification). Given a learned dictionary $D \in \mathbb{R}^{p \times N}$, which has N p-*dimensional* basis elements, the goal of sparse representation is to select a few basis elements for best representing the input signal $x \in \mathbb{R}^p$. Mathematically, it can be formulated as the following *sparse coding* problem:

$$\alpha^* = \arg\min_{\alpha} \|x - D\alpha\|_2^2 + \lambda \|\alpha\|_1, \tag{9.4}$$

where $\alpha^* \in \mathbb{R}^N$ is called *sparse representation* or *sparse code* of x with respect to the dictionary D, $\|\alpha\|_1$ is the L1 norm of α, and λ is a parameter that controls the sparsity of α^* or the number of nonzero entries in α^*. The larger λ is, the sparser α^* is and the fewer nonzero entries α^* has.

Sparse representation-based classification (SRC) [29] was recently proposed and has been widely used for face recognition. In SRC, to classify a new sample, all training samples from different classes are used to represent it in a competitive manner, and the class label is determined by choosing the class that best reconstructs it. Specifically, the training samples belonging to the same class are first column-wisely grouped into *sub-dictionaries*, which are further combined to form a *global dictionary* $D \in \mathbb{R}^{p \times N}$:

$$D = [D_1, \cdots, D_i, \cdots, D_K] \tag{9.5}$$

$$= [d_{1,1}, d_{1,2}, \cdots, d_{i,j}, \cdots, d_{K,N_K}], \tag{9.6}$$

where D_i is the sub-dictionary of class i, $d_{i,j}$ is the *j*th training sample of class i, K is the total number of classes, N_K is the total number of training samples in class K, and N is the total number of training samples equal to $\sum_{i=1}^{K} N_i$. To classify a new sample $x \in \mathbb{R}^p$, its sparse code $\alpha^* \in \mathbb{R}^N$ is first computed with respect to the global dictionary D according to Eq. (9.4). Then the residue with respect to each class is calculated:

$$r_i = x - D_i \alpha_i^*, \quad i \in \{1, \cdots, K\}, \tag{9.7}$$

where $r_i \in \mathbb{R}^p$ is the residue with respect to class i and α_i^* carries entries of α^* corresponding to the indices of columns in D belonging to D_i. Finally, the signal x is classified to the class with the minimum L2 residue norm.

To segment the prostate in daily treatment images, sparse representation-based classification is used to enhance the prostate in CT images by pixel-wise classification in order to overcome the poor contrast of the prostate images. Then, based on the classification results, the previously segmented prostates of the same patient can be aligned onto the current image for multi-atlas-based segmentation [21]. Since our segmentation is guided by the classification, the segmentation accuracy highly depends on the classification performance. However, the conventional SRC suffers from two main limitations when applied to the pixel-wise classification. First, the conventional SRC cannot be directly adapted to the large-scale problem where the size of training samples is huge. Second, when training samples of different classes are highly correlated, the classification performance of the conventional SRC is limited. To overcome these limitations, especially for the purpose of segmentation, we propose four extensions to the SRC, which are elaborated in the following paragraphs one by one.

9.3.5 Discriminant Sub-dictionary Learning

In pixel-wise classification, it is common for different classes to have similar training samples. In such a case, the performance of SRC is limited. Discriminant sub-dictionary learning aims to learn sub-dictionaries as distinct as possible. Here, we propose to combine feature selection with the dictionary learning method as a way to learn discriminant sub-dictionaries. First, a feature selection technique is used to select discriminant features so that the output training samples of different classes are as distinct as possible. A dictionary learning method is subsequently adopted to learn a compact representation of these discriminant training samples in order to make the size of the sub-dictionary feasible.

In the context of prostate segmentation, each voxel needs to be identified as prostate or background. Considering that each training sample is represented by a feature vector, which can be intensity or image features, the selection of features discriminant between prostate and background classes aims to best distinguish prostate voxels from background voxels. In this work, feature ranking, based on the Fisher separation criterion (FSC) [30], is adopted to select those discriminant features. Specifically, for each feature f, we compute its FSC score as $|\mu_1 - \mu_2|/\sqrt{v_1 + v_2}$, where μ_1 and μ_2, v_1 and v_2 are the sample means and variances of feature f in prostate and background classes, respectively. Features with high FSC scores are considered discriminant and, thus, are selected while features with low FSC scores are discarded in the final feature-based representation.

After feature selection, due to the large size of training samples in pixel-wise classification, it is practically infeasible to directly use them to form sub-dictionaries. For storage and computational efficiency, it is necessary to adopt a dictionary

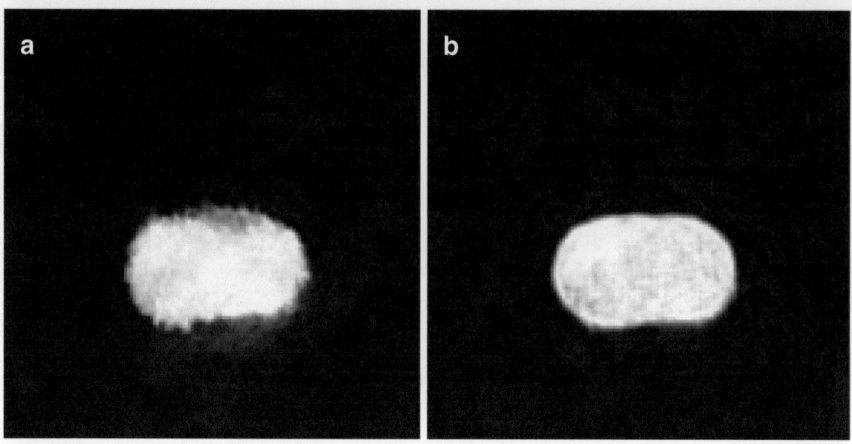

Fig. 9.6 (a) Zigzag prostate boundary caused by the L1-regularized sparse coding and (b) smooth prostate boundary by adopting the elastic net in SRC

learning method to learn a compact representation of those discriminant training samples for each class. In this work, we use the K-means clustering method that preserves the discriminant characteristics of training samples.

Once sub-dictionaries of different classes are learned, their columns are first normalized to the unit norm and then put together to form the global dictionary according to Eq. (9.5) for classification.

9.3.6 Elastic Net

Due to the fuzziness of the prostate boundary in CT images, prostate and background voxels drawn near the prostate boundary can be quite similar, which inevitably introduces highly correlated elements between sub-dictionaries. As the sparse coding only selects one of the highly correlated elements due to its sparsity nature, samples with similar features can have distinct sparse codes and, thus, be classified into different classes because of small noises. As a result, the classification of prostate boundary voxels in the new treatment image is instable, which causes a zigzag and unclear boundary in the classification map (Fig. 9.6).

To address this problem, we replace the traditional L1-regularized sparse coding with the elastic net [31], which compromises between sparsity and stability. Instead of using only the L1 constraint to regularize the least squares problem, the elastic net balances between the L1 constraint and the L2 constraint:

$$\alpha^* = \arg\min_{\alpha} \|x - D\alpha\|_2^2 + \lambda_1 \|\alpha\|_1 + \frac{\lambda_2}{2} \|\alpha\|_2^2 \tag{9.8}$$

As we know [31], the solution of the L2-regularized least squares problem is stable. Thus, adding L2 regularization helps stabilize the sparse code. In this work, we propose to use the elastic net to replace the traditional L1-regularized sparse coding in pixel-wise classification where there exist highly correlated elements in sub-dictionaries of different classes. Practically, we found that boundary-smoothing effects can be achieved by stabilizing the sparse code. Figure 9.6 visually compares the classification results of the L1-regularized sparse coding and the elastic net.

9.4 Residue-Based Linear Regression

In the traditional SRC, residue norms with respect to each class are compared, and the new sample is classified to the class with the minimum residue norm. In such cases, residues of different features are equally treated. While it is reasonable when features are of the same type and importance, it is not desirable in other cases. Usually each voxel is represented by the combination of different types of features; thus, the discriminabilities of individual features are different and their contributions to classification are also different. Therefore, equally weighting them in determining the class label limits the classification performance. Besides, the traditional SRC is a hard classification method, which only assigns a class label to the new sample. In contrast, soft classification provides more quantitative information, especially in the decision margin where the class membership is unclear. Based on these observations, we propose to learn a linear regression model to predict the class probability based on the residues, which extends SRC from hard classification to soft classification. Specifically, in the training stage, after the discriminant sub-dictionaries are learned, we can compute the residue vectors of each training sample with respect to the prostate and background classes. These two residue vectors are then concatenated into a long vector, which is used as features in the linear regression model to estimate the class probability. In the testing stage, given a testing voxel, we first compute the two residue vectors, and then use the learned linear regression model to estimate its class probability.

By incorporating the residue-based linear regression into SRC, full residual information is used, instead of just using their norm. Besides, individual features are weighted by their contributions in predicting the class probability. Compared with the traditional SRC, we found the classification performance can be increased by using residue-based linear regression.

9.5 Iterative SRC

Segmentation using classification methods which overlook spatial regularization is often criticized because each pixel is independently processed and, thereby, can be easily misclassified. Recently, Tu proposed the auto-context model [32], which uses context information to iteratively refine the classification results. Specifically, at each iteration, previous classification results at context locations are extracted as context features to assist the classification in the current iteration. Each pixel can be represented by a feature vector that contains both its original features and the context features, which are updated iteratively. As the classification iterates, these context features become more discriminative and, thus, more helpful in the classification. As a result, the classification probability map becomes clearer and clearer. Inspired by this idea, we incorporate the context information into SRC and propose the iterative SRC.

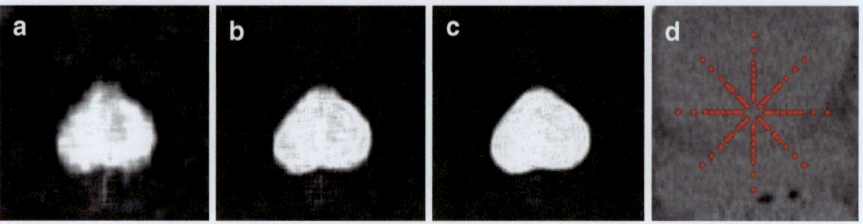

Fig. 9.7 (**a–c**) Show the classification results in the first, second, and third iteration, respectively. *Red points* in (**d**) show the context locations of the center pixel in the image dictionary learning [25]

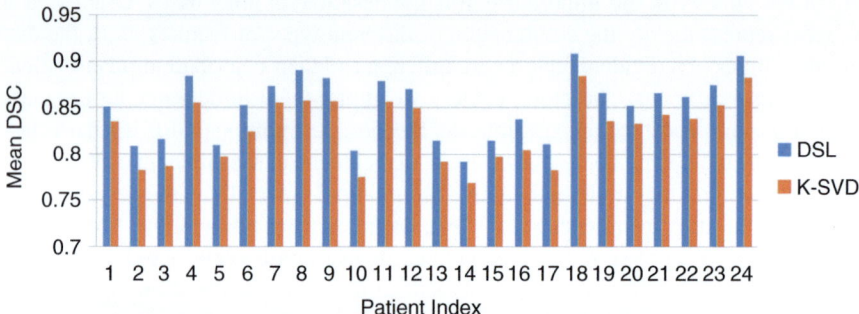

Fig. 9.8 Quantitative comparison between discriminant sub-dictionary learning (*DSL*) and K-SVD dictionary learning [25].

In the iterative SRC, initially we start with a uniform probability map since no classification has been performed. Due to the lack of discriminability, these context features are filtered out by discriminant sub-dictionary learning, which means the context features are not included in the first classification iteration. In the later classification iterations, the context features are iteratively updated and start to encode more and more accurate class probability information about its surrounding pixels, which can be considered as effective high-level features. As a result, more context features are identified as the topmost discriminant features in the discriminant sub-dictionary learning and, thus, selected to guide the refinement of the classification results. Figure 9.7 shows typical classification probability maps at different iterations, which clearly justifies the effectiveness of the iterative SRC.

9.6 Experimental Results

The evaluation of this method is based on 330 CT images from 24 patients. Each patient has more than 9 daily CT scans. The axial image size is 512×512 with voxel size 1×1 mm. The inter-slice distance is 3 mm. The manual segmentation results provided by a clinical expert are available for each CT image to serve as the ground truth. The Dice similarity coefficient (DSC) [33], as a widely adopted segmentation measure, is used again for evaluating our prostate segmentation method. To show the effectiveness of our method, we evaluate each component independently.

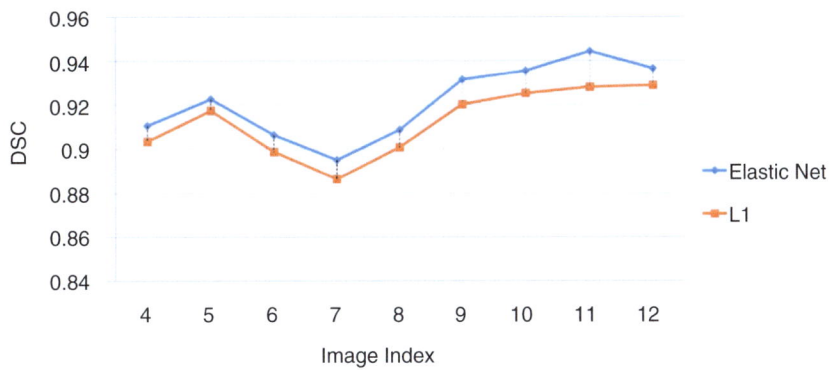

Fig. 9.9 Quantitative comparison between L1 sparse coding and elastic net

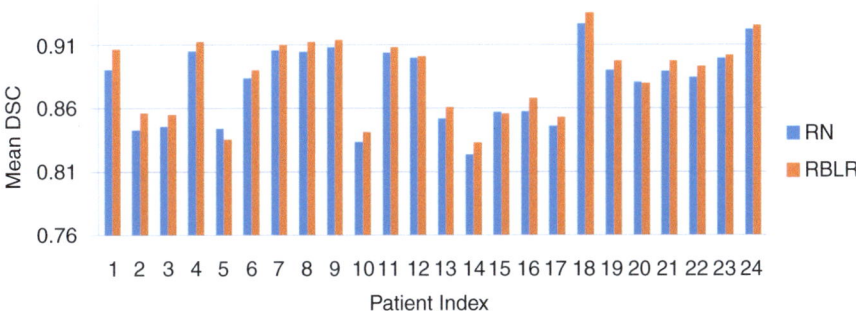

Fig. 9.10 Quantitative comparison between residue-norm-based classification (*RN*) and residue-based linear regression (*RBLR*) with 1-iteration

Figure 9.8 quantitatively compares our discriminant sub-dictionary learning with the K-SVD dictionary learning method [25] for learning the sub-dictionaries used in the SRC. For fair comparison, we also adopted feature selection for K-SVD. To exclude the influence from other components. Here, we did not use other extensions, including elastic net, residue-based linear regression, and iterative SRC. Clearly, we can see from Fig. 9.8 that our discriminant sub-dictionary learning (DSL) performs better in terms of the segmentation accuracy.

Figure 9.9 compares the elastic net and the traditional L1 sparse coding in the SRC using 12 images of one patient. The visual comparison of the respective classification response maps is shown in Fig. 9.6. Both quantitative and qualitative results indicate that the elastic net is better suited to this application than the traditional L1 sparse coding. Figure 9.10 gives the quantitative comparison between the residue-norm-based hard classification and the proposed residue-based linear regression. By weighting all features differently according to their contributions to the classification, residue-based linear regression further boosts the classification performance, compared with the traditional residue-norm-based hard classification. Finally, we compared the effectiveness of context features in our framework. As we can see in Fig. 9.11, the segmentation accuracy is generally improved when using more iterations.

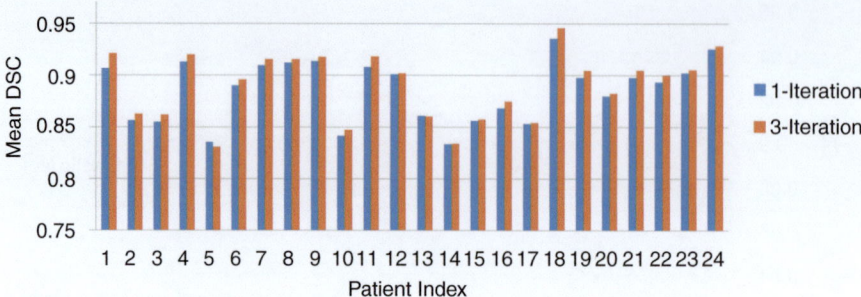

Fig. 9.11 Quantitative comparison of auto-context scheme with 1 and 3 iterations

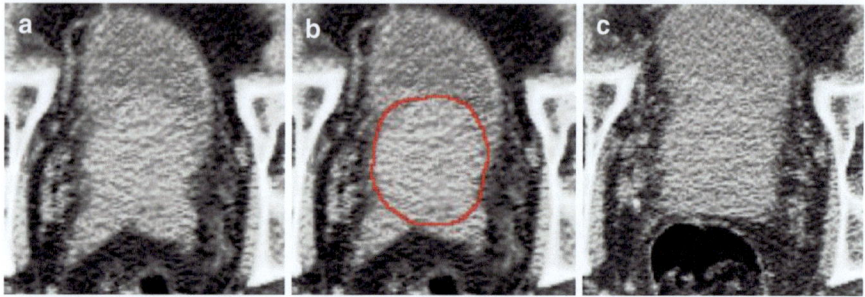

Fig. 9.12 (**a**) An image slice obtained from a patient on certain treatment day, with the prostate boundary superimposed in (**b**). (**c**) An image slice obtained from the same patient on other treatment day. Note the significant image appearance difference due to the existence of bowel gas

9.6.1 Sparse Label Propagation for Automatic Prostate Segmentation in Daily Treatment CTs

In this section, we will introduce a multi-atlas-based sparse label propagation method for automatic prostate localization in daily treatment CT images. There are two main challenges to segmenting the prostate in treatment CT images: (1) low image contrast between the prostate and the surrounding tissues and (2) the prostate motion and image appearance across different treatment days which can be large, even for the same patient. These two challenges, illustrated in Fig. 9.12, will be addressed by the sparse label propagation method.

In the field of medical image analysis, multi-atlas-based image segmentation is a widely used method for automatic organ segmentation. Here, atlases usually denote training images with segmentation ground truths. It mainly consists of two stages, namely, the registration stage and the label fusion stage. In the registration stage, each atlas is registered to the target image to be segmented using simple rigid and affine transformations or complex diffeomorphic transformations. In the label fusion stage, each registered atlas is aggregated to provide the final segmentation result of the target image, the most straightforward aggregation scheme is majority voting. More advanced aggregation schemes are also derived, such as STAPLE [34] and nonlocal mean-based label propagation [35, 36]. The proposed method is based on nonlocal mean label propagation, which provides the following advantages over other label fusion techniques: (1) The nonlocal mean searching strategy relaxes the degree of correspondence accuracy requirement, such that linear registration algorithms can be used during the registration step and avoid the computational intensive nonlinear registration process, and (2) it allows many-to-one correspondences to identify a set of good candidate voxels in the atlases to use during the label fusion step.

The conventional label propagation method [35, 36] uses a 3D patch centered at each voxel as the voxel's anatomical features. Good candidate voxels are determined as voxels with high patch similarity to the to-be-labeled voxel in the target image (we name it as reference voxel below). These candidate voxels are used to vote for the label of the reference voxel. However, there are two major limitations in the conventional label propagation framework when applied to the prostate CT image segmentation. First, in [35, 36], the patch-based representation is constructed by using only the voxel intensity information, which may not be able to effectively distinguish voxels belonging to the prostate from non-prostate regions due to low image contrast in the prostate CT images. Second, the weight of each candidate voxel for label propagation is determined by directly comparing the similarity between the patch-based representation of each candidate voxel with that of the reference voxel in the target image, which may not be robust against outlier candidate voxels and, thus, increase the risk of misclassification.

The new method introduced in this section has the following advantages, compared to the conventional label propagation method: (1) To deal with the low image contrast problem in prostate CT images, a new patch-based representation is derived in the discriminative feature space with logistic sparse Lasso. The derived patch-based representation can capture salient features to effectively distinguish prostate voxels from non-prostate voxels and can, thus, serve as effective anatomical features for each voxel. (2) For each reference voxel in the new prostate CT, its new patch-based features are reconstructed by sparse representation of the patch-based features of candidate voxels in the previous images (i.e., training images) of the same patient. The reconstruction weights estimated by sparse representation are then used for label propagation. Due to the robustness property of sparse representation against outliers, the segmentation accuracy can be further improved. (3) A hierarchical segmentation strategy is proposed for first segmenting voxels with high segmentation confidence in the new treatment image, and then using their segmentations to provide useful context information for aiding the segmentation of other low-confidence voxels, which are more difficult to segment.

9.7 Patch-Based Representation in the Discriminative Feature Space

We propose a patch-based representation in the high-dimensional discriminative feature space. We denote the current available M training images from the same patient as $I_1(x),\cdots,I_M(x)$ (i.e., including the planning image and previously

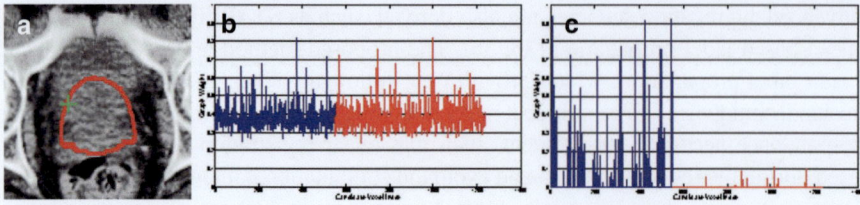

Fig. 9.13 (**a**) A prostate CT image slice, with the reference voxel highlighted by the *green cross* and also the prostate boundary of the segmentation ground truth highlighted by the *red contour*. Note that the reference voxel is in the prostate region but close to the prostate boundary. (**b**) Graph weights between the reference voxel (*green cross*) and candidate voxels from training images estimated by pair-wise Euclidean distance. (**c**) The same graph weights as (**b**) estimated by the proposed sparse label propagation strategy

Table 9.4 The average Fisher score calculated from 330 prostate images with different voxel features: the higher the score, the better the discriminant power of the feature is

Features	Intensity	Intensity patch	Patch in feature space	Patch in feature space + logistic Lasso
Fisher score	2.15	5.83	8.89	12.14

segmented treatment images). As illustrated in Fig. 9.13, each previously segmented treatment image is first rigidly aligned to the planning image, based on the pelvic bone structures similar to [37], for removing the whole-body rigid motion. Each training image $I_i(x)(i=1,\cdots,M)$ is convolved with a set of feature extraction kernels $\psi^j(x)(j=1,\cdots,n)$ to produce different feature maps $F_i^j(x)$:

$$F_i^j(x) = I_i(x) * \psi^j(x)(j=1,\ldots,n), \quad (9.9)$$

where n denotes the number of kernels for extracting features. Here, 14 Haar wavelet [38], 9 histogram-of-oriented-gradient (HOG) [39], and 30 local-binary-pattern (LBP) [40] kernels are adopted to extract features. Based on the feature $F_i^j(x)$ calculated by Eq. (9.9), we can obtain a patch-based representation of each voxel x. Suppose that a $K \times K$ patch is adopted, and then each voxel x has a $K \times K \times n$ dimensional anatomical features, denoted as $f(x)$.

Then, feature selection is performed on $f(x)$ in order to reduce the noise effect and feature redundancy. Logistic regression serves as a good choice for performing this task. Moreover, the aim of feature selection is to select a small subset of the most informative features as anatomical features, which can be well accomplished by enforcing the sparsity constraint during the logistic regression process:

$$J(\beta,b) = \sum_{c=1}^{P} \log\left(1+\exp\left(-L_c\left(\beta^T f(x_c)+b\right)\right)\right)+\lambda\|\beta\|_1 \quad (9.10)$$

where β is the sparse coefficient vector, $\|\cdot\|_1$ is the L$_1$ norm, b is the intercept scalar, and λ is the regularization parameter. x_c is a training sample drawn from the training images $(c=1,\cdots,P)$, with label $L_c = 1$ if x_c belongs to the prostate and $L_c = -1$ otherwise. The optimal solution (β^{opt} and b^{opt}) to minimize Eq. (9.10) can be estimated by the Nesterov's method. The final selected features are those with nonzero entries in β^{opt}. Here we denote the final features of each voxel x as $a(x)$ (Table 9.4).

9.8 Hierarchical Prostate Segmentation in New Treatment Images with Sparse Label Propagation

The general nonlocal mean label propagation principle can be summarized by Equation (9.11):

$$S_{MA}(x) = \frac{\sum_{i=1}^{M}\sum_{y\in\Omega} w_i(x,y) S_i(y)}{\sum_{i=1}^{M}\sum_{y\in\Omega} w_i(x,y)} \quad (9.11)$$

where Ω denotes the image domain and S_{MA} denotes the prostate probability map of the target image I_{new} estimated by multi-atlas-based labeling. $S_i(y)=1$ if y belongs to the prostate region in I_i, and $S_i(y)=0$ otherwise. $w_i(x,y)$ is the graph weight between voxel x and y in I_i. Typically, the searching range is spatially confined to the neighborhood of voxel x, instead of the whole image domain Ω. We denote the neighborhood of a voxel x in image I_i as $N_i(x)$.

Therefore, the key for nonlocal mean-based label propagation is how to define the graph weight $w_i(x,y)$. The most straightforward solution is to use the Euclidean distance between the features of x and y. However, the graph weight defined in this way may not be able to effectively identify the most representative candidate voxels in atlases to estimate the prostate probability of a reference voxel, especially when the reference voxel is located near the prostate boundary, which is the most difficult region to segment correctly.

Figure 9.13a shows a prostate CT image slice with the reference voxel highlighted by the green cross (i.e., a voxel belonging to the prostate region but close to the prostate boundary). The prostate boundary of the ground truth is highlighted by the red contour. Figure 9.13b shows the graph weights associated with the reference voxel and the candidate voxels from the training images, estimated by the Euclidean distance. Without loss of generality, we use blue to highlight the graph weights corresponding to the prostate sample voxels and red to highlight the graph weights corresponding to the non-prostate sample voxels.

It can be observed from Fig. 9.13b that, with the conventional label propagation strategy, lots of non-prostate sample voxels are also assigned with large weights during the label propagation step, which significantly increases the risk of misclassification.

Motivated by the superior discriminant power of sparse representation, we propose to enforce the sparsity constraint in the conventional label propagation framework to resolve this issue. More specifically, we estimate the sparse graph weights based on Lasso to reconstruct the patch-based features of each voxel x in I_{new} by using the features of neighboring voxels in the training images. To do this, we first organize all features $a(y)$ of y in $N_i(x)$ as columns in a matrix $\mathbf{A}$, which is also called a dictionary. Then, we can estimate the corresponding sparse coefficient vector θ_x of a voxel x by minimizing Eq. (9.12):

$$J(\theta_x) = \frac{1}{2}\left|a_{I_{new}}(x) - A * \theta_x\right|_2^2 + \lambda\|\theta_x\|_1, \quad (9.12)$$

and the graph weight $w_i(x, y)$ can be set to the corresponding element in the optimal solution of Eq. (9.12).

Figure 9.13c shows the resulting graph weights estimated by the proposed sparse label propagation strategy. It can be observed from Fig. 9.13c that, by enforcing the sparsity constraint, candidate voxels assigned with large graph weights are identified as prostate, while candidate voxels from non-prostate regions are mostly assigned with zero or very small graph weights. These results show the advantage of the proposed sparse label propagation strategy.

To improve the robustness of sparse label propagation, a hierarchical segmentation scheme is also used. Specifically, the prostate probability map is estimated in an iterative manner using the results of the previous iteration to extract context features for prostate segmentation.

9.9 Experimental Results

The segmentation accuracy of the proposed method is systematically evaluated on a 3D prostate CT image database with 24 patients. Here, each patient has more than 10 treatment images, and 24 patients totally have 330 images. Each image is collected on a Siemens Somatom or a Primatom CT-on-rails scanner with an in-plane image size of 512×512, a voxel size of 1×1 mm^2, and an inter-slice thickness of 3 mm.

Figure 9.14 shows typical segmentation results with the proposed method. The average Dice ratio obtained on 24 patients with different segmentation strategies is also plotted in Fig. 9.15. It can be observed that the new method introduced in this section achieves the highest segmentation accuracy among the other methods under comparison, which illustrates the robustness and effectiveness of the new method [8].

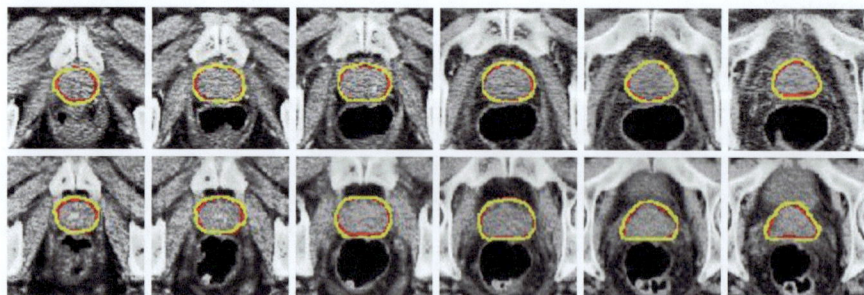

Fig. 9.14 Typical segmentation performance of the proposed method on two patients. Each *row* represents a patient

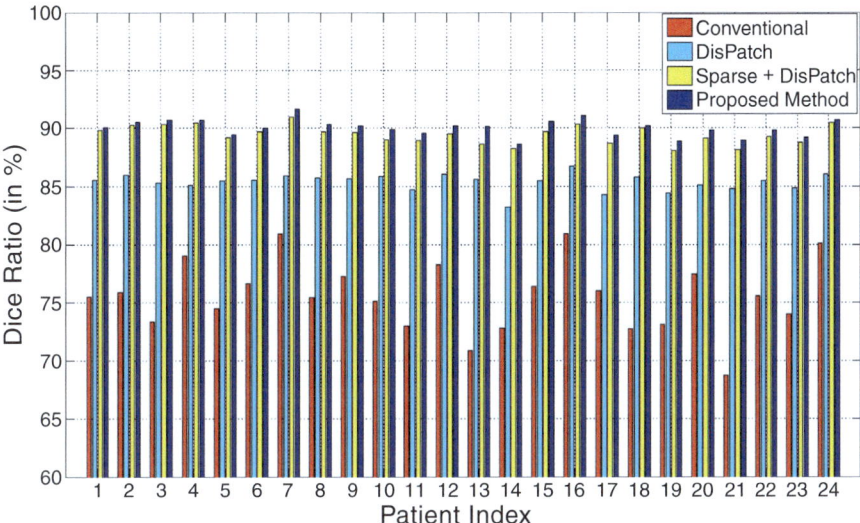

Fig. 9.15 Average Dice ratios obtained from 24 patients with different segmentation strategies

9.9.1 Prostate Segmentation in CT Images via Spatial-Constrained Multitask Feature Selection

In this section, we will introduce an application of the spatial-constrained multitask feature selection technique for prostate segmentation in CT images. Previous learning-based methods [7, 37] first collect the voxels from certain slices and then conduct both the feature selection and the subsequent prostate-likelihood estimation for all voxels in those selected slices jointly. However, different local regions usually prefer choosing different features to better discriminate between their prostate and non-prostate voxels, as indicated by a typical example in Fig. 9.16. In this example, we extracted features (i.e., Haar wavelet, HoG, LBP) for three different local regions and then applied Lasso (a supervised feature selection technique as introduced in [41]) for the respective features' selection. From the results shown in Fig. 9.16, we can see that the selected features from three local regions are completely different, demonstrating the necessity of selecting the respective features for each local region.

Here, we present a novel local learning strategy: Partition each 2D slice into several nonoverlapping local blocks, and then select their respective local features to predict the prostate likelihood for each local block. This will be achieved by our proposed Spatial-COnstrained Transductive Lasso (SCOTO) and support vector regression (SVR), respectively, which will be detailed below. The major difference between the previous learning-based methods and our method is explained in Fig. 9.17.

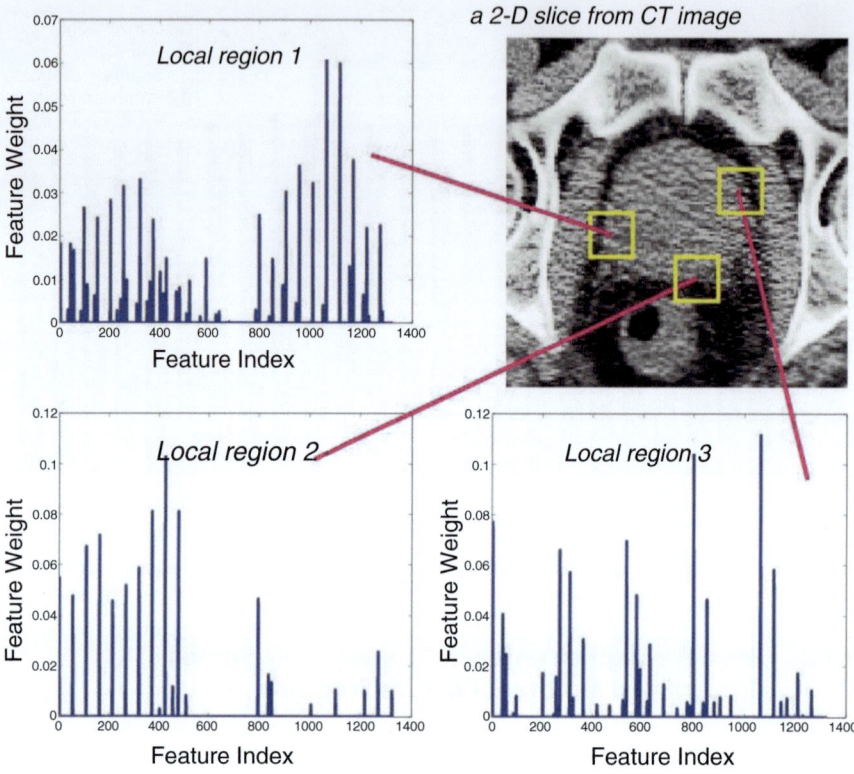

Fig. 9.16 A typical example showing the importance of selecting different features for different local regions (i.e., three *yellow rectangles*)

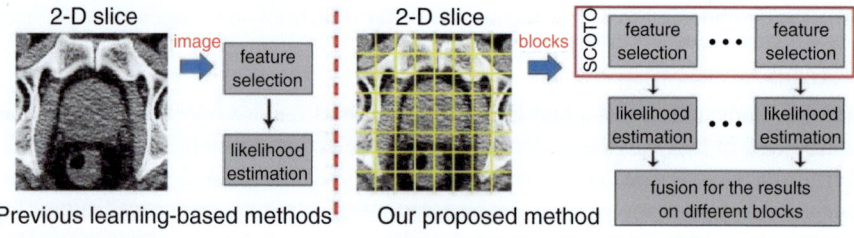

Fig. 9.17 The difference between the previous learning-based methods and our proposed method. Specifically, our method adopts a local feature selection and prostate-likelihood estimation strategy

Note that, before segmentation on the current treatment image, the physician only needs to spend a few seconds to specify just the first and last slices of the prostate region in the CT image. With this minimal user interaction, the segmentation results can be significantly improved, compared with the fully automatic methods [7, 37].

9.10 Implementation

Our proposed method is mainly composed of two steps: (1) the prostate-likelihood estimation step and (2) the multi-atlas-based label fusion step. In the prostate-likelihood estimation step: *First*, all previous and current treatment images are rigidly aligned to the planning image based on the pelvic bone structures. Then, we extract the ROI regions according to the prostate center in the planning image. *Second*, for the current treatment image, a physician is required to specify the first and last slices of the prostate in the CT images. By combining the voxels in the specified slices with the voxels sampled from the planning and previous treatment images according to the previous segmentation results, we can extract 2D low-level features (LBP [40], HoG [39], and Haar wavelets [38]) for all of these voxels separately from their original CT images. Then, each 2D slice will be partitioned into several nonoverlapping blocks. The proposed SCOTO is applied for joint feature selection for all blocks, and SVR is further adopted to predict the 2D prostate-likelihood map for all the voxels in the current slice. *Finally*, the predicted 2D prostate-likelihood map of each individual slice will be merged into a 3D prostate-likelihood map.

In the multi-atlas-based label fusion step, to make full use of the prostate shape information, all manually segmented prostate regions in both the planning and previous treatment images of the same patient will be rigidly aligned to the estimated 3D prostate-likelihood map of the current treatment image. Then, majority voting will be applied to fuse the labels from all different aligned images and obtain the final segmentation result.

9.11 Prostate-Likelihood Estimation via SCOTO

The planning image and its corresponding manual segmentation result are denoted as $\mathbf{I}_p$ and $\mathbf{G}_p$, respectively. The nth treatment image, which is the current treatment image, is denoted as $\mathbf{I}_n$. The previous treatment images and their corresponding manual segmentation results are denoted as $\mathbf{I}_1, \cdots, \mathbf{I}_{n-1}$ and $\mathbf{G}_1, \cdots, \mathbf{G}_{n-1}$, respectively. Also, the final 3D prostate-likelihood map and its segmentation result for the current treatment image $\mathbf{I}_n$ by adopting the proposed method are denoted as $\mathbf{M}_n$ and $\mathbf{S}_n$, respectively.

For each slice, we first partition the slice into nonoverlapping $N_x \times N_y$ blocks as shown in Fig. 9.17. Then, for the ith block, we use $l_i \in \mathbb{R}$ and $u_i \in \mathbb{R}$ to denote the numbers of training voxels and testing voxels, respectively. $N \in \mathbb{R} \left(N = N_x \times N_y \right)$ denotes the total number of blocks in the current slice. $\mathbf{y}_i \in \mathbb{R}^{l_i + u_i}$ and $\mathbf{F}_i \in \mathbb{R}^{(l_i + u_i) \times d}$ denote the ground-truth label and feature matrix for all the training and testing voxels, respectively. Without loss of generality, all the training voxels are listed before the testing voxels in both $\mathbf{y}_i$ and $\mathbf{F}_i$. d denotes the number of features. It is noteworthy that the labels of testing voxels in $\mathbf{y}_i$ are set to 0. Also, in $\mathbf{y}_i$, the labels of training voxels are set to 1 if they belong to the prostate, and set to 0 if they belong to the background.

Mathematically, the objective function of SCOTO can be formulated as follows:

$$\min_{\beta_1,\cdots,\beta_N} \left\{ \begin{array}{l} \sum_{i=1}^{N}\left[\left\|\mathbf{J}_i\left(\mathbf{y}_i - \mathbf{F}_i\boldsymbol{\beta}_i\right)\right\|_2^2 + \lambda_S\|\boldsymbol{\beta}_i\|_1 + \dfrac{\lambda_L}{(l_i+u_i)^2}\boldsymbol{\beta}_i^{\mathrm{T}}\mathbf{F}_i^{\mathrm{T}}\mathcal{L}_i\mathbf{F}_i\boldsymbol{\beta}_i\right] \\ + \dfrac{\lambda_E}{|H(i)|}\sum_{i=1}^{N}\sum_{j \in H(i)}\|\boldsymbol{\beta}_i - \boldsymbol{\beta}_j\|_2^2 \end{array} \right\}, \quad (9.13)$$

where $\boldsymbol{\beta}_1,\cdots,\boldsymbol{\beta}_N\left(\boldsymbol{\beta}_i \in \mathbb{R}^d\right)$ are the parameters to learn, which indicates the weights of individual features for each block. $\lambda_S, \lambda_L, \lambda_E \in \mathbb{R}$ are the parameters to control the corresponding terms. $\mathbf{J}_i \in \mathbb{R}^{(l_i+u_i) \times (l_i+u_i)}$, which is used to indicate the training voxels, is a diagonal matrix defined as $\mathbf{J}_i = \mathrm{diag}\left[\overbrace{1/l_i,\cdots,1/l_i}^{l_i},\overbrace{0,\cdots,0}^{u_i}\right]$. $\mathcal{L}_i \in \mathbb{R}^{(l_i+u_i) \times (l_i+u_i)}$ is the graph Laplacian with the same definition as that in [42]. $H(i)$ denotes the neighbors of the ith block, and $|H(i)|$ is the cardinality of $H(i)$.

In Eq. (9.13), the first term with three subterms focuses on each individual block: The 1st subterm indicates the reconstruction error, the 2nd subterm imposes the sparsity constraint with the $L1$ norm, and the 3rd subterm is the graph Laplacian imposing the manifold constraint on both the training and testing voxels since a large amount of testing voxels can be well used for training. The second term is the smoothness term on the neighboring blocks, so that the neighboring blocks are encouraged to choose similar features since they usually have similar appearance.

After using SCOTO for feature selection, for the ith block, the features, which correspond to the entries in $\boldsymbol{\beta}_i$ that are larger than 0, will be selected. So we can finally obtain the new feature matrices $\mathbf{F}_i^{'}\left(i=1,\cdots,N\right)$ by selecting the columns in $\mathbf{F}_i$ corresponding to the selected features.

9.12 Prostate-Likelihood Estimation

With the obtained new feature matrices $\mathbf{F}_i^{'}\left(i=1,\cdots,N\right)$, we can estimate the prostate likelihood for each block. For each individual block, we apply SVR, which is a conventional regression method, to predict the prostate likelihood for all the voxels in each block. Specifically, the SVR model is first trained by voxels in $\mathbf{F}_i^{'}$ as well as available labels in $\mathbf{y}_i$ and then tested on the u_i testing voxels in the ith block for prostate prediction. All the predicted likelihood will be finally normalized into [0,1].

9.13 Experimental Results

Here, we use three common evaluation metrics: the Dice ratio, the true-positive fraction (TPF), and the centroid distance (CD) for evaluation.

Table 9.5 Comparison of experimental results among different feature selection methods, with the best results marked by bold font

Level	Methods	Dice (mean ± std)	TPF (mean ± std)	CD (mean ± std) (x/y/z) (mm)
Image-level feature selection	Lasso$_S$	0.874 ± 0.083	0.869 ± 0.107	0.71 ± 0.56/0.80 ± 0.61/0.67 ± 0.53
	tLasso$_S$	0.917 ± 0.053	0.899 ± 0.084	0.54 ± 0.37/0.50 ± 0.38/0.40 ± 0.33
Block-level feature selection	mRMR	0.893 ± 0.033	0.912 ± 0.047	0.50 ± 0.34/0.72 ± 0.41/0.36 ± 0.33
	Lasso$_B$	0.922 ± 0.039	0.909 ± 0.042	0.47 ± 0.39/0.47 ± 0.37/0.33 ± 0.34
	tLasso$_B$	0.932 ± 0.036	0.919 ± 0.040	0.37 ± **0.17**/0.41 ± 0.35/0.32 ± 0.33
	Fused Lasso	0.928 ± 0.047	0.906 ± 0.043	0.34 ± 0.37/0.42 ± 0.38/0.34 ± 0.51
	SCOTO	**0.941 ± 0.030**	**0.924 ± 0.037**	**0.25** ± 0.18/**0.30 ± 0.22/0.27 ± 0.29**

Table 9.6 The results of mean Dice ratio and median TPF, compared with the related methods, with the best results marked by bold font

	Patient no.	Image no.	Method	Mean DSC	Median TPF
Other datasets	3	40	Davis et al. [22]	0.820	N/A
	13	185	Chen et al. [13]	N/A	0.840
Same dataset	24	330	Feng et al. [45]	0.893	N/A
			Liao et al. [37]	0.899	N/A
			Shi et al. [43]	0.920	0.901
			Our method	**0.941**	**0.932**

We introduce several related feature selection methods for comparison, which include Lasso$_S$ and tLasso$_S$ (by applying Lasso [41] and tLasso [43] on slice-level feature selection, respectively), Lasso$_B$ and tLasso$_B$ (by applying Lasso and tLasso on block-level feature selection, respectively), mRMR [19], and fused Lasso [44].

Table 9.5 lists the segmentation accuracies obtained by the different feature selection schemes, and the best results are marked by the bold fonts. We found that SCOTO can achieve superior performance over the related methods. Specifically, we also found that (1) the block-level methods are better than the slice-level ones, which validates our assumption that different local regions prefer choosing different features; (2) manifold constraint is useful for improving the results (by comparing tLasso$S$ with Lasso$_S$, and tLasso$_B$ with Lasso$_B$); and (3) the spatial-constraint smoothness term leads to better results (by comparing SCOTO with tLasso$_B$).

To further evaluate the performance of the proposed method, the results of several state-of-the-art methods are illustrated for comparison (see Table 9.6), which include deformable-model-based methods [13, 45], registration-based methods [22, 37], and learning-based methods [7, 43]. From the results listed in Table 9.6, we can find that the proposed method outperforms the related methods in terms of higher mean Dice ratio and median TPF. Also, we illustrate in Fig. 9.18 several typical segmented examples as well as a prostate-likelihood map, with the red curves denoting the manual segmentation results by the physician and the yellow curves denoting the segmentation results by the proposed method.

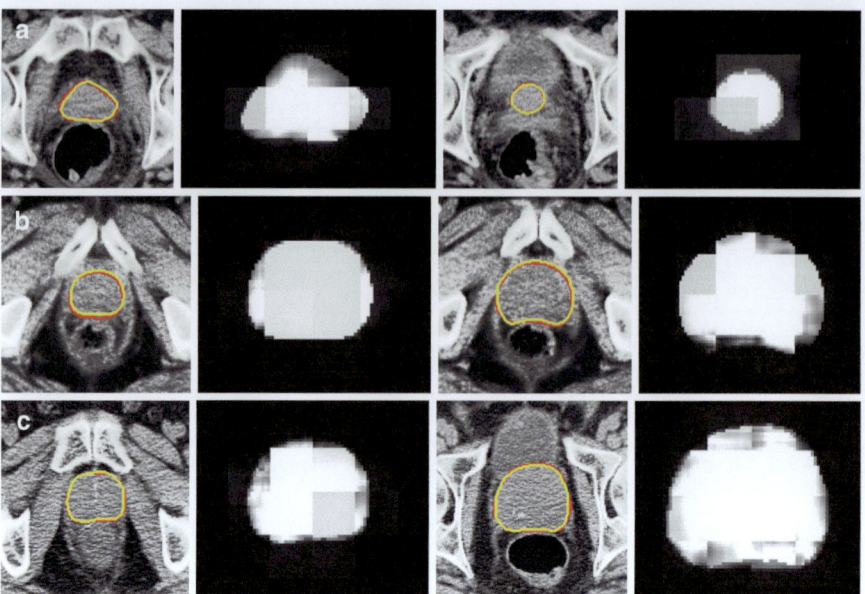

Fig. 9.18 Typical segmentation results and prostate-likelihood maps by the proposed method. (**a**) Typical results of the 14th image of patient 3, with Dice ratio of 0.898. (**b**) Typical results of the 10th image of patient 11, with Dice ratio of 0.929. (**c**) Typical results of the 8th image of patient 24, with Dice ratio of 0.924. *Red* and *yellow contours* indicate the manual and automatic segmentations, respectively

9.13.1 Distributed Discriminant Dictionary (DDD) Learning for MR Prostate Segmentation

As introduced in the previous sections, the sparse representation technique has been successfully employed for prostate segmentation in CT images. In this section, its capability of segmenting prostate MR images, which is important for biopsy and the optimization of the radiotherapy dose [9, 17, 18], is further investigated. Specifically, both image appearance and organ shape are proposed to be modeled by sparse representation and then integrated into the deformable segmentation framework for prostate segmentation, as shown schematically in Fig. 9.19. Note that, instead of imposing Gaussian distribution on the appearance and shape distribution, a dictionary learning method is employed here for building appearance and shape models in a *nonparametric* fashion. Specifically, a distributed discriminative dictionary (DDD) learning-based appearance model and ensemble learning of classifiers is first built to attract the deformable model toward the object boundary. Then, a sparse shape constraint (SSC)-based shape model [20] is adopted to ensure the shape regularity of the deformed model.

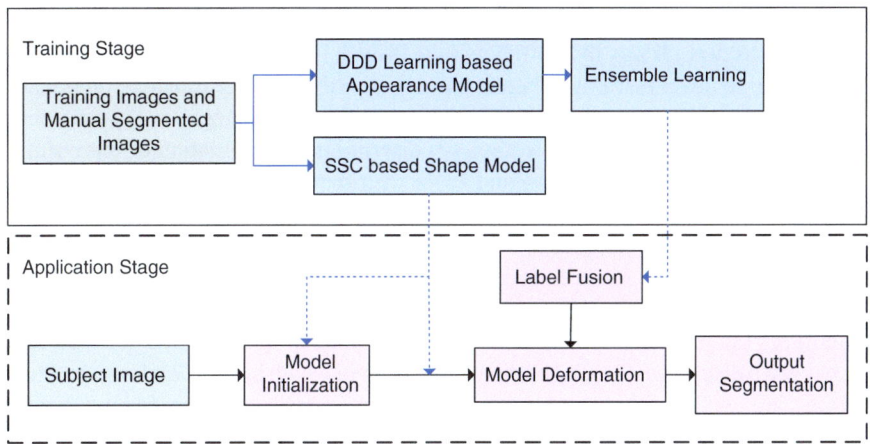

Fig. 9.19 The schematic description of proposed deformable segmentation framework

9.14 Distributed Discriminative Dictionary (DDD) Learning-Based Appearance Model

In the sparse representation theory, data is modeled by a linear combination of a few elements, called atoms. Each atom is selected from an over-complete dictionary, in which the number of atoms usually exceeds the dimension of the data space. Given a dictionary $D \in \mathbb{R}^{M \times Q}$, which has Q atoms (each with M dimensions), the goal of sparse representation for a testing sample $f \in \mathbb{R}^M$ is to select a small number of atoms from D to best represent f. Mathematically, the sparse representation problem can be formulated as the following minimization problem:

$$\alpha = \mathrm{argmin}_\alpha \|f - D\alpha\|_2^2 + \beta_0 \|\alpha\|_1 \tag{9.14}$$

Here, $\alpha \in \mathbb{R}^Q$ is a coefficient vector including the linear coefficients for the atoms in the dictionary D. $\|a\|_1$ is an l_1 norm on α for guaranteeing the sparsity of α. β_0 is the parameter that controls the number of nonzero elements (or sparsity) in α. The number of nonzero elements in α decreases with the increase of the value of β_0. By solving Eq. (9.14), the testing sample f can be reconstructed by $D\alpha$.

For *sparse representation-based classification* (SRC), the prostate and non-prostate sub-dictionaries $\{D_{PR}, D_{NPR}\}$ are jointly used to represent a new testing sample $f \in \mathbb{R}^M$. This sample is labeled as the class that best reconstructs it through sparse representation. By combining the sub-dictionaries to form a single global dictionary $D = [D_{PR}, D_{NPR}] \in \mathbb{R}^{M \times (Q_{PR} + Q_{NPR})}$, the sparse code of the new sample f can be solved as $\alpha \in \mathbb{R}^{Q_{PR} + Q_{NPR}}$, according to Eq. (9.14). Here $\alpha = [\alpha_{PR}^T, \alpha_{NPR}^T]^T$, where

$\{\boldsymbol{\alpha}_{\text{PR}}, \boldsymbol{\alpha}_{\text{NPR}}\}$ carries the elements of $\boldsymbol{\alpha}$, corresponding to the indices of the columns belonging to $\{\boldsymbol{D}_{\text{PR}}, \boldsymbol{D}_{\text{NPR}}\}$ in $\boldsymbol{D}$. To boost the discriminative power of the above dictionary pair, a novel learning scheme, namely, distributed discriminative dictionary (DDD) learning, is proposed, which involves three novel strategies: (1) sparse dictionary learning with feature selection, (2) discriminative integration of representation residuals by LDA learning, and (3) a distributed learning strategy for local subsurfaces with consistent appearance.

First, the minimal-redundancy-maximal-relevance algorithm [19] is employed to build a discriminative feature space. Compared to other feature selection methods [46] that only select individual features with the highest discrimination, mRMR minimizes the redundancy of the selected features as well. Thus, the selected features span a discriminative and compact subspace in which prostate and non-prostate tissues are well separated. After mRMR feature selection, the feature vector $\boldsymbol{f} \in \mathbb{R}^M$ of each training sample is now represented by a reduced feature vector $\hat{\boldsymbol{f}} \in \mathbb{R}^{M'}$, which includes only the set of selected features. Since the dictionary learning is constrained in a discriminative space, the learned sub-dictionaries will contain discriminative information. Consequently, $\hat{\boldsymbol{D}}_{\text{PR}} \in \mathbb{R}^{M' \times \hat{Q}_{\text{PR}}}$ and $\hat{\boldsymbol{D}}_{\text{NPR}} \in \mathbb{R}^{M' \times \hat{Q}_{\text{NPR}}}$ encode *distinctive* appearance characteristics, which can be used to classify prostate and non-prostate tissues. By solving the sparse representation problem of Eq. (9.14) using the combined dictionary $\hat{\boldsymbol{D}} = \left[\hat{\boldsymbol{D}}_{\text{PR}}, \hat{\boldsymbol{D}}_{\text{NPR}} \right]$, the reconstruction residual $\hat{\boldsymbol{r}} = \left[\hat{\boldsymbol{r}}_{\text{PR}}^{\text{T}}, \hat{\boldsymbol{r}}_{\text{PR}}^{\text{T}} \right]^{\text{T}}$ can be computed from the sparse coefficients $\hat{\boldsymbol{\alpha}}_{\text{PR}}$ and $\hat{\boldsymbol{\alpha}}_{\text{NPR}}$ as $\hat{\boldsymbol{r}}_{\text{PR}} = \hat{\boldsymbol{f}} - \hat{\boldsymbol{D}}_{\text{PR}} \hat{\boldsymbol{\alpha}}_{\text{PR}}$ and $\hat{\boldsymbol{r}}_{\text{NPR}} = \hat{\boldsymbol{f}} - \hat{\boldsymbol{D}}_{\text{NPR}} \hat{\boldsymbol{\alpha}}_{\text{NPR}}$.

Second, a linear classifier in the above residual space is learned by Fisher-LDA. Thus, by combining the discriminative dictionary learning and Fisher-LDA residual integration, the prostate-likelihood map for the proposed deformable model can be reformulated as $h_i = \text{sigmoid}\left(\boldsymbol{\omega}^{\text{T}} \begin{pmatrix} \hat{\boldsymbol{r}}_{\text{PR}} \\ \hat{\boldsymbol{r}}_{\text{NPR}} \end{pmatrix} - \delta \right)$, where $\text{sigmoid}(\cdot)$ denotes the sigmoid function. The parameters of the classifier, $\boldsymbol{\omega} \in \mathbb{R}^{2M'}$ and δ, are calculated as $\boldsymbol{\omega} = \text{argmax}_{\boldsymbol{\omega}} \dfrac{\boldsymbol{\omega}^T \Gamma_B \boldsymbol{\omega}}{\boldsymbol{\omega}^T \Gamma_W \boldsymbol{\omega}}$ and $\delta = \boldsymbol{\omega}^T \cdot \dfrac{\mu_{\text{PR}} + \mu_{\text{NPR}}}{2.0}$, where Γ_B and Γ_W are the interclass and the intra-class scatter matrices in the residual space $\hat{\boldsymbol{r}}$. μ_{PR} and μ_{NPR} denote average prostate residuals $\hat{\boldsymbol{r}}_{\text{PR}}$ and average non-prostate residuals $\hat{\boldsymbol{r}}_{\text{NPR}}$, respectively. In this way, elements in the residual vectors $\begin{pmatrix} \hat{\boldsymbol{r}}_{\text{PR}} \\ \hat{\boldsymbol{r}}_{\text{NPR}} \end{pmatrix}$ are assigned with different weights (by the corresponding elements in $\boldsymbol{\omega}$) for optimally separating the prostate from non-prostate tissues, which further improves the discriminative power of standard dictionary learning.

Third, a "divide-and-conquer" learning strategy is designed, in which the global surface is partitioned into a set of subsurfaces with consistent appearance. Discriminant dictionary learning is applied on these distributed subsurfaces to further improve the performance of tissue differentiation. Specifically, the deformable model is divided into L subsurfaces corresponding to L local regions along the prostate boundary. Each subsurface $l \in \{1,...,L\}$ can be attached by a pair of distributed sub-dictionaries, $\widehat{D}_{PR}^{l}$ and $\widehat{D}_{NPR}^{l}$, learned from samples extracted around the lth subsurface. Then, based on the sparse coefficients, $\widehat{\alpha}_{PR}^{l}$ and $\widehat{\alpha}_{NPR}^{l}$, the reconstruction residual $\widehat{r}^{l} = \left[\left(\widehat{r}_{PR}^{l} \right)^{T}, \left(\widehat{r}_{NPR}^{l} \right)^{T} \right]^{T}$ for a testing sample $\widehat{f}$, can be computed by the mapping functions $\widehat{r}_{PR}^{l} = \widehat{f} - \widehat{D}_{PR}^{l} \widehat{\alpha}_{PR}^{l}$ and $\widehat{r}_{NPR}^{l} = \widehat{f} - \widehat{D}_{NPR}^{l} \widehat{\alpha}_{NPR}^{l}$.

Thus, the prostate likelihood of the ith vertex, estimated by distributed dictionaries at the lth subsurface, can be formulated as:

$$h_i^l = \text{sigmoid}\left(\omega^{l^T} \begin{pmatrix} \widehat{r}_{PR}^l \\ \widehat{r}_{NPR}^l \end{pmatrix} - \delta^l \right) \quad (9.15)$$

These local tissue scores $\{h_i^l\}_{l=1}^{L}$ are used as appearance cues to guide subsurfaces of the deformable model onto the prostate boundary during deformable segmentation. For overlapping regions between two neighboring subsurfaces, the tissue scores are estimated by the minimum distance criteria: each voxel is labeled by the subsurface with the closest central point.

According to the above DDD learning method, one dictionary for each subsurface can be learned. During this training stage, typically a subset of voxels is randomly selected to serve as training data, due to the large number of voxels around each subsurface. However, this approach may lead to a low-accuracy classifier if the sampled voxels are not representative. To relieve this phenomenon and increase the robustness of sparse representation-based classification, the idea of bagging [47] is further adopted in the DDD learning.

9.15 Sparse Shape Constraint (SSC)-Based Shape Model

To build the shape prior in the deformable model by sparse learning techniques, a recently proposed method, called the sparse shape composition method [48], is employed. Specifically, by denoting D_s as a large shape repository that includes the shape instances of training subjects, the approximation of an input shape vector v by D_s is formulated as the following optimization problem in the SSC method:

$$\left(\alpha_s^{opt}, \psi^{opt}, e^{opt} \right) = \text{argmin}_{\alpha_s, \psi, e} \left\| \psi(v) - D_s \alpha_s - e \right\|^2 + \beta_1 \left\| \alpha_s \right\|_1 + \beta_2 \left\| e \right\|_1 \quad (9.16)$$

Here, ψ is an affine transformation matrix, which aligns surface vector v to the mean shape vector. α_s denotes the sparsity coefficient for linear combination, and e compensates the large residual errors caused by a few mispositioned vertices. Minimization of Eq. (9.16) is a two-step iteration scheme. At each iteration, the affine transformation ψ is first estimated. Then, based on the current estimated ψ, Eq. (9.16) can be solved as a sparse representation problem. These two steps are iteratively performed until convergence.

With the help of SSC, the current surface vector v can be easily represented by inverse affine transformation of its sparse linear representation $\psi^{opt-1}\left(D_s \alpha_s^{opt}\right)$, which can be regarded as the shape prior regularization. Finally, the deformable model can be evolved to the object boundary iteratively, under the appearance model based on the DDD learning, the nonparametric shape model based on the SSC learning, as well as the smooth constraint [49].

9.16 Experimental Results

The proposed method was evaluated on both internal and public datasets. The internal dataset contains 75 T2-weighted MR images with ground-truth segmentations provided by a clinical collaborator. The public dataset "the MICCAI 2012 challenge data" contains 50 T2-weighted MR images with corresponding ground-truth segmentations.

For the internal dataset, three other state-of-the-art prostate segmentation methods are compared to demonstrate the effectiveness of the proposed deformable model in T2-weighted MR prostate segmentation, including ASM and two multi-atlas-based methods [50]. Table 9.7 reports the mean and standard deviation of DSC, sensitivity, PPV, and ASD between automatic segmentations and manual segmentations for the proposed method and the three other methods. It should be noted that the proposed 75 images include 30 images used in Liao's method [51] and 66 images used in [50]. According to Table 9.7, the proposed method achieves the best performance among all methods under comparison.

To validate the performance of the proposed method on segmenting different zones of the prostate, Fig. 9.20 shows the segmentation results of the apex, base, and central slices, with comparison to segmentation via the ASM method. As can be seen, even though the appearance and shape is much more complicated on the base and apex regions than in the central region, the proposed method still achieves more accurate classification results. The average DSC of the proposed segmentation method for the apex, central, and base regions of the prostate is 84.9 %, 93.6 %, and 81.8 %, respectively, compared to 59.2 %, 83.3 %, and 58.7 % obtained by the ASM method. Besides, the proposed prostate segmentation method achieves a median

Table 9.7 Mean value and standard deviation (std) of DSC, sensitivity, PPV, and ASD between automatic segmentations and manual segmentations for the proposed method and three other methods on the internal dataset

Method	Image no.	DSC (in %)	Sensitivity (in %)	PPV (in %)	ASD (in mm)
ASM	75	74.5±11.3	75.7±18.3	79.1±11.8	4.16±3.64
Liao et al. [51]	30	86.7±2.2	NA	NA	1.90±1.60
Liao et al. [50]	66	88.3±2.6	NA	NA	1.8±0.9
Proposed	75	**89.1±3.6**	**89.9±7.0**	**89.0±6.2**	**1.67±0.61**

NA in the table denotes that the corresponding measurement was not reported in the literature. The best performance of each measurement is shown in bold lettering

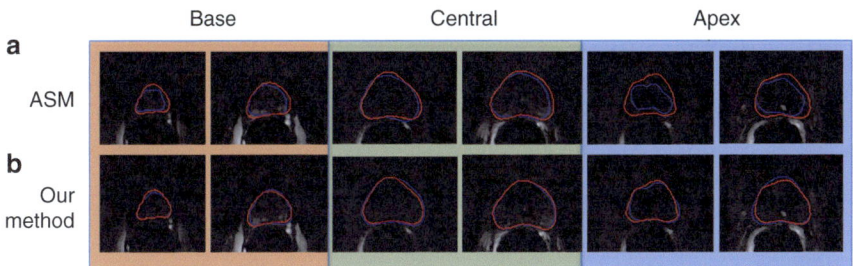

Fig. 9.20 Typical segmentation results for prostate base, central, and apex regions of two patients produced by (**a**) ASM and (**b**) the proposed deformable model. (**a**) demonstrates the segmentation results for ASM, and (**b**) demonstrates the segmentation results for proposed deformable model. The three main columns show the segmentation results for the apex, central, and base regions of the two patients, respectively. *Red contours* indicate the manual segmentations, and *blue contours* indicate the automatic segmentations

Table 9.8 Mean value and standard deviation (std) of DSC, sensitivity, PPV, and ASD between automatic segmentations and manual segmentations for the proposed method and four other methods on the public dataset

Method	Image no.	DSC (in %)	Sensitivity (in %)	PPV (in %)	ASD (in mm)
PASM [52]	50	77.0 ± 23.0	NA	NA	4.10 ± 7.81
AAM [53]	50	81.0 ± 12.0	NA	NA	NA
Martin et al. [54]	50	84.0 ± NA	**87.0 ± NA**	84.0 ± NA	2.41 ± NA
Birkbeck et al. [55]	50	86.0 ± NA	NA	NA	**1.91 ± NA**
Proposed	50	**87.4 ± 3.8**	82.6 ± 7.2	**93.3 ± 3.5**	1.92 ± 0.90

NA in the table means that the corresponding measurement was not reported in the literature. The best performance of each measurement is shown in bold lettering

DSC of 86.7 %, 94.2 %, and 84.2 % on the apex, central, and base regions of the prostate, respectively, which are much higher than the median scores obtained by the ASM method (67.5 %, 86.6 %, and 65.5 %, respectively).

Next, the proposed deformable model with DDD learning is further evaluated on the public MICCAI 2012 challenge database. Comparing with four other state-of-the-art prostate segmentation methods [52–55], Table 9.8 reports the means and standard deviations of DSC, sensitivity, PPV, and ASD between automatic segmentations and manual segmentations for the proposed method. Since all mentioned methods in Table 9.8 were evaluated on the same dataset, the comparisons are informative to show that the proposed method achieves the best performance among all methods under comparison.

9.17 Summary

In this chapter, we have discussed several recent machine learning methods for assisting radiation treatment of prostate cancer, including (1) boosting-based landmark detection method for fast prostate localization, (2) sparse representation-based classification methods for CT and MR prostate segmentation, (3) sparse label propagation for CT prostate localization, and (4) a semiautomatic prostate segmentation method that uses group sparsity for joint feature selection in neighboring local regions. Evaluated on both internal large datasets and a public dataset, the recently developed machine learning approaches have shown better robustness and accuracy, compared with the traditional intensity-based segmentation methods.

References

1. Weiss E, Hess CF. The impact of gross tumor volume (GTV) and clinical target volume (CTV) definition on the total accuracy in radiotherapy theoretical aspects and practical experiences. Strahlenther Onkol. 2003;179(1):21–30.
2. Brouwer CL, et al. 3D Variation in delineation of head and neck organs at risk. Radiat Oncol. 2012;7:32.
3. Sharp G, et al. Vision 20/20: perspectives on automated image segmentation for radiotherapy. Med Phys. 2014;41(5):050902.
4. Rohlfing T, et al. Quo vadis, atlas-based segmentation? In: Handbook of biomedical image analysis. USA: Springer; 2005. p. 435–86.
5. Heimann T, Meinzer HP. Statistical shape models for 3D medical image segmentation: a review. Med Image Anal. 2009;13(4):543–63.
6. Geremia E, et al. Spatial decision forests for MS lesion segmentation in multi-channel magnetic resonance images. Neuroimage. 2011;57(2):378–90.
7. Li W, et al. Learning image context for segmentation of the prostate in CT-guided radiotherapy. Phys Med Biol. 2012;57(5):1283–308.
8. Criminisi A, Shotton J, Konukoglu E. Decision forests: a unified framework for classification, regression, density estimation, manifold learning and semi-supervised learning. Found Trends Comput Graph Vis. 2012;7(2–3):81–227.
9. Shukla-Dave A, Hricak H. Role of MRI in prostate cancer detection. NMR Biomed. 2014;27(1):16–24.
10. Freedman D, et al. Model-based segmentation of medical imagery by matching distributions. IEEE Trans Med Imaging. 2005;24(3):281–92.
11. Costa MJ, et al. Automatic segmentation of bladder and prostate using coupled 3D deformable models. Med Image Comput Comput Assist Interv. 2007;10(Pt 1):252–60.
12. Foskey M, et al. Large deformation three-dimensional image registration in image-guided radiation therapy. Phys Med Biol. 2005;50(24):5869.
13. Chen S, Lovelock DM, Radke RJ. Segmenting the prostate and rectum in CT imagery using anatomical constraints. Med Image Anal. 2011;15(1):1–11.
14. Haas B, et al. Automatic segmentation of thoracic and pelvic CT images for radiotherapy planning using implicit anatomic knowledge and organ-specific segmentation strategies. Phys Med Biol. 2008;53(6):1751.
15. Ghosh P, Mitchell M. Segmentation of medical images using a genetic algorithm. In: Proceedings of the 8th annual conference on Genetic and evolutionary computation. Seattle:ACM; 2006. p. 1171–8.
16. Viola P, Jones MJ. Robust real-time face detection. Int J Comput Vis. 2004;57(2):137–54.
17. Zhan Y, Dewan M, Harder M, Krishnan A, Zhou XS. Robust automatic knee MR slice positioning through redundant and hierarchical anatomy detection. IEEE Trans Med Imaging. 2011;30(12):2087–100.
18. Zhan Y, Zhou XS, Peng Z, Krishnan A. Active Scheduling of Organ Detection and Segmentation in Whole-Body Medical Images. In: Metaxas D et al., editors. Medical Image Computing and Computer-Assisted Intervention – MICCAI 2008. Berlin/Heidelberg: Springer; 2008. p. 313–21.

19. Peng H, Fulmi L, Ding C. Feature selection based on mutual information criteria of max-dependency, max-relevance, and min-redundancy. IEEE Trans Pattern Anal Mach Intell. 2005;27(8):1226–38.
20. Zhang S, Zhan Y, Metaxas DN. Deformable segmentation via sparse representation and dictionary learning. Med Image Anal. 2012;16(7):1385–96.
21. Gao Y, Zhang Y, Shen D. Incremental learning with selective memory (ILSM): towards fast prostate localization for image guided radiotherapy. IEEE Trans Med Imaging. 2014;33(2):518–34.
22. Davis BC, et al. Automatic segmentation of intra-treatment CT images for adaptive radiation therapy of the prostate. Med Image Comput Comput Assist Interv. 2005;8(Pt 1):442–50.
23. Garrigues P, Olshausen B. Group sparse coding with a laplacian scale mixture prior. Adv Neural Inf Process Syst. 2010;23:1–9.
24. Krause A, Cevher V. Submodular dictionary selection for sparse representation. In: ICML 2010: proceedings of the 27th international conference on Machine learning. Haifa: Omnipress; 2010.
25. Aharon M, Elad M, Bruckstein A. K-SVD: an algorithm for designing overcomplete dictionaries for sparse representation. IEEE Trans Signal Process. 2006;54(11):4311–22.
26. Huang J, Yang M. Fast sparse representation with prototypes. In: Computer Vision and Pattern Recognition (CVPR), 2010 IEEE conference on. San Francisco, CA; 2010.
27. Jiang Z, Lin Z, Davis LS. Learning a discriminative dictionary for sparse coding via label consistent K-SVD. In: Computer Vision and Pattern Recognition (CVPR), 2011 IEEE conference on. Providence, RI; 2011.
28. Baraniuk R, et al. Applications of sparse representation and compressive sensing. Proc IEEE. 2010;98(6):906–9.
29. Wright J, et al. Robust face recognition via sparse representation. IEEE Trans Pattern Anal Mach Intell. 2009;31(2):210–27.
30. Elisseeff IGA. An introduction to variable and feature selection. J Mach Learn Res. 2003;3:1157–82.
31. Zou H, Hastie T. Regularization and variable selection via the Elastic Net. J Royal Stat Soc B. 2005;67:301–20.
32. Tu Z, Bai X. Auto-context and its application to high-level vision tasks and 3D brain image segmentation. IEEE Trans Pattern Anal Mach Intell. 2010;32(10):1744–57.
33. Dice LR. Measures of the amount of ecologic association between species. Ecology. 1945;26(3):297–302.
34. Warfield SK, Zou KH, Wells WM. Simultaneous truth and performance level estimation (STAPLE): an algorithm for the validation of image segmentation. IEEE Trans Med Imaging. 2004;23(7):903–21.
35. Coupé P, et al. Patch-based segmentation using expert priors: application to hippocampus and ventricle segmentation. Neuroimage. 2011;54(2):940–54.
36. Rousseau F, Habas PA, Studholme C. A supervised patch-based approach for human brain labeling. IEEE Trans Med Imaging. 2011;30(10):1852–62.
37. Liao S, Shen D. A learning based hierarchical framework for automatic prostate localization in CT images. In: Madabhushi A et al., editors. Prostate cancer imaging. Image analysis and image-guided interventions. Berlin/Heidelberg: Springer; 2011. p. 1–9.
38. Mallat SG. A theory for multiresolution signal decomposition: the wavelet representation. IEEE Trans Pattern Anal Mach Intell. 1989;11(7):674–93.
39. Dalal N, Triggs B. Histograms of oriented gradients for human detection. 2005.
40. Ojala T, Pietikainen M, Maenpaa T. Multiresolution gray-scale and rotation invariant texture classification with local binary patterns. IEEE Trans Pattern Anal Mach Intell. 2002;24(7):971–87.
41. Tibshirani R. Regression shrinkage and selection via the lasso: a retrospective. J Royal Stat Soc B Stat Methodol. 2011;73(3):273–82.
42. Belkin M, Niyogi P, Sindhwani V. Manifold regularization: a geometric framework for learning from labeled and unlabeled examples. J Mach Learn Res. 2006;7:2399–434.

43. Shi Y, et al. Transductive prostate segmentation for CT image guided radiotherapy. In: Wang F et al., editors. Machine learning in medical imaging. Berlin/Heidelberg: Springer; 2012. p. 1–9.
44. Tibshirani R, et al. Sparsity and smoothness via the fused lasso. J Royal Stat Soc B Stat Methodol. 2005;67(1):91–108.
45. Feng Q, et al. Segmenting CT prostate images using population and patient-specific statistics for radiotherapy. In: Proceedings of the sixth IEEE international conference on symposium on biomedical imaging: From Nano to Macro. Boston: IEEE Press; 2009. p. 282–5.
46. Jain A, Zongker D. Feature selection: evaluation, application, and small sample performance. IEEE Transactions Pattern Anal Mach Intell. 1997;19(2):153–8.
47. Bühlmann P. Bagging, boosting and ensemble methods. In: Gentle JE, Härdle WK, Mori Y, editors. Handbook of computational statistics. Berlin/Heidelberg: Springer; 2012. p. 985–1022.
48. Zhang S, et al. Towards robust and effective shape modeling: sparse shape composition. Med Image Anal. 2012;16(1):265–77.
49. Shen D, Ip HHS. A Hopfield neural network for adaptive image segmentation: an active surface paradigm. Pattern Recognit Lett. 1997;18(1):37–48.
50. Liao S, et al. Automatic prostate MR image segmentation with sparse label propagation and domain-specific manifold regularization. In: Gee J et al., editors. Information processing in medical imaging. Berlin/Heidelberg: Springer; 2013. p. 511–23.
51. Liao S, et al. Representation learning: a unified deep learning framework for automatic prostate MR segmentation. In: Mori K et al., editors. Medical image computing and computer-assisted intervention – MICCAI 2013. Berlin/Heidelberg: Springer; 2013. p. 254–61.
52. Kirschner M, Jung F, Wesarg S. Automatic prostate segmentation in MR images with a probabilistic active shape model. In: PRostate MR Image SEgmentation, PROMISE 2012. Nice: Electronic Publication; 2012. p. 28–35.
53. Maan B, van der Heijden F. Prostate MR image segmentation using 3D active appearance models. In: PRostate MR Image SEgmentation, PROMISE 2012. Nice: Electronic Publication; 2012. p. 44–51.
54. Martin S, Troccaz J, Daanen V. Automated segmentation of the prostate in 3D MR images using a probabilistic atlas and a spatially constrained deformable model. Med Phys. 2010;37(4):1579–90.
55. Birkbeck N, Zhang J, Zhou SK. Region-specific hierarchical segmentation of MR prostate using discriminative learning. In: The PRostate MR Image SEgmentation, PROMISE 2012. Nice: Electronic Publication; 2012.

Knowledge-Based Treatment Planning   10

Issam El Naqa

Abstract

Prior information about patient status and previously archived treatment plans, particularly if performed by expert clinicians, could be used to inform the treating team of a current pending case. This notion of using prior treatment planning information constitutes the underlying principle of the so-called knowledge-based treatment planning (KBTP). In this chapter, we will discuss KBTP and provide some examples highlighting its current status, the role of machine learning, and its potential for decision support in radiotherapy.

10.1 Introduction

Radiotherapy planning is a laborious computer-aided process that aims to provide a blueprint for the treatment that would be followed meticulously and precisely over several weeks. It involves the determination of treatment parameters that would be considered optimal in the management of a patient's cancer. These parameters include target volume, dose-limiting normal tissue structures, treatment volume, dose prescription, dose fractionation, dose distribution, positioning of the patient, treatment machine settings, and adjuvant therapies [1, 2].

Conventionally, this process would involve acquiring patient image data by CT/PET/MR scans (most typically fully 3D computed tomography (CT) scans). Then, the physician outlines the tumor and important normal structures on a computer (contouring), based on the CT scan using a specific set of guidelines such as the International

---

I. El Naqa
Department of Oncology, McGill University, Montreal, QC, Canada

Department of Radiation Oncology, University of Michigan, Ann Arbor, USA
e-mail: issam.elnaqa@mcgill.ca; ielnaqa@med.umich.edu

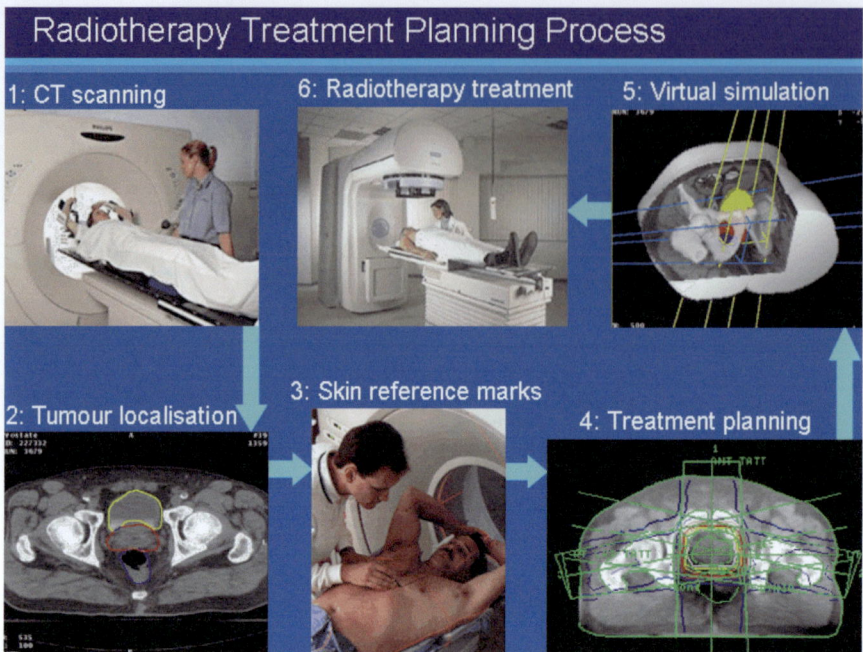

Fig. 10.1 Summary of a typical treatment planning process (The image is courtesy M. Lewis from imPACT)

Commission on Radiation Units and Measurements (ICRU) Reports 50, 62, 71, and 83 [3]. Afterwards, the dosimetrist sometimes with the aid of a physicist would come up with proper beam arrangements and weighting to meet the requirements of attending physician using forward (3D conformal radiotherapy (3D-CRT)) or inverse planning (intensity-modulated radiotherapy (IMRT)) strategies. The planning is verified prior to delivery by estimating the treatment dose distributions with prescribed doses in the treatment planning system (TPS) using dose calculation algorithms as shown in Fig. 10.1.

More recently, there has been interest in using prior treatment planning information, referred to as knowledge-based treatment planning (KBTP), possibly generated by experts to aid daily practice. The motivation for such an approach lies in reducing current complexity and time spent on generating a new treatment plan from each incoming patient. Clinicians typically have their own set of manual templates that they often use, which may give rise to inconsistency and increased patient risk [4–6]. It is believed that such a standardization process based on KBTP can help enhance consistency, efficiency, and plan quality.

10.2 Framework for Knowledge-Based Treatment Planning

The development of a framework for KBTP would require accumulation of information from past experiences related to patient, disease, imaging, treatment setup, dose, etc. A depiction of such system is shown in Fig. 10.2, in which a retrieval

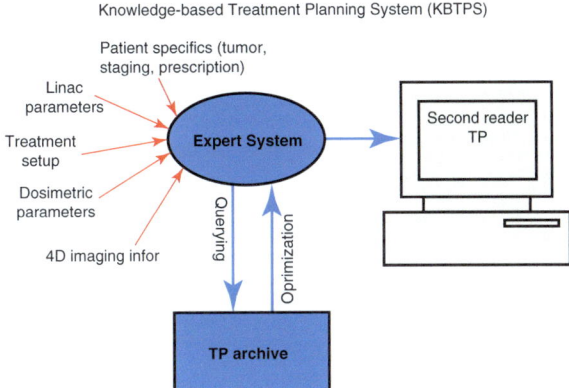

Fig. 10.2 Schematic of a hypothetical KBTPS, in which the user builds a query using features related to patient, disease, imaging, treatment setup, dose, etc for the treatment plan (TP). The database returns a set of similar treatment plans that the user could select from to optimize and compare with the current one according to the query

system allows the user to query "similar" cases from their archive and propagate such information to optimize the current plan at hand.

10.3 Clinical Applications

The implementation of knowledge-based approaches has taken several interesting directions as quality control of treatment and as a process to generate new IMRT plans as briefly discussed in the following.

10.3.1 Treatment Assessment Tools

Zhang et al. developed a machine learning approach for predicting normal tissue complications from dose-volume planning constraints without the need for explicit plan computation and demonstrated their method in cases of prostate and head and neck cancers [7]. Moore et al. developed and evaluated a model for predicting the organ-at-risk (OAR) dose from its overlap with the PTV and the prescription dose, and they demonstrated that the dose received by the parotid gland or the rectum could be indeed reduced using their model [5]. This work was extended to predict achievable dose-volume histograms (DVHs) using skewed Gaussian distributions from individual patient anatomy as shown in Fig. 10.3 [8].

10.3.2 IMRT Planning

As a demonstration of the feasibility of applying KBTP to the generation of IMRT plans in prostate cancer, Chanyavanich et al. presented a semiautomated method based on mutual information to identify similar patient cases by matching 2D beam's eye-view projections of contours [4]. This approach was further evaluated on a larger pool of patients, and reported reductions were significantly lower for the rectum with 40 % of cases; the KBTP plan had better DVHs for rectum and bladder;

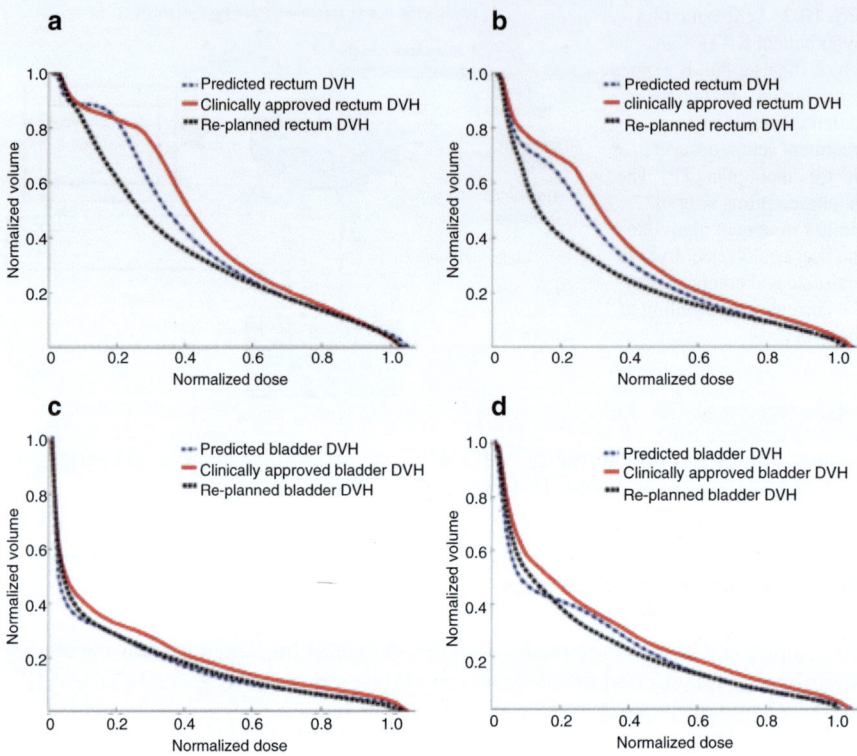

Fig. 10.3 Clinically approved DVHs compared to refined predicted DVHs and replanned DVHs for the (**a**, **b**) rectum and (**c**, **d**) bladder

in 54 % of cases, the comparison was equivocal; in 6 % of cases, it was inferior for both bladder and rectum as shown in Fig. 10.4 [6].

10.4 Role of Machine Learning

The framework of KBTP in Fig. 10.2 represents an ideal scenario for applying techniques of machine learning or expert systems. As an example of the latter, Petrovic et al. presented a method based on case-based reasoning (CBR) system to generate dose plans for prostate cancer patients. The proposed CBR system applied a modified Dempster-Shafer approach to fuse dose plans suggested by the most similar cases retrieved from the archive database [9]. The Dempster-Shafer theory (DST) allows for combining evidence information from different sources [10]. To mimic the continuous learning characteristic of oncologists, the weights corresponding of the features used in the retrieval process are updated after generating a new

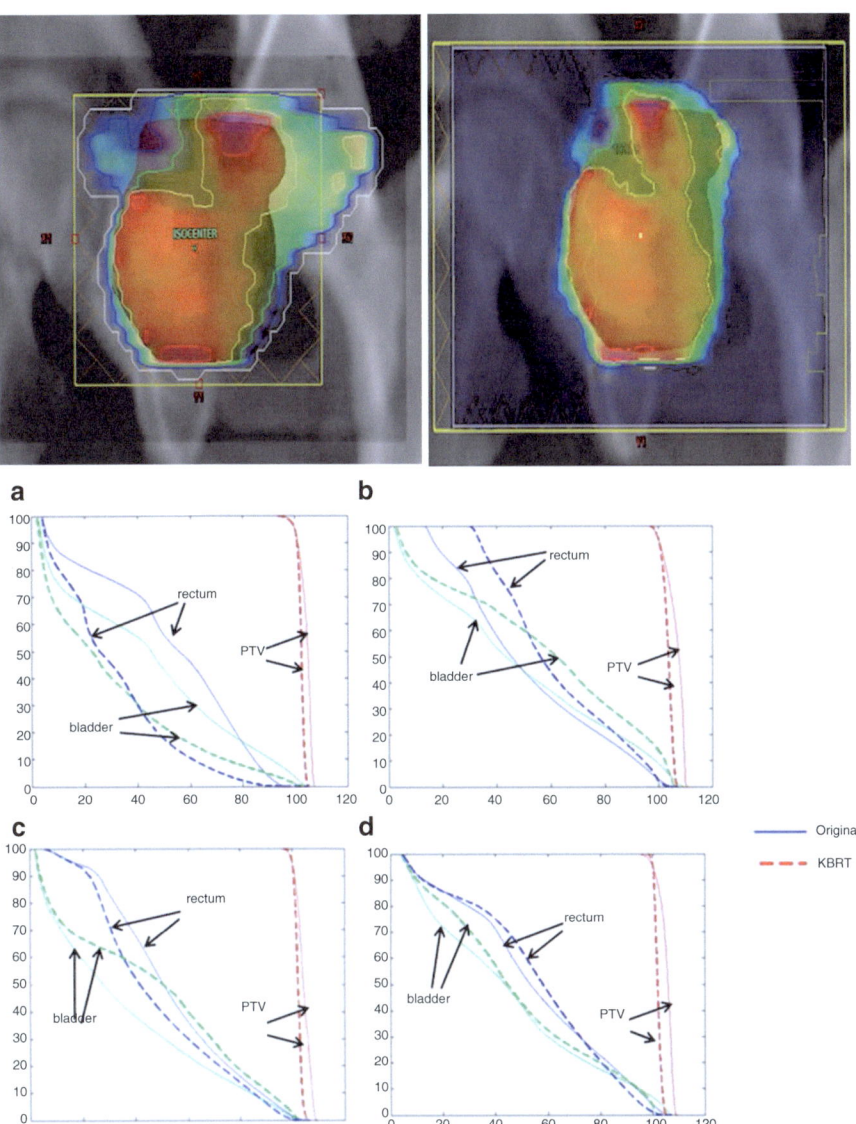

Fig. 10.4 Examples of application of knowledge-based radiation therapy (KBRT) from IMRT planning. *Top* Example of match case's fluence map for 1 beam orientation overlaid on query case's planning target volume before deformation (*left*) and after deformation (*right*). *Bottom* (**a–d**) Comparison categories between the original and plans with different scenarios. The percentage reflects the time the KBRT provides better performance

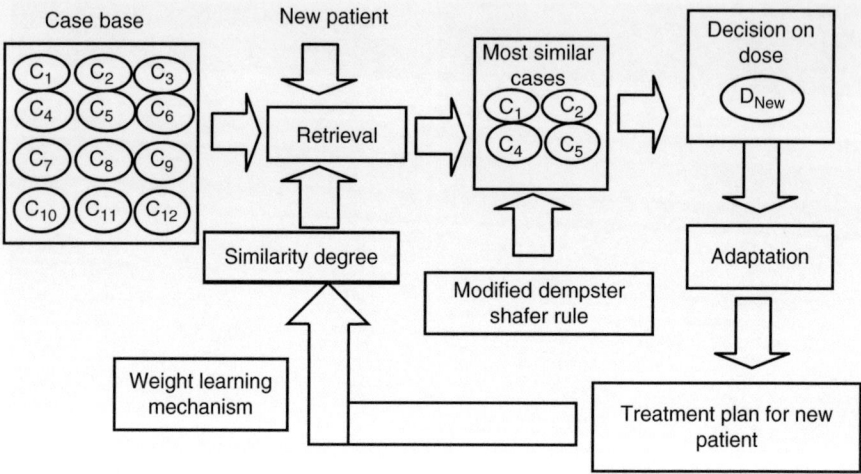

Fig. 10.5 Architecture of a case-based reasoning system for prostate planning

treatment plan; a depiction of the system is shown in Fig. 10.5 [9]. Features related to clinical stage and geometry of prostate are used to retrieve the most similar cases from the archive. Similarity metric is based on using fuzzy sets and information is combined from these cases using a modified DST rule.

Conclusions

In this chapter, we have presented the framework of knowledge-based treatment planning for radiotherapy, in which an archive is queried for "similar" cases and information is propagated to optimize a current treatment plan. We presented different applications of KBTP in radiotherapy from quality control/assurance, assessment tool, to generating complex IMRT plans. In addition, we presented a case of using expert systems to combine decision information. KBTP is a promising area in radiotherapy that lends itself naturally to machine learning application with potential to improve decision support system.

References

1. Videtic GMM, Woody N, Vassil AD. Handbook of treatment planning in radiation oncology. 2nd ed. New York: Demos Medical; 2015.
2. Khan FM. Treatment planning in radiation oncology. 2nd ed. Philadelphia: Lippincott Williams & Wilkins; 2007.
3. Hodapp N. The ICRU Report 83: prescribing, recording and reporting photon-beam intensity-modulated radiation therapy (IMRT). Strahlenther Onkol Organ der Deutschen Rontgengesellschaft. 2012;188:97–9.

4. Chanyavanich V, Das SK, Lee WR, Lo JY. Knowledge-based IMRT treatment planning for prostate cancer. Med Phys. 2011;38:2515–22.
5. Moore KL, Brame RS, Low DA, Mutic S. Experience-based quality control of clinical intensity-modulated radiotherapy planning. Int J Radiat Oncol Biol Phys. 2011;81:545–51.
6. Good D, Lo J, Lee WR, Wu QJ, Yin FF, Das SK. A knowledge-based approach to improving and homogenizing intensity modulated radiation therapy planning quality among treatment centers: an example application to prostate cancer planning. Int J Radiat Oncol Biol Phys. 2013;87:176–81.
7. Zhang HH, D'Souza WD, Shi L, Meyer RR. Modeling plan-related clinical complications using machine learning tools in a multiplan IMRT framework. Int J Radiat Oncol Biol Phys. 2009;74:1617–26.
8. Appenzoller LM, Michalski JM, Thorstad WL, Mutic S, Moore KL. Predicting dose-volume histograms for organs-at-risk in IMRT planning. Med Phys. 2012;39:7446–61.
9. Petrovic S, Mishra N, Sundar S. A novel case based reasoning approach to radiotherapy planning. Expert Syst Appl. 2011;38:10759–69.
10. Yager RR, Liu L. Classic works of the Dempster-Shafer theory of belief functions. Berlin/New York: Springer; 2008.

Part IV

Machine Learning Delivery and Motion Management

Artificial Neural Networks to Emulate and Compensate Breathing Motion During Radiation Therapy

11

Martin J. Murphy

Abstract

A number of treatment sites for external-beam radiation therapy, such as lung, breast, pancreatic, and liver cancers, move as the patient breathes, which compromises the precision of their irradiation. Modern radiation treatment modalities attempt to deal with this by adapting the radiation delivery to the respiratory motion as it occurs. This requires system control processes that can detect and anticipate respiratory movement patterns on a patient-by-patient basis in real time. Because breathing can be very idiosyncratic, this problem is a good candidate for machine learning algorithms that can be trained to model individual breathing patterns. Neural networks have proven quite effective in this capacity. This chapter describes the nature of the motion-compensated treatment problem and the issues in using a neural network to handle it.

11.1 Background

A number of treatment sites for external-beam radiation therapy, such as lung, breast, pancreatic, and liver cancers, move as the patient breathes, which compromises the precision of their irradiation. For the purpose of this chapter, we will use lung tumors as the paradigm to represent this motion problem.

To achieve the best likelihood of effective beam coverage for a treatment target that moves during respiration, there are four basic approaches: (1) inhibit the movement via breath holding or physical restraints, (2) enlarge the therapy beam field so that the tumor never moves outside of it (the margin approach), (3) turn the beam on

---

M.J. Murphy
Department of Radiation Oncology, Virginia Commonwealth University,
401 College Street, P.O. Box 980058, Richmond, VA 23298-0058, USA
e-mail: MMurphy@mcvh-vcu.edu

only when the tumor is at or near the beam isocenter (the gating approach), and (4) move the beam or the patient synchronously with breathing so that the beam stays continuously aligned with the tumor (the tracking approach) [30]. In the tracking approach, the beam can be realigned by moving the linear accelerator (LINAC) itself [1, 5, 42, 43, 45] or shifting the multileaf collimator (MLC) aperture [7, 19, 24, 25, 26, 27, 34, 39, 41, 48] or, in the case of a charged-particle beam, magnetically steering the beam [3]. Alternatively, the patient can be moved by shifting the couch, so that the tumor remains at a fixed beam isocenter [23, 36, 38, 46, 47]. Gating and tracking are the two approaches that call for adapting to tumor motion in real time.

There are two fundamental problems in adapting to tumor motion: (1) determining the precise tumor position at any given time and (2) making a synchronized adaptive response to maintain beam/tumor alignment. Tumor position can either be measured directly via imaging or other detection methods, or it can be inferred by measuring respiratory movement that is reliably correlated with the tumor movement and can act as a surrogate for it [2, 10, 14, 15, 18, 20, 35, 50, 52–54]. In this chapter we are particularly interested in the problem of inferring the tumor movement from some kind of surrogate respiratory signal.

No adaptive response to movement can occur instantaneously, so it is necessary to compensate for delays between localization of the tumor and adjustment of the beam timing or alignment. This comes down to predicting the future tumor position (or its surrogate respiratory signal) by an amount equal to the response delay time so that the adaptation is synchronized to the tumor's actual position.

Figures 11.1 and 11.2 illustrate two representative patients' breathing patterns, as measured by an optical marker placed on the chest. Figures 11.3 and 11.4 show how a sequence of measurements of surface breathing movement (via the marker) can be correlated with the tumor's actual position, measured via X-ray fluoroscopic imaging. These four figures combine to demonstrate the complications presented by the

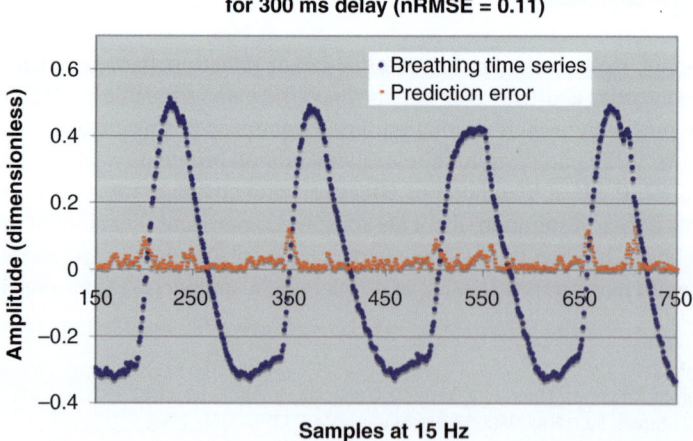

Fig. 11.1 An example of regular breathing

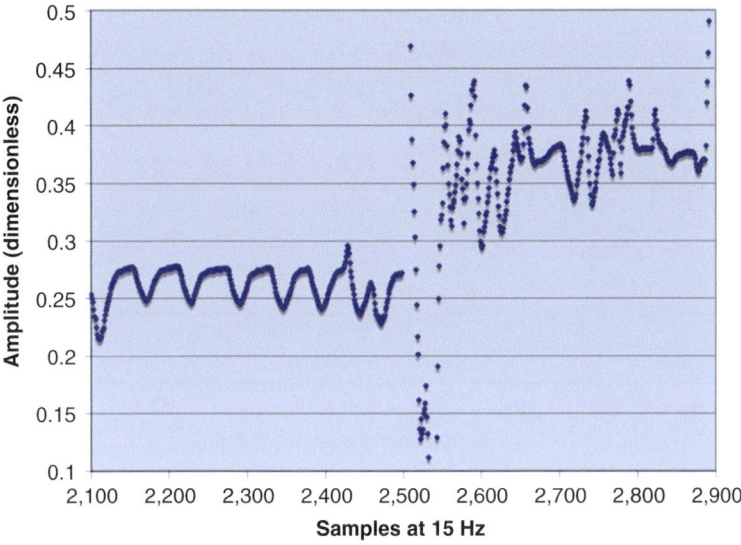

Fig. 11.2 An example of highly irregular breathing

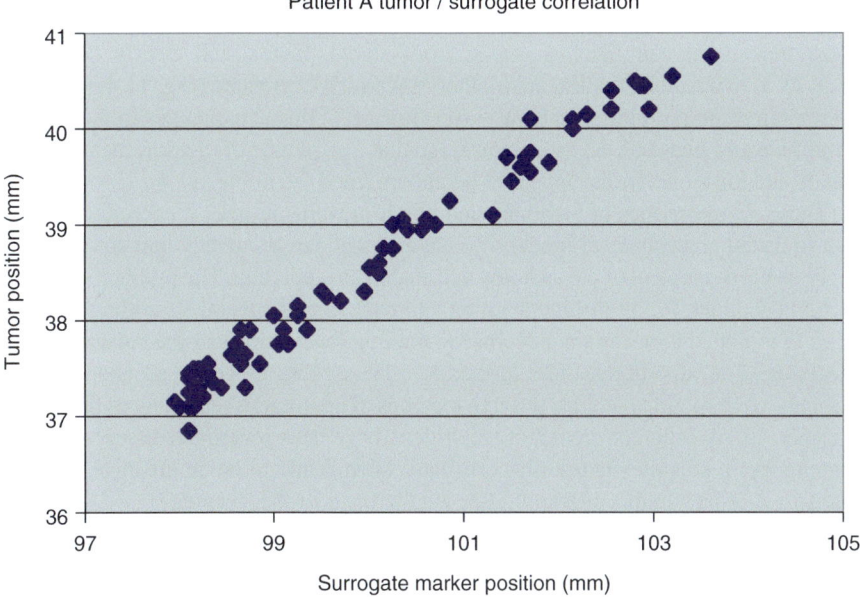

Fig. 11.3 A sequence of measurements of tumor position and chest marker position, showing a tight correlation over time [35]

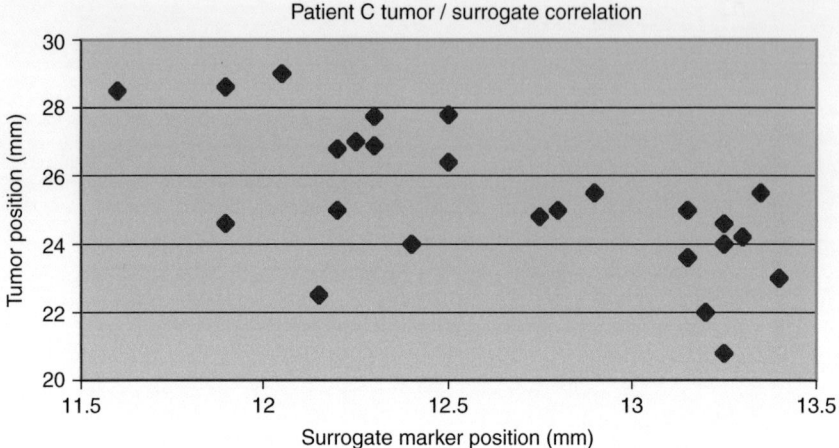

Fig. 11.4 An example of tumor and chest marker positions that do not maintain a tight correlation over time [35]

breathing prediction and correlation problem. Although superficially regular (as in Fig. 11.1), normal breathing is not strictly periodic, but changes amplitude and period over time [22, 49]. In extreme cases, the breathing pattern can be highly irregular to the point of appearing chaotic (Fig. 11.2). The relationship between, e.g., tumor and chest movement can likewise range from stable, linear, and tightly correlated (Fig. 11.3) to unstable, nonlinear, or otherwise poorly correlated (Fig. 11.4) [35]. The tumor/surrogate correlation can vary over time (e.g., through changes in the relative amplitude and phase of the movements), so that a sequence of measurements of surrogate and tumor positions appear to be uncorrelated (as in Fig. 11.4).

These characteristics of breathing and tumor movement make it exceedingly difficult to devise a mechanical model of breathing that can accurately and continuously describe the movement of the anatomy and enable its prediction. The problem is instead a good candidate for a machine learning approach, using general algorithms that can learn to imitate the movement patterns via training on examples of the patient's actual breathing. The algorithms must furthermore be capable of continual adaptation to changes in the motion patterns, through methods of continuous retraining as the patient breathes. Many different prediction algorithms have been investigated (see, e.g., [6]). Among them, adaptive neural networks have been found to be an effective machine learning approach to this problem. They are the focus of this chapter.

11.2 Using an Artificial Neural Network (ANN) to Model and Predict Breathing Motion

The basic mechanism for maintaining beam alignment with a moving tumor is illustrated schematically in Figs. 11.5 and 11.6. Figure 11.5 is an "open loop" control architecture that is appropriate for either a gating or a beam tracking scheme. The

11 Artificial Neural Networks to Emulate and Compensate Breathing Motion

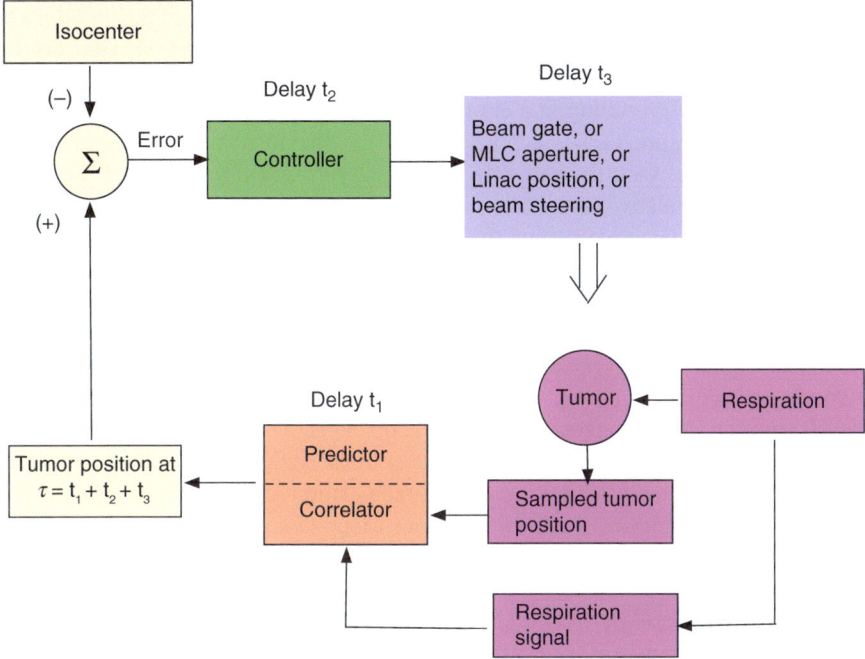

Fig. 11.5 An open control loop architecture for maintaining beam and tumor alignment

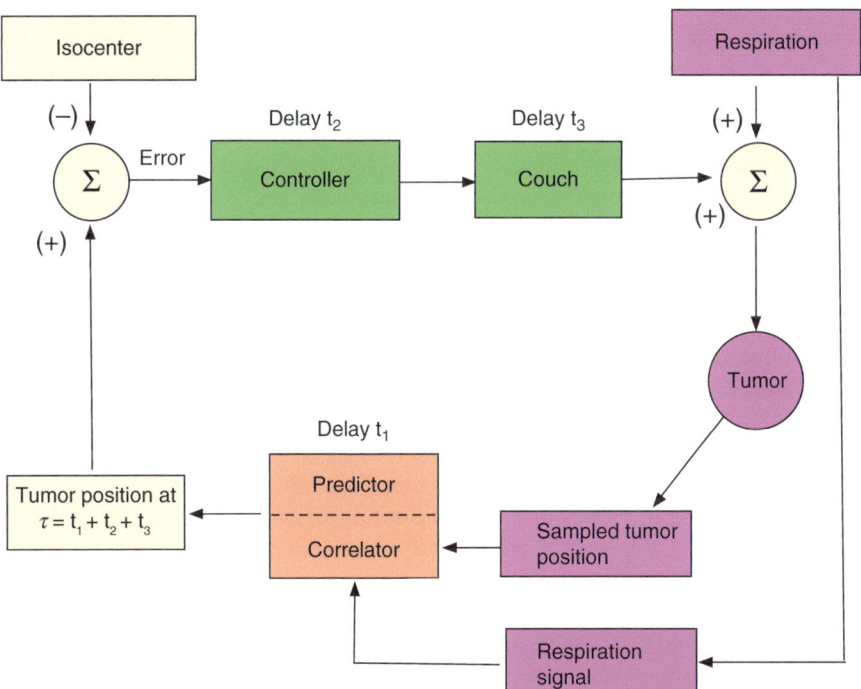

Fig. 11.6 A closed-loop beam alignment architecture

tumor moves solely under the influence of patient movement (e.g., breathing). Respiration and/or tumor position sensors provide the input to the loop. The corrective signal propagates through various system components, each of which takes some time to react, resulting in a cumulative delay before the beam responds with the correction. Figure 11.6 is a "closed-loop" architecture in which the system's response combines with the patient's anatomical movement to influence the position of the target relative to the beam isocenter and thus the input to the loop. This is required for an adaptive system that moves the couch and patient relative to the beam as the tumor moves, so as to keep the tumor at a fixed position (set point) in space. In this case, respiration and couch shifts combine to move the tumor. In both architectures, the tumor position can be established either by following a surrogate breathing signal that correlates with tumor motion or by directly observing the tumor's position or both.

The subject of this chapter is the control loop element identified in Figs. 11.5 and 11.6 as the "correlator/predictor". This element receives as input some measurement of breathing and provides the anticipated position of the tumor as input to the beam or couch controller. To allow for control loop delays, the "correlator/predictor" must emulate the patient's breathing in order to predict the future respiratory signal and/or tumor position.

An artificial neural network (ANN) is a trainable machine learning algorithm. One form that is very useful for predicting a signal amplitude has the basic architecture shown in Fig. 11.7. In this kind of application, we have some measured signal $S(t)$ as input and a future instance of that signal $S(t+n)$ as the output target. The job of the ANN is to make an estimate $S'(t+n)$ of the future target signal from samples of the input signal. The input layer of the network is provided with discrete measurement samples from the past signal history, the hidden layers compute weighted combinations of the input data, and the output layer delivers an estimate of the target signal at a future time. In Fig. 11.7 the target signal is a future sample of the input signal, in which case the network is trained to imitate the input signal so that it can predict its future behavior. When the target signal finally arrives at time $t+n$, the prediction $S'(t+n)$ is compared to it, an error is computed, and this error is used to adjust the network weights so as to produce a more accurate prediction of the next sample. Figure 11.8 shows a configuration to use the input signal $S(t)$ to predict a different signal $P(t)$ that is correlated in some way with the input signal. In this case the network is trained to predict the correlated target signal from the input. The target prediction might be for the present moment or some future time.

In our breathing prediction problem, we identify the input data with a sequence of discrete measurements of the patient's breathing. This could be as simple as the time history of the amplitude of a single breathing signal, such as a moving marker [29] or spirometer signal [15, 54], or it could comprise simultaneous measurements of multiple breathing signals [53]. If we are only interested in predicting breathing movement to compensate for a treatment system's lag time, then the target signal would be a future instance of the patient's measured breathing, and the network's output would be an estimate of that future instance. If we are

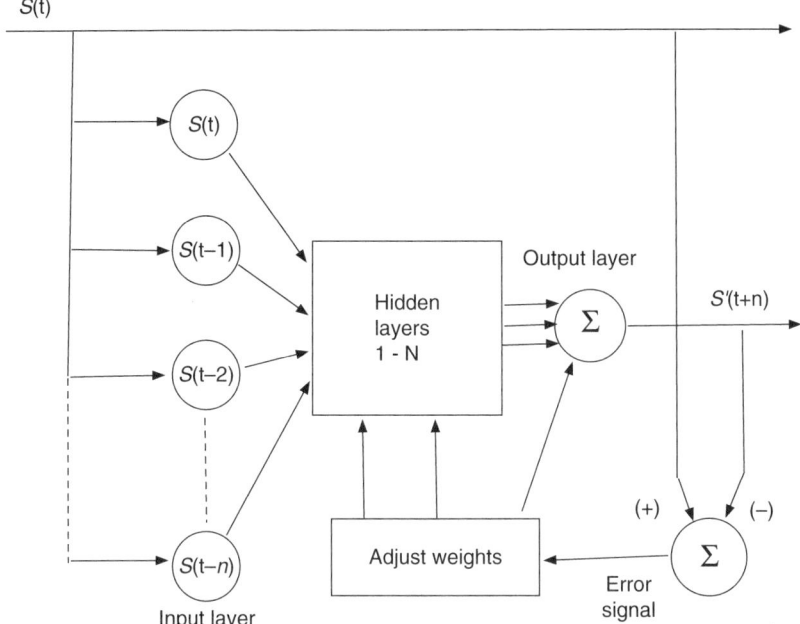

Fig. 11.7 An artificial neural network architecture to predict a signal amplitude $S(t)$

interested in deducing the tumor position from the measured breathing signal, then the input signal would be a breathing surrogate measurement, and the target data would be a measurement of the tumor's spatial position at some particular time. It could be the tumor position at the present time, in which case the ANN makes a spatial correlation between the tumor and breathing motions, or it could be the future position of the tumor, in which case the network performs both a correlation and a temporal prediction to arrive at a good estimate of the tumor location.

11.3 Neural Network Architectures for Correlation and Prediction

11.3.1 The Single Neuron, or Linear Filter

We can introduce the basic computational components of an artificial neural network for correlation and prediction by considering a simple network configured to predict the future amplitude of a single breathing signal sampled at discrete time intervals. It begins with a single neuron, as shown in Fig. 11.9 (This has historically been known as a linear perceptron). The input is the amplitude history of the measured signal $S(t)$, sampled at n intervals of τ seconds. For breathing, which has a period of a few seconds for most people, τ might be on the order of 100 ms. We take

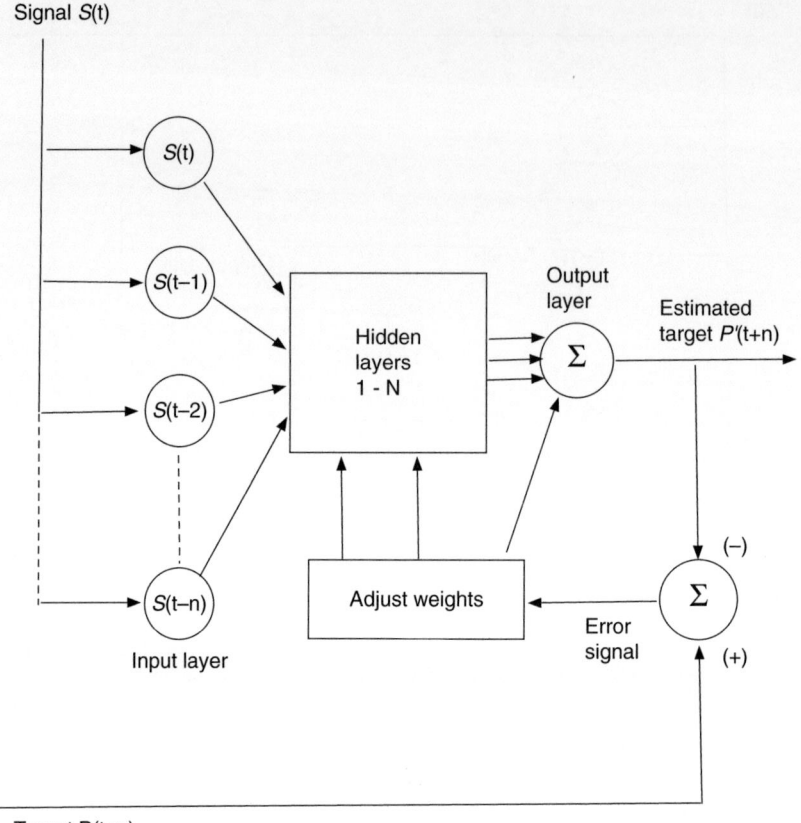

Fig. 11.8 An ANN configured to predict a different signal $P(t+n)$ that is correlated with the input $S(t)$

the N most recent samples. Each sample is multiplied by a weight w_i and the N samples are summed:

$$S'(t) = {}^N\sum_{i=1} w_i S(t-i\tau) \qquad (11.1)$$

If we stop here, we have a simple linear filter, where $S'(t)$ is the filter's estimate of the signal amplitude at the present time, based on the previous N samples. $S'(t)$ is compared to $S(t)$ and the error is used to adjust the weights until the difference is minimized. If we want it to predict S at some future time $t + \Delta t$, rather than the present, we wait Δt seconds for the actual signal to arrive, compare it to S' to find the error, and adjust the weights accordingly.

The linear filter (i.e., a single neuron) in Fig. 11.9 and Eq. 11.1 can do a reasonable job of predicting breathing, provided that the pattern isn't too changeable or irregular [31]. It provides a starting point to introduce several basic elements in the development of ANNs for prediction and correlation.

The weights are initially optimized in the training stage. For a basic signal prediction filter, this typically consists of presenting the filter with prerecorded signal

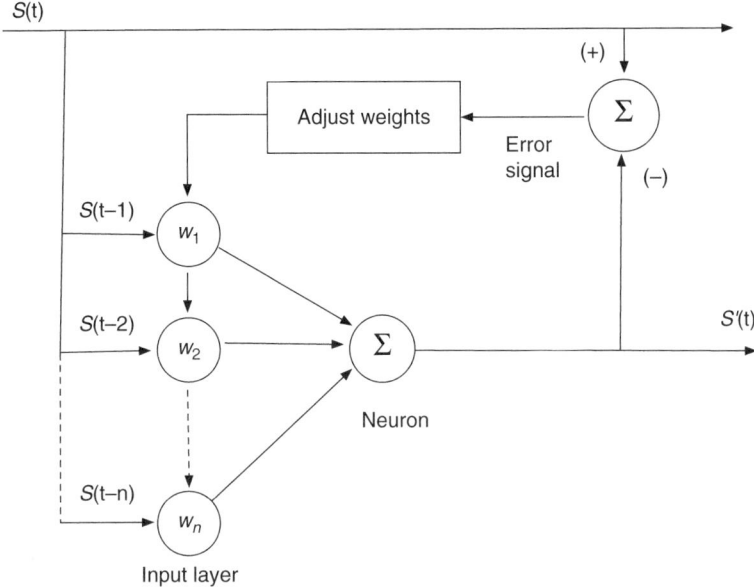

Fig. 11.9 A simple linear filter for prediction

histories that are representative of the signal that one ultimately wants to predict. For example, if one wants a filter customized to emulate and predict a particular patient's breathing, one begins by recording a segment of the patient's breathing signal. This is presented to the filter incrementally via a sliding window that is N samples wide. The filter gets a set of N samples up to a time t at the inputs, makes a prediction for $t + \Delta t$, compares the prediction to its target (which is the recorded signal at $t + \Delta t$), adjusts the weights, steps forward one sample, and repeats the process. This is an example of supervised sequential training. Sequential training has the advantage that, as the filter is presented with new breathing data that it hasn't seen before, it can continue the process, retraining continuously to adapt to new breathing patterns.

The initial training process must be done in such a way that it doesn't "see" future samples in the training stage before they would actually arrive in real time.

The simplest training algorithm for a linear neuron is the LMS (least mean square) method. Let S_i be the vector of N input samples from the ith training signal history, let W_i be the vector of N weights assigned to the inputs, and let ε_i be the difference between the predicted and target signal sample. The updated weight vector is

$$W_{i+1} = W_i + \alpha \varepsilon_i S_i \qquad (11.2)$$

where α is a parameter that determines the speed of convergence. In the case of sequential training, each training signal history S_i is simply the previous signal history advanced by one sample.

There are numerous other algorithms to update the weights. For a more comprehensive review of ANN training algorithms, see, for example, Haykin [13] or Haykin [12] or another introductory textbook on neural networks.

11.3.2 The Basic Feedforward Artificial Neural Network for Prediction

Soon after the single-neuron perceptron was proposed as a primitive machine learning algorithm for pattern recognition, Minsky proved that it, and any linear combination of neurons performing the function of Eq. 11.1, could only do linear discrimination and was incapable of performing even a simple exclusive-or function [28]. This led to the development of nonlinear networks of neurons for more complex pattern recognition and signal processing. Figure 11.10 is a schematic of the simplest nonlinear neural network – a feedforward network with one hidden layer – for signal prediction. The inputs are distributed in parallel to two or more neurons like the one in Fig. 11.9 (the simple linear filter). These make up the "hidden" layer. (The layer is "hidden" because it can't be reached directly from the outside.) The output x of each neuron is passed through a nonlinear "activation" function f(x) (the sigmoid function in Eq. 11.3 is the most commonly used), weighted, and

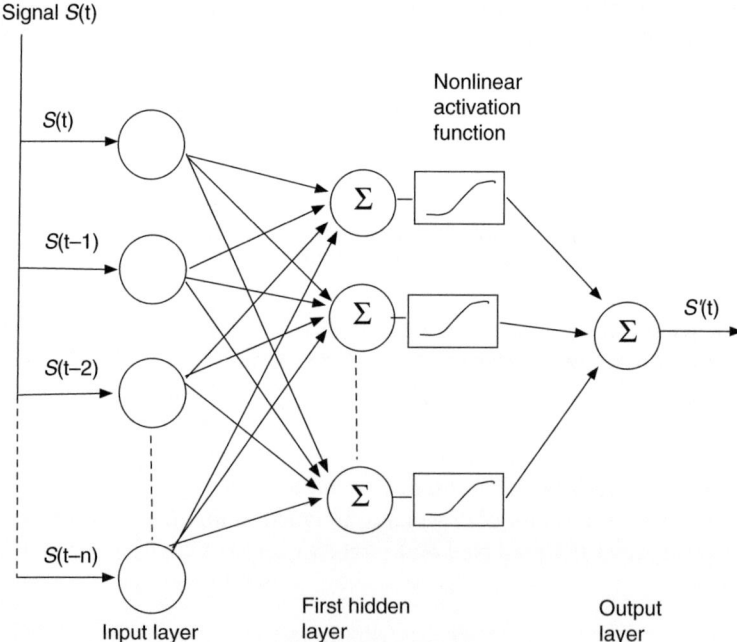

Fig. 11.10 A basic feedforward network with one hidden layer of neurons and a single neuron in the output layer

summed with the others in the output neuron, which delivers the final signal estimate.

$$y = f(x) = 1/(1+e^{-x}) \tag{11.3}$$

$$df/dx = y(1-y) \tag{11.4}$$

The activation function must be nonlinear; otherwise, the network is reducible to a single linear neuron and nothing is gained.

11.3.3 Training the Feedforward Network

Each input to each neuron in the hidden and output layers in Fig. 11.10 has an independently variable weight. However, the weights in the hidden layer are "blocked" from the output signal error by the nonlinear activation function. This prevents a simple linear generalization of the LMS algorithm in Eq. 11.2. The problem is solved by the method of error back propagation.

Although the basics of back propagation can be found in any textbook on neural networks (e.g., [12]), there is some advantage to providing them here, using the simple two-layer network in Fig. 11.10 as the architecture. Let layer 1 be the hidden layer and layer 2 be the output layer (in this case just one neuron). Let the index i apply to the data samples and j to the number of neurons in layer 1 (and also the equal number of input weights to layer 2). Let $W_{1,j}$ be the vector of weights for the jth neuron in layer 1 (with components $w_{1,ji}$) and W_2 be the weight vector for the output (layer 2) neuron (with components $w_{2,j}$). The outputs of the layer 1 neurons are $x_{1,j}$ before activation and $y_{1,j}$ after activation. The error in the predicted output signal is ε.

In the forward pass, the delta is calculated for layer 2:

$$\Delta_2 = \varepsilon.$$

In the backward pass, the deltas for layer 1 propagate through the derivative of the transfer function:

$$\Delta_{1,j} = \left[y_{1,j}(1 - y_{1,j}) \right] \left[\Delta_2 w_{2,j} \right].$$

The incremental changes to the weights in the two layers are then calculated (in this example via LMS):

$$\delta w_{2,j} = \alpha \Delta_2 y_{1,j}$$
$$\delta w_{1,ji} = \alpha \Delta_{1,j} S_i.$$

In addition to LMS, there are a number of other algorithms that can be used to update the weights [12, 13]. Regardless of which one is used, there are some general principles to be followed to get the best results. The first step is to initialize all of the weights. The usual practice is to choose them randomly, because this gets the neurons acting independently. However, there is always some chance that a random initialization will come up with an unfavorable filter that performs badly. This can be avoided by performing the random initialization and subsequent training multiple times while testing each fully-trained filter on an independent validation signal. The set of weights that does the best job of predicting the validation signal becomes the optimal filter for application to test signals. The validation signal can be any part of the prerecorded signal that wasn't used for training.

It is also generally the case that a single pass through the training data will not result in optimal convergence of the weights. It is therefore customary to run through the training data repeatedly, starting each subsequent training pass at the weights from the prior training pass. Each pass is called an *epoch*. However, there is the risk of *overtraining* the filter after too many epochs. In this case the filter becomes completely optimized to emulate the training data but cannot generalize effectively to signals it has not yet seen. This can be avoided by testing each epoch of trained filter on the validation data and terminating the training when the filter's performance on the validation set is clearly worse than its performance on the training data.

The feedforward breathing prediction network in Fig. 11.10 can be generalized to perform temporal prediction and position correlation by comparing its output to some measure of tumor position, as in Fig. 11.8.

11.3.4 The Recurrent Network

A recurrent network is a closed-loop feedback architecture in which signals from the hidden and output layers are fed back to previous hidden layers and/or to the input layer. This architecture is inspired by the observation that the human brain is a recurrent network of neurons. In Fig. 11.11, a simple recurrent network for prediction feeds the previous $m-1$ predictions back to the input layer at each time step $S(t)$ of the input signal. The output signals are held back by the prediction interval τ before they are supplied to the input, so that the error between $S(t)$ and $S'(t)$ can be computed and used to update the weights. The hoped-for advantage is that the raw input data from the (potentially noisy) measurements is supplemented by filtered data from the outputs that will smooth out the network's response. A recurrent network can be trained in the same way as a feedforward network, e.g., via back propagation.

11.3.5 Using a Kalman Filter to Predict/Correct as Part of the Training Loop

Consider a system that is being observed via periodic data samples. Suppose each data sample fluctuates randomly due to the behavior of the system itself (plant noise) and uncertainty in the measurements (measurement noise). If the system's evolving state is governed by a linear function, then the best estimate of the next

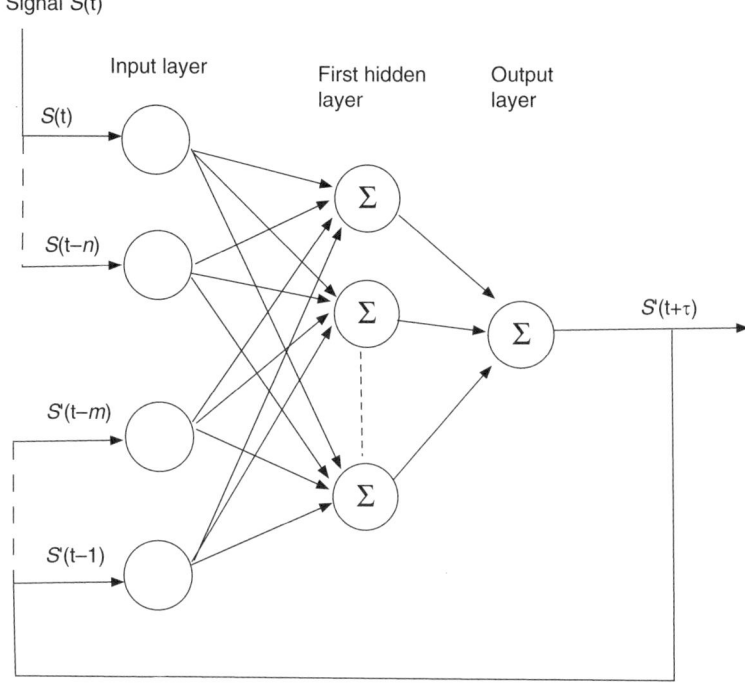

Fig. 11.11 A basic recurrent network

sample is provided by the Kalman filter predictor, which is a continuously-updating algorithm that takes its present estimate of the system's state, makes a prediction of the next signal sample, combines it with the next available data measurement, and calculates a correction to update the state of the system, which is then recirculated via a prediction/correction loop. Such a filter continuously adapts to the evolution of the system.

A breathing signal has variability that can be divided between two sources: irregularity in the actual breathing (plant noise) and errors in the observations (measurement noise). This has inspired studies to predict breathing with a Kalman filter. However, the breathing system is nonlinear, and consequently the Kalman filter must be generalized to an extended Kalman filter (EKF). The extended Kalman filter attempts to linearize the observations (typically via a Taylor expansion) so that the basic Kalman prediction/correction algorithm can be used. Unfortunately, this has proved problematic, and the performance of an EKF for breathing has generally not been as favorable as other methods.

Looking back to Fig. 11.7, one sees that the weights are updated from the most recent error signal. These error signals also incorporate plant and measurement noise, which suggests that an extended Kalman filter can be used to train an ANN [51]. In this application it would be used to calculate (predict) each successive

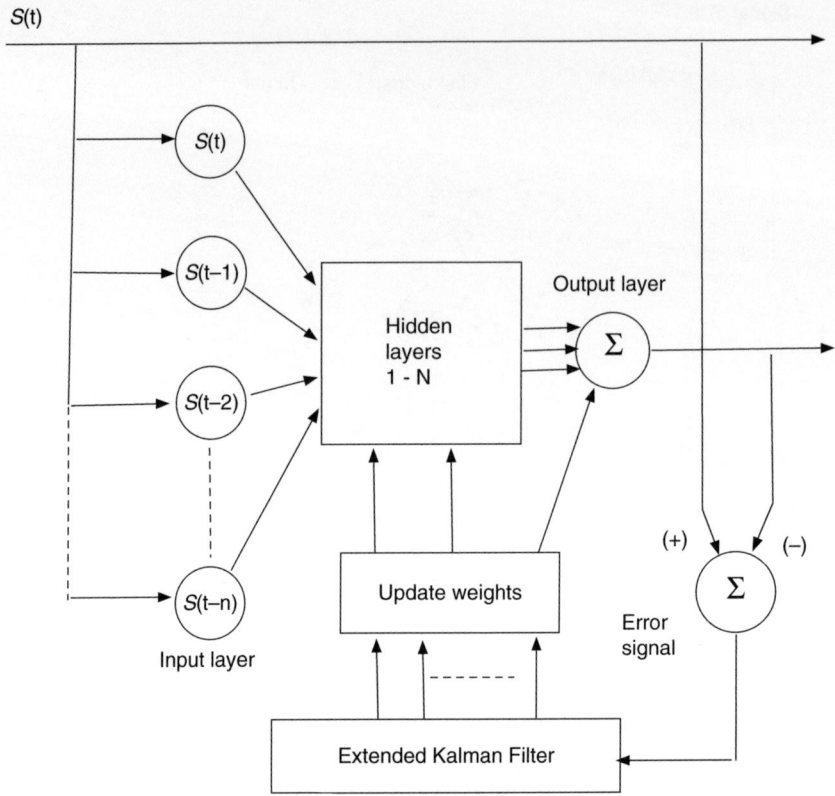

Fig. 11.12 A recurrent network employing an extended Kalman filter to compute the updates to the weights

update to the weights, and thus the state of the ANN, rather than model and update the breathing state itself. This could combine the advantages of both the ANN and the EKF. Figure 11.12 illustrates the strategy. This breathing prediction architecture has been studied by, e.g., Lee et al. [21], who present the details for computing the EKF prediction/correction of the network weights.

A recurrent EKF-ANN with p outputs describes the system state with a vector of s neuron weights, which requires an error covariance matrix of size s^2 and computational complexity of order $O(ps^2)$. This can become demanding when there are a large number of inputs to the network. However, it is possible to decouple the individual weights in the EKF stage, so that the error covariance matrix becomes block diagonal and the computational complexity is reduced to order $O(ps)$ [21, 37].

11.3.6 A Network with Multiple Breathing Signal Inputs

The discussion of ANNs for breathing prediction and correlation has so far used the simple case of a single one-dimensional breathing signal $S(t)$ supplied as input.

Fig. 11.13 An array of optically-tracked LED markers to measure breathing

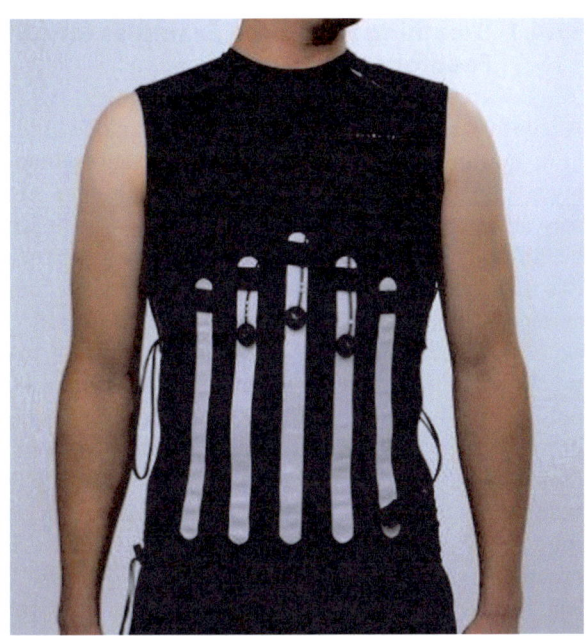

In a clinical setting one can often have multiple sensors, each measuring up to three spatial degrees of freedom in movement. Figure 11.13 illustrates such a situation, in which the CyberKnife (Accuray Incorporated, Sunnyvale CA) utilizes an array of optically-tracked infrared emitters distributed on the patient's chest and abdomen to record breathing. The breathing data are correlated with periodic X-ray measurements of tumor position to provide a targeting signal to the linear accelerator, which makes compensating corrections to the treatment beam's direction [42, 43]. This has the advantage of multiple redundant measurements to reduce the influence of measurement noise and the capacity to determine during training whether the patient is a chest or abdominal breather.

The most basic generalization to m breathing signal sources is simply to make up an input layer that provides n taps of each signal, for a total of nm input nodes. However, for a breathing patient, the m sets of input samples will be correlated with one another. In the EKF-ANN this correlation will be reflected in the error covariances. This can be dealt with by coupling the Kalman filters for each signal channel (while keeping the weights decoupled, as above). This has been studied by Lee et al. [21]. Alternatively, one can make a principal component analysis (PCA) of the signals to obtain an input vector of maximally-uncorrelated data.

11.4 Performance of Neural Networks to Predict Tumor Motion

The problem of predicting breathing with an artificial neural network has been studied by a number of researchers (e.g., [4, 11, 16, 17, 21, 29, 32, 33, 40, 44]).

11.4.1 Breathing Prediction Examples for a Simple Feedforward Network

A feedforward network can have more than one hidden layer, each of which can have multiple neurons. It can also have more than one neuron in the output layer (cf Fig. 11.14). The output of each hidden neuron is passed through the activation function before it is summed by the neurons in the next layer. It has been found, however, that a feedforward network with just one hidden layer of two neurons, and one output neuron, can predict breathing more or less as well as more complicated layered architectures [4, 16, 29]. We can therefore use such a simple network to learn some important things about basic breathing prediction. The following examples of feedforward ANN results for breathing prediction were all obtained with a single breathing amplitude (displacement of a chest marker) for the input signal, two neurons in the hidden layer, a sigmoid activation function, and one output neuron (for the future signal amplitude). After initial training via LMS, the network was updated (adapted) each time a new breathing data point became available. To quantify the accuracy of breathing prediction, the dimensionless quantity of normalized root-mean-square error (Eq. 11.5) was used to compare the predicted (P_i) and actual (D_i)

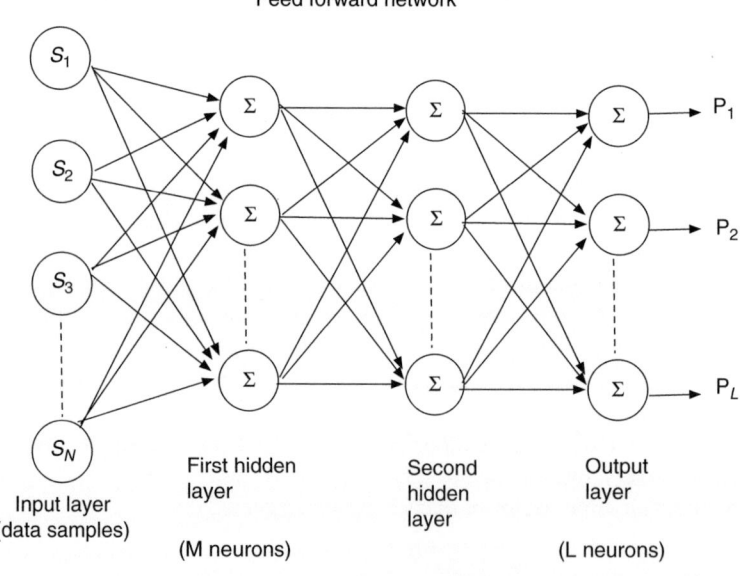

Fig. 11.14 A general feedforward network with multiple layers containing multiple neurons

future amplitudes, for prediction horizons (i.e., lag times) ranging from 100 to 500 ms.

$$n\text{RMSE} = \left[\sum_i (D_i - P_i)^2 / \sum_i (D_i - \mu)^2 \right]^{1/2} \quad (11.5)$$

Here μ is the mean of all of the observations.

There are several parameters to determine when designing the feedforward neural network prediction filter – the length (in seconds) of the input signal history and the number of samples in that history (i.e., the sampling rate), the number of training epochs, and the training rate α in the LMS updating rule (Eq. 11.2). Without going into detail about the testing of the network, which is reported in detail in Murphy and Pokhrel [33], it suffices to say that the performance of the network in predicting a variety of different breathing examples was explored by varying each of these network parameters, to find the values that provided the best results. One obvious question to ask is whether a single network setup can do a reasonable job of predicting a wide range of breathing patterns or if the filter setup needs to be optimized to each individual patient. To answer this question, the filter setup was first optimized for each patient breathing history, and its accuracy was noted. Then a globally-optimal sampling length and rate, number of training epochs, and training rate were identified in the results and used to configure a standardized filter, which was then tested against all of the individual patient histories. Figure 11.15 shows the results [33].

Patients 1–14 were randomly selected from a cohort treated for lung cancer and displayed a wide range of breathing patterns; patients 15–27 were healthy volunteers coached to regularize their breathing via audiovisual feedback [8, 9]. The

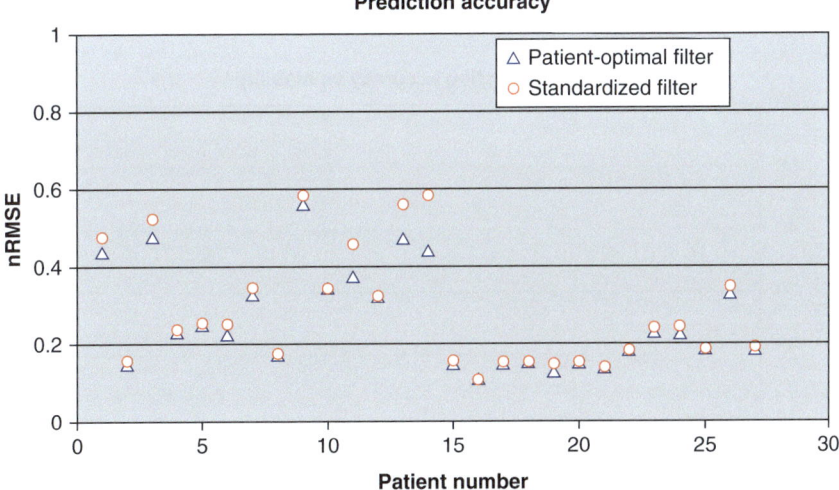

Fig. 11.15 Prediction accuracy of an ANN customized to each patient compared to the accuracy of an ANN with a fixed configuration [33]

standardized filter did essentially as well as the personalized filter for the healthy coached patients and continued to do reasonably well even for the most erratic lung cancer patients. This offers encouragement that it is not necessary to go through an involved filter optimization process for each patient.

The accuracy of any predictive filter can be expected to diminish if the breathing pattern changes over time, simply because the filter must retrain itself to adapt to the changes, and that takes time. This can be demonstrated by calculating a breathing regularity measure and then looking at prediction accuracy as a function of that measure.

For a finite length of continuous patient breathing signal $S(t)$, the autocorrelation coefficient $C(\tau)$ is defined as the cross-correlation integral of $S(t)$ with itself, at delay time τ:

$$C(\tau) = \int S(t) S(t-\tau) dt \qquad (11.6)$$

For a stationary periodic signal, the average value of $C(\tau)$ versus τ will be approximately constant, while for a nonstationary (time-changing) signal the average of $C(\tau)$ will become smaller with increasing τ. We can characterize the stability of the signal by the inverse of the rate at which the average correlation coefficient decays with τ. Call this the correlation decay time. To compute the decay time, a 60 s window was set at a point in the breathing time series, and $C(\tau)$ was computed for $0 < \tau < 60$ s. The peak values of the positive half cycle of the autocorrelation coefficient were plotted in a semilog scale as a function of τ. The inverse slope of the graph gave the decay time for that particular position of the breathing signal window. A rapidly changing signal will have a short decay time; a slowly changing signal will have a long decay time; a perfectly stationary signal will have an infinite decay time.

Figure 11.16 shows the prediction accuracy of the ANN filter as a function of the breathing signal's decay time (from [33]). As expected, rapidly evolving breathing patterns are harder to predict, no matter how well the filter is designed.

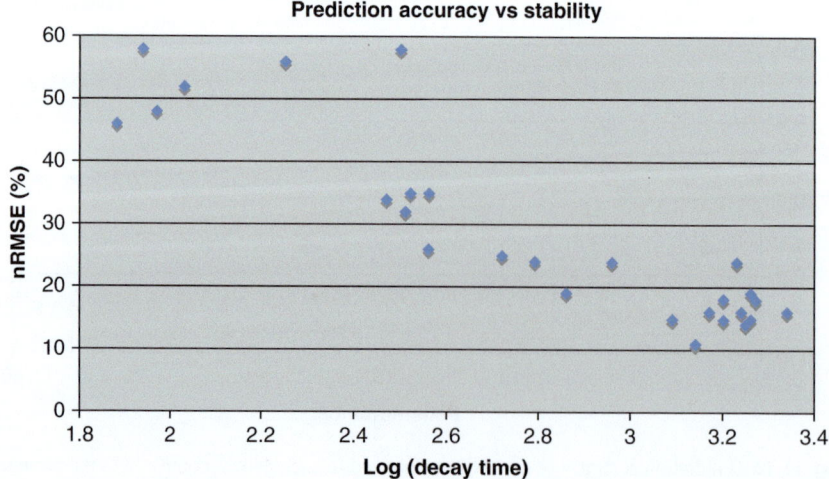

Fig. 11.16 The prediction accuracy of a neural network filter as a function of the stability of the breathing signal, as characterized by the decay time of its autocorrelation. Shorter decay times correspond to more rapidly changing breathing patterns (From [33])

11.5 Summary

Adaptive breathing compensation during radiation therapy requires a means to predict tumor movement either directly from imaging data or indirectly from surrogate breathing data. Although breathing appears superficially regular in most individuals, it is actually variable in period and amplitude. Furthermore, the relative movement of different parts of the anatomy under the influence of breathing can change over time, making it difficult to associate tumor movement with other surrogate movements. Machine learning algorithms offer an attractive way to emulate these complicated behaviors without recourse to biomechanical modeling. They are intrinsically capable of conforming to individual breathing patterns and adapting in real time to changes in breathing behavior.

The artificial neural network is a simple machine learning algorithm that has been shown to be effective at predicting breathing behavior. It offers a clear advantage over a basic linear adaptive filter without much additional computational burden [31]. More usefully, it has been found by numerous researchers that an acceptable level of prediction accuracy can be achieved with a very simple network architecture and that adding feedback loops or more layers with more neurons often provides little or no further improvement [4, 16, 29]. Furthermore, it is not generally necessary to customize the network architecture to each individual patient [33]. This is most clearly the case when steps are taken to regularize an individual's breathing through training and feedback [33].

While the latencies of various motion-adaptive therapy devices can be (and have been) systematically reduced, so that temporal prediction becomes less important in a tumor tracking system, the problem of tracking the tumor's motion from surrogate breathing signals remains. This application of ANNs has not been studied as well as temporal prediction and invites further investigation.

References

1. Adler JR, Murphy MJ, Chang S, Hancock S. Image-guided robotic radiosurgery. Neurosurgery. 1999;44:1299–306.
2. Ahn S, Yi B, Suh Y, Kim J, Lee S, Shin S, Choi E. A feasibility study on the prediction of tumour location in the lung from skin motion. Br J Radiol. 2004;77:588–96.
3. Bert C, Saito N, Schmidt A, Chaudhri N, Schardt D, Rietzel E. Target motion tracking with a scanned particle beam. Med Phys. 2007;34(12):4768–71.
4. Davuluri P, Hobson RS, Murphy MJ, Najarian K. Performance comparison of Volterra predictor and neural network for breathing prediction. In: First international conference on biosciences. Cancun, Mexico: IEEE. 2010. p. 6–10.
5. Depuydt T, Verellen D, Hass O, Gevaert T, Linthout N, Duchateau M, Tournel K, Reynders T, Leysen K, Hoogeman M, Storme G, De Ridder M. Geometric accuracy of a novel gimbals based radiation therapy tumor tracking system. Radiother Oncol. 2011;98(3):365–72.
6. Ernst F, Schweikard A. Robotic LINAC tracking based on correlation and prediction. In: Murphy MJ, editor. Motion adaptation in radiation therapy. New York, NY: Taylor and Francis; 2012.
7. Falk M, Munck AF, Rosenchöld P, Keall P, Catell H, Cho BC, Poulson P, Povsner S, Sawant A, Zimmerman J, Korreman S. Real-time dynamic MLC tracking for inversely optimised arc therapy. Radiother Oncol. 2010;94:218–23.
8. George R, Chung TD, Vedam SS, Ramakrishnan V, Mohan R, Weiss E, Keall PJ. Audio-visual biofeedback for respiratory-gated radiotherapy: impact of audio instruction and audio-visual

biofeedback on respiratory-gated radiotherapy. Int J Radiat Oncol Biol Phys. 2006;65(3): 924–33.
9. George R, Suh Y, Murphy M, Williamson J, Weiss E, Deall P. On the accuracy of a moving average algorithm for target tracking during radiation therapy treatment delivery. Med Phys. 2008;35(6):2356–65.
10. Gierga DP, Brewer J, Sharp GC, Betke M, Willett CG, Chen GTY. The correlation between internal and external markers for abdominal tumors: implications for respiratory gating. Int J Radiat Oncol Biol Phys. 2005;61(5):1551–8.
11. Goodband JH, Haas OCL, Mills JA. A comparison of neural network approaches for on-line prediction in IGRT. Med Phys. 2008;35(3):1113–22.
12. Haykin S. Neural networks and learning machines. 3rd ed. London: Pearson; 2009.
13. Haykin S. Kalman filtering and neural networks. New York: Wiley Interscience; 2001.
14. Hoisak JDP, Sixel KE, Tirona R, Cheung PCF, Pignol J-P. Correlation of lung tumor motion with external surrogate indicators of respiration. Int J Radiat Oncol Biol Phys. 2004; 60(4):1298–306.
15. Hoisak JDP, Sixel KE, Tirona R, Cheung PCF, Pignol J-P. Prediction of lung tumour position based on spirometry and on abdominal displacement: accuracy and reproducibility. Radiother Oncol. 2006;78(3):339–46.
16. Isaksson M, Jalden J, Murphy MJ. On using an adaptive neural network to predict lung tumor motion during respiration for radiotherapy applications. Med Phys. 2005;32(12):3801–9.
17. Kakar M, Mystrom H, Aarup LR, Nottrup TJ, Olsen DR. Respiratory motion prediction by using the adaptive neuro fuzzy inference system (ANFIS). Phys Med Biol. 2005;50:4721–8.
18. Kanoulas E, Aslam JA, Sharp GC, Berbeco RI, Nishioka S, Shirato H, Jiang SB. Derivation of the tumor position from external respiratory surrogates with periodical updating of the internal/external correlation. Phys Med Biol. 2007;52(17):5443–56.
19. Keall PJ, Cattell H, Pokhrel D, Dieterich S, Wong K, Murphy MJ, Vedam SS, Wijesooriya K, Mohan R. Geometric accuracy of a system for real time target tracking with a dynamic MLC. Int J Radiat Oncol Biol Phys. 2006;65(5):1579–84.
20. Koch N, Liu HH, Starkschall G, Jacobson M, Forster KM, Liao Z, Komaki R, Stevens CW. Evaluation of internal lung motion for respiratory-gated radiotherapy using MRI: part I– correlating internal lung motion with skin fiducial motion. Int J Radiat Oncol Biol Phys. 2004;60(5):1459–72.
21. Lee SJ, Motai Y, Murphy M. Respiratory motion estimation with hybrid implementation of extended Kalman filter. IEEE Trans Ind Electron. 2012;59(11):4421–32.
22. Liang P, Pandit JJ, Robbins PA. Non-stationarity of breath-by-breath ventilation and approaches to modeling the phenomenon. In: Semple SJG, Adams L, Whipp BJ, editors. Modeling and control of ventilation. New York: Plenum; 1995. p. 117–21.
23. Malinowski K, D'Soua WD. Couch-based target alignment. In: Murphy MJ, editor. Motion adaptation in radiation therapy. New York, NY: Taylor and Francis; 2012.
24. McQuaid D, Webb S. IMRT delivery to a moving target by dynamic MLC tracking: delivery for targets moving in two dimensions in the beam's-eye view. Phys Med Biol. 2006;51: 4819–39.
25. McQuaid D, Webb S. Target-tracking deliveries using conventional multileaf collimators planned with 4D direct-aperture optimization. Phys Med Biol. 2008;53:4013–29.
26. McQuaid D, Partridge M, Symonds Tayler R, Evans PM, Webb S. Target-tracking deliveries on an Elekta linac: a feasibility study. Phys Med Biol. 2009;54:3563–78.
27. McQuaid D, Webb S. Fundamentals of tracking with a linac MLC. In: Murphy MJ, editor. Motion adaptation in radiation therapy. New York, NY: Taylor and Francis; 2012.
28. Minsky M, Papert S. Perceptrons. Cambridge: MIT Press; 1969.
29. Murphy MJ, Jalden J, Isaksson M. Adaptive filtering to predict lung tumor breathing motion free breathing. In: Proceedings of the 16th international congress on computer-assisted radiology and surgery. Paris; 2002. p. 539–44.
30. Murphy MJ. Tracking moving organs in real time. In: Chen and Bortfield, editors. Seminars in radiation oncology, vol 14 (1). 2004. p. 91–100.

31. Murphy MJ, Dieterich S. Comparative performance of linear and nonlinear neural networks to predict irregular breathing. Phys Med Biol. 2006;51:5903–14.
32. Murphy MJ. Using neural networks to predict breathing motion. In: Seventh international congress on machine learning applications. San Diego, CA: IEEE. 2008. p. 528–32.
33. Murphy MJ, Pokhrel D. Optimization and evaluation of an adaptive neural network filter to predict respiratory motion. Med Phys. 2009;36(1):40–7.
34. Neicu T, Shirato H, Seppenwoolde Y. Synchronized moving aperture radiation therapy (SMART); average tumour trajectory for lung patients. Phys Med Biol. 2003;48:587–98.
35. Ozhasoglu C, Murphy MJ. Issues in respiratory motion compensation during external-beam radiotherapy. Int J Radiat Oncol Biol Phys. 2002;52:1389–99.
36. Podder TK, Buzurovic I, Galvin JM, Yu Y. Dynamics-based decentralized control of robotic couch and multi-leaf collimators for tracking tumor motion. IEEE Int Conf Robot Automat. 2008;19(23):2496–502.
37. Puskorius GV, Feldkamp LA. Neurocontrol of nonlinear dynamical systems with Kalman filter trained recurrent networks. IEEE Trans Neural Netw. 1994;5(2):279–97.
38. Qiu P, D'Souza WD, McAvoy TJ, Liu KJR. Inferential modeling and predictive feedback control in real-time motion compensation using the treatment couch during radiotherapy. Phys Med Biol. 2007;52:5831–54.
39. Rangaraj D, Papiez L. Synchronized delivery of DMLC intensity modulated radiation therapy for stationary and moving targets. Med Phys. 2005;32:1802–17.
40. Ren Q, Nishioka S, Shirato H, Berbeco RI. Adaptive prediction of respiratory motion for motion compensation radiotherapy. Phys Med Biol. 2007;52(22):6651–61.
41. Sawant A, Venkat R, Srivastava V, Carlson D, Povzner S, Cattell H, Keall P. Management of three-dimensional intrafraction motion through real-time DMLC tracking. Med Phys. 2008;35:2050–61.
42. Schweikard A, Glosser G, Bodduluri M, Murphy MJ, Adler JR. Robotic motion compensation for respiratory movement during radiosurgery. Comput Aided Surg. 2000;5:263–77.
43. Schweikard A, Shiomi H, Adler J. Respiration tracking in radiosurgery. Med Phys. 2004;31(1):2738–41.
44. Sharp GC, Jiang SB, Shimizu S, Shirato H. Prediction of respiratory tumour motion for real-time image-guided radiotherapy. Phys Med Biol. 2004;49(3):425–40.
45. Solberg TD, Medin PM, Ramirez E, Ding C, Foster RD, Yordy J. Commissioning and initial stereotactic ablative radiotherapy experience with Vero. J Appl Clin Med Phys. 2014;15(2):205–25.
46. D'Souza WD, Naqvi SA, Uu CX. Real-time intra-fraction motion tracking using the treatment couch: a feasibility study. Phys Med Biol. 2005;50:4021–33.
47. D'Souza WD, McAvoy TJ. An analysis of the treatment couch and control system dynamics for respiration-induced motion compensation. Med Phys. 2006;33(12):4701–9.
48. Tacke MB, Nill S, Krauss A, Oelfke U. Real-time tumor tracking: automatic compensation of target motion using the Siemens 160 MLC. Med Phys. 2010;37:753–61.
49. Tobin MJ, Mador MJ, Guenther SM, Lodato RF, Sackner MA. Variability of resting respiratory drive and timing in healthy subject. J Appl Physiol. 1988;65:309–17.
50. Tsunashima Y, Sakae T, Shioyama Y, et al. Correlation between the respiratory waveform measured using a respiratory sensor and 3D tumor motion in gated radiotherapy. Int J Radiat Oncol Biol Phys. 2004;60(3):951–8.
51. Williams RJ. Training recurrent networks using the extended Kalman filter. Int Joint Conf Neural Netw. 1992;4:241–6.
52. Wu H, et al. Gating based on internal/external signals with dynamic correlation updates. Phys Med Biol. 2008;53(24):7137–50.
53. Yan H, Yin F-F, Zhu G-P, Ajlouni M, Kim JH. Adaptive prediction of internal target motion using external marker motion: a technical study. Phys Med Biol. 2006;51(1):31–44.
54. Zhang T, et al. Application of the spirometer in respiratory gated radiotherapy. Med Phys. 2003;30(12):3165–71.

Image-Based Motion Correction

12

Ruijiang Li

Abstract

This chapter will discuss dedicated machine learning techniques for motion management using imaging information. We will cover a wide range of well-established machine learning techniques, including principal component analysis, linear discriminant analysis, artificial neural networks, and support vector machine, etc. Motion management techniques including both respiratory gating and real-time tumor tracking will be discussed. In this chapter, we will demonstrate how to utilize domain-specific knowledge and prior imaging information to achieve more accurate and robust motion management in radiotherapy. Finally, future research directions in the clinical applications of machine learning for motion management will be discussed.

12.1 Introduction

Radiation therapy is a major modality for treating cancer patients. Studies have shown that an increased radiation dose to the tumor will lead to improved local control and survival rates [16, 18, 19, 24]. However, in many anatomic sites, e.g., lung, liver, and pancreas, the tumor can move significantly with respiration, up to ~2–3 cm [17, 21]. The respiratory tumor motion has been a major challenge in radiotherapy to deliver sufficient radiation dose without causing secondary cancer or severe radiation damage to the surrounding healthy tissue.

Motion-adaptive radiotherapy explicitly accounts for the tumor motion during radiation dose delivery, in which respiratory gating and tumor tracking are two

---

R. Li, PhD
Department of Radiation Oncology, Stanford University,
875 Blake Wilbur Drive, Stanford, CA 94305-5847, USA
e-mail: rli2@stanford.edu

promising approaches [6, 8]. Respiratory gating limits radiation exposure to a portion of the breathing cycle when the tumor is in a predefined gating window [5]. Tumor tracking, on the other hand, allows continuous radiation dose delivery by dynamically adjusting the radiation beam so that it follows the tumor movement in real time [7]. For either technique to be effective, accurate measurement of the respiration signal is required.

Conventional methods for respiration measurement are either invasive or unreliable. Methods based on fiducial markers require an invasive implantation procedure and involves serious medical risks to the patient, e.g., pneumothorax for lung cancer patients [4]. In addition, the fiducial markers may drift relative to the tumor, which will lead to erroneous results if only the markers are tracked. On the other hand, measurement of external respiration surrogates using infrared reflective marker, spirometer, pressure belt, etc. generally lacks sufficient accuracy to infer the tumor position, because the relations between internal tumor and external surrogate may change over time, either intra- or inter-fractionally [22, 23].

Accurate, noninvasive methods that are based on direct measurement of internal patient anatomy and the tumor are critically needed, in order to realize the full potential of motion-adaptive radiotherapy. The onboard x-ray imaging system is widely available on modern linacs and provides effective means of imaging internal anatomy. However, this brings considerable challenges when fiducial markers are not present in radiographic images, because it is often very difficult for humans to visualize the tumor directly in projection images. Various machine learning techniques have been applied to solve this challenging problem.

In this chapter, we will summarize the recent advances in the application of machine learning techniques for motion management in radiotherapy. Techniques developed for both respiratory gating and real-time tumor tracking will be discussed. We will demonstrate how to utilize domain-specific knowledge and prior imaging information to achieve more accurate and robust motion management in radiotherapy. Finally, we point out some future research directions that may further improve the accuracy of tumor localization.

12.2 Respiratory Gating Based on Fluoroscopic Images

The initial efforts on image-based respiratory gating have been mainly focused on template matching [1, 3]. In this approach, a set of representative reference templates are first generated which corresponds to the treatment positions of the target in the gating window using fluoroscopic images acquired during the patient setup. The similarity scores between the reference templates and the incoming fluoroscopic images acquired during treatment delivery are calculated and then converted into gating signals. Template matching does not fully utilize the information, in particular, those images outside the gating window.

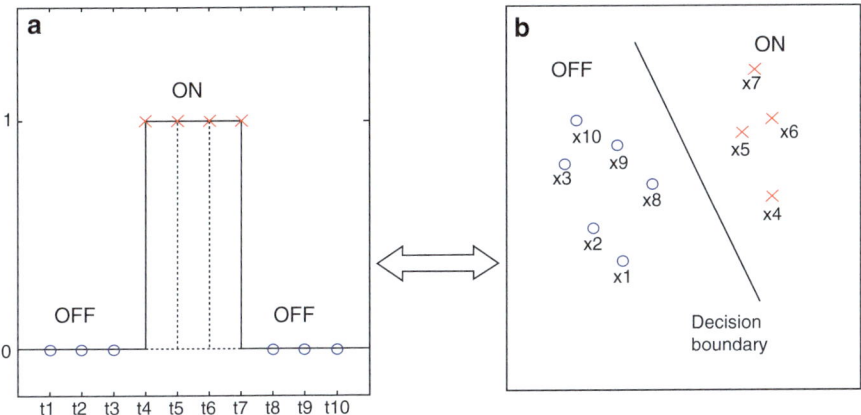

Fig. 12.1 Cast the gating problem (**a**) into a binary classification problem (**b**) (Reprint from Cui et al. [2])

From a machine learning perspective, respiratory gating can be formulated as a binary classification problem (Fig. 12.1), because the output of the gating system is binary, either beam on or beam off. The advantage of the machine learning approach over conventional template matching is that it utilizes both positive examples (inside the gating window) and negative examples (outside the gating window) in generating the gating signal.

One major issue with learning from images is the large dimensionality of the raw data. Even if a smaller region of interest (ROI) is used, each ROI image typically has a dimensionality of 10,000. In addition, high-dimensional data also requires increased computational cost, which creates problems for real-time applications such as respiratory gating. Thus, it is crucial to apply dimensionality reduction techniques to images before any learning procedure is performed.

Cui et al. [2] presented the first study using machine learning approaches for respiratory gating. In their approach, the fluoroscopic images acquired during patient setup are first transformed into a lower dimensional space using principal component analysis (PCA) for training purposes. These samples with class label are used to train a classifier based on support vector machine (SVM). After the optimal classifier is determined, new images are acquired during treatment delivery, which are projected to the same PCA feature space, and passed to the SVM classifier obtained in the training session. The output of the classifier is the predicted label of the new image, which determines whether to turn the beam on or off at any given time. When tested on five sequences of fluoroscopic images from five lung cancer patients, the SVM classifier was found to be slightly more accurate on average (1–3 %) than the template matching method, and the average duty cycle is 4–6 % longer.

In a follow-up study, Lin et al. [15] performed more comprehensive evaluations of different combinations of dimensionality reduction and classification techniques. They investigated four nonlinear dimensionality reduction techniques, including locally linear embedding (LLE), local tangent space alignment (LTSA), Laplacian eigenmap (LAP), and diffusion maps (DMAP). For classification, a three-layer artificial neural network (ANN) was used in addition to SVM. Performance was evaluated on ten fluoroscopic image sequences of nine lung cancer patients. It was found that among all combinations of dimensionality reduction techniques and classification methods, PCA combined with either ANN or SVM achieved a better performance than the other nonlinear manifold learning methods. ANN when combined with PCA achieved a better performance than SVM, with 96 % classification accuracy and 90 % recall rate, although the target coverage is similar at 98 % for the two classification methods. Furthermore, the running time for both ANN and SVM with PCA is around 6.7 ms on a Dual Core CPU, within tolerance for real-time applications. Overall, ANN combined with PCA was found to be a better candidate than other combinations for real-time gated radiotherapy.

In the above previous works, PCA was used as a dimensionality reduction technique to preprocess the data. In [12] the generalized linear discriminant analysis (GLDA) was applied to the respiratory gating problem. The fundamental difference from conventional dimensionality reduction techniques is that GLDA explicitly takes into account the label information available in the training set and therefore is efficient for discrimination among classes. On average, GLDA was demonstrated to perform similarly with PCA trained with SVM at high nominal duty cycles and outperform PCA in terms of classification accuracy (CA) and target coverage (TC) at lower nominal duty cycle (20 %). A major advantage of GLDA is its robustness, while CA and TC using PCA can be reduced by up to 10 % depending on the data dimensionality. With only 1-dimensional feature vectors, GLDA is much more computationally efficient than PCA. Therefore, GLDA is an effective and efficient method for respiratory gating with markerless fluoroscopic images.

12.3 Real-Time Tumor Tracking Based on Fluoroscopic Images

Since the output of a real-time tumor tracking system is a continuous variable, it can be formulated as a regression problem from a machine learning perspective. Lin et al. [14] proposed to use learning algorithm for tumor tracking in fluoroscopic images, based on the observation that the motion of some anatomic features in the images may be well correlated to the tumor motion (Fig. 12.2). The correlation between the tumor position and the motion pattern of surrogates can be captured by regression analysis techniques. The proposed algorithm consists of four main

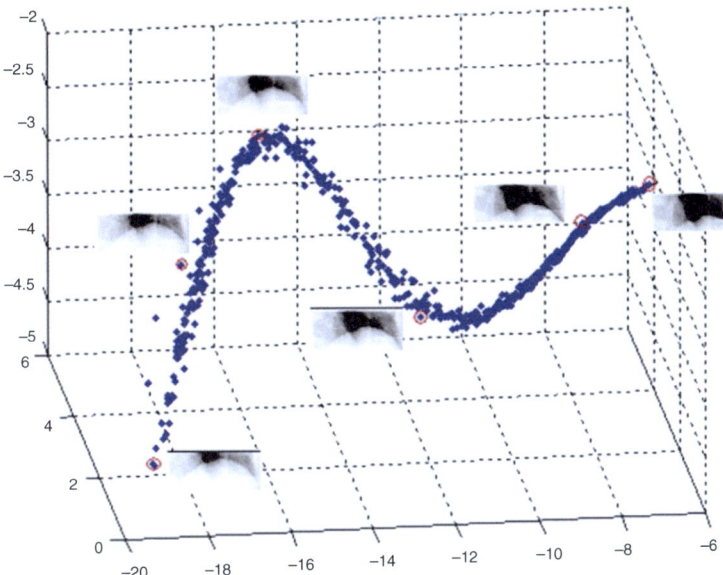

Fig. 12.2 3D embedding of the diaphragm ROI images using PCA. Representative images are shown next to circled points at different location in the 3D PCA space, representing different positions of the diaphragm (Reprint from Lin et al. [14])

steps: [1] selecting surrogate regions of interest (ROIs), [2] extracting spatiotemporal patterns from the surrogate ROIs using PCA, [3] establishing regression between the tumor position and the spatiotemporal patterns, and [4] predicting the tumor location using the established regression model. In a clinical setting, the first three steps would be performed using training image data before the treatment, while the final step would be performed in real time using the image data acquired during treatment delivery.

They evaluated several regression techniques for tracking purposes, including linear regression, second-order polynomial regression, ANN, and SVM. The experimental results based on fluoroscopic sequences of 10 lung cancer patients demonstrate a mean tracking error of 1.1 mm and a maximum error at a 95 % confidence level of 2.3 mm for the proposed tracking algorithm. The results suggest that the machine learning approaches are promising for real-time tumor tracking. However, these methods have to be fully validated before their clinical use. In particular, PCA is sensitive to the tumor size and position, so if the tumor changes size or relative position with respect to the chosen surrogates, the regression model needs to be re-evaluated. This suggests that a separate training data set may be required for each treatment fraction for the learning technique to work well.

Li and Sharp [13] proposed a fluoroscopic fiducial tracking method that exploits the spatial relationship among the multiple implanted fiducials. The spatial relationships between multiple implanted markers are modeled as Gaussian distributions of their pairwise distances over time. The means and standard deviations of these distances are learned from training sequences, and pairwise distances that deviate from these learned distributions are assigned a low spatial matching score. The spatial constraints are incorporated in two different algorithms: a stochastic tracking method and a detection-based method. In the stochastic method, hypotheses of the "true" fiducial position are sampled from a pre-trained respiration motion model. Each hypothesis is assigned an importance value based on image matching score and spatial matching score. Learning the parameters of the motion model is needed in addition to learning the distribution parameters of the pairwise distances in the proposed stochastic tracking approach. In the detection-based method, a set of possible marker locations are identified by using a template matching-based fiducial detector. The best location is obtained by optimizing the image matching score and spatial matching score through non-serial dynamic programming. The proposed method was evaluated using a retrospective study of 16 fluoroscopic videos of liver cancer patients with implanted fiducials. On the patient data sets, the detection-based method gave the smallest error (0.39 ± 0.19 mm). The stochastic method performed well (0.58 ± 0.39 mm) when the patient breathed consistently; the average error increased to 1.55 mm when the patient breathed differently across sessions.

12.4 Real-Time Tumor Tracking via Volumetric Imaging Based on a Single X-Ray Image

Li et al. [9] have recently made a breakthrough in reconstructing volumetric images and localizing lung tumors in real time using a single x-ray projection image. The method is based on an accurate patient-specific lung motion model and uses the CT images acquired during simulation as the reference anatomy. For lung cancer patients, a respiration-correlated 4DCT is typically acquired for treatment simulation purposes. Deformable image registration (DIR) is performed between a reference CT image and all other CT images, and a set of displacement vector fields (DVFs) will be obtained, which basically tells how each voxel/point in the lung moves, or its 3D motion trajectory. Given the dense DVFs, a patient-specific lung motion model is built based on PCA. The PCA motion model is accurate, efficient, and flexible and imposes implicit regularization on its representation of the lung motion [11]. As a result, a few scalar variables (i.e., PCA coefficients) are sufficient in order to accurately derive the dynamic lung motion for a given patient. Therefore, limited information, e.g., a single x-ray projection, can be used to reconstruct the volumetric image of the patient anatomy, in which the PCA coefficients are optimized such that the projection of the reconstructed volumetric image corresponding

Fig. 12.3 Tumor localization results (*dots*) for two patients with irregular breathing. The solid lines are ground truths. Only the direction with the largest motion (axial) is shown. The average error is 1.9 and 0.9 mm for the two patients, respectively

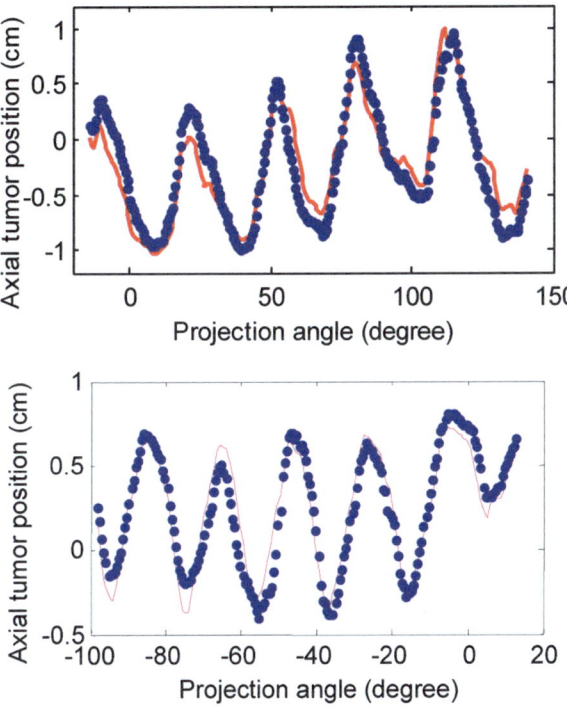

to the new DVF matches with the measured x-ray projection. Once the optimal DVF has been found, the 3D tumor location relative to the reference position defined in 4DCT can be determined. The algorithm was implemented on graphic processing unit (GPU). The average computation time for image reconstruction and 3D tumor localization from an x-ray projection ranges between 0.2 and 0.3 s on the C1060 GPU card.

The algorithm was evaluated on five lung cancer patients for tumor tracking accuracy [10]. The raw cone beam x-ray images during the CBCT scan for treatment setup purposes were used retrospectively to test the algorithm. For all patients in this study, there were no implanted fiducial makers. To evaluate the tracking accuracy, the tumor was manually marked by the clinician in the largest continuous set of projections in which the tumor was visible. All five patients had somewhat irregular breathing during the CBCT scans. Figure 12.3 shows the localization results for two patients. The average tumor localization error is <2 mm for all five patients [20]. Figure 12.4 shows the raw x-ray projections and the coronal and sagittal views of the reconstructed images at two breathing phases. Overall, the algorithm gives realistic and consistent anatomy during respiration, including tumor, diaphragm, and bronchial and vascular structures.

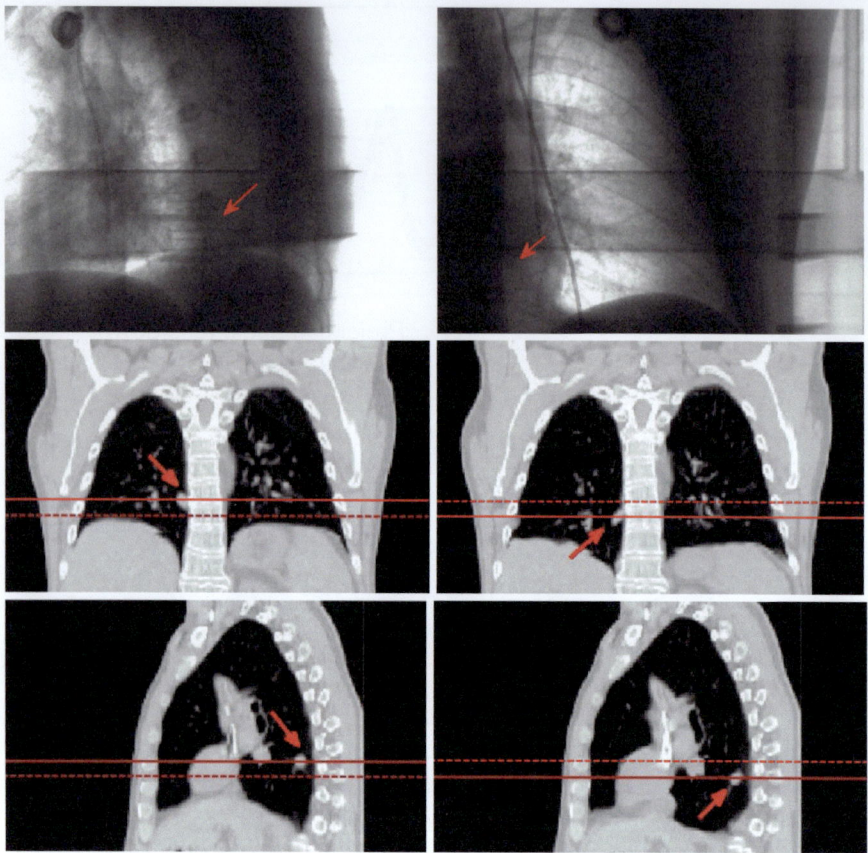

Fig. 12.4 Real-time image reconstruction and tumor localization results. *Left column*: cone beam x-ray projection image (*top*) and the coronal and sagittal views of the reconstructed image at an EOE phase (*middle and bottom*). *Right column*: same as left column, except at an EOI phase. *Red arrow* indicates the tumor

12.5 Summary and Future Directions

A variety of well-established machine learning approaches have been applied to respiratory gating and real-time tumor tracking, such as PCA, LDA, ANN, SVM, etc. The application of machine learning in image-based motion management has been shown to be promising, with gating accuracy of above 95 % and tracking errors of ~2 mm. However, these results were achieved often in well-controlled, favorable conditions. For example, most analysis was done in the anterior-posterior direction, in which the tumor is more clearly visualized. Also, many require a separate training data set on each treatment fraction in order to achieve an optimal performance, which may not be practical in clinical settings. These logistic issues need to be taken into account for their practical use. For these techniques to be adopted in the clinic, much more validation needs to be conducted. In particular, methods that are more robust, reliable, and require minimum human input are critically needed.

References

1. Berbeco RI, Mostafavi H, Sharp GC, Jiang SB. Towards fluoroscopic respiratory gating for lung tumours without radiopaque markers. Phys Med Biol. 2005;50:4481–90.
2. Cui Y, Dy JG, Alexander B, Jiang SB. Fluoroscopic gating without implanted fiducial markers for lung cancer radiotherapy based on support vector machines. Phys Med Biol. 2008;53:N315–27.
3. Cui Y, Dy JG, Sharp GC, Alexander B, Jiang SB. Robust fluoroscopic respiratory gating for lung cancer radiotherapy without implanted fiducial markers. Phys Med Biol. 2007;52:741–55.
4. Geraghty PR, Kee ST, McFarlane G, Razavi MK, Sze DY, Dake MD. CT-guided transthoracic needle aspiration biopsy of pulmonary nodules: needle size and pneumothorax rate. Radiology. 2003;229:475–81.
5. Jiang SB. Radiotherapy of mobile tumors. Semin Radiat Oncol. 2006;16:239–48.
6. Keall P. 4-dimensional computed tomography imaging and treatment planning. Semin Radiat Oncol. 2004;14:81–90.
7. Keall PJ, Cattell H, Pokhrel D, Dieterich S, Wong KH, Murphy MJ, Vedam SS, Wijesooriya K, Mohan R. Geometric accuracy of a real-time target tracking system with dynamic multileaf collimator tracking system. Int J Radiat Oncol Biol Phys. 2006;65:1579–84.
8. Keall PJ, Mageras GS, Balter JM, Emery RS, Forster KM, Jiang SB, Kapatoes JM, Low DA, Murphy MJ, Murray BR, Ramsey CR, Van Herk MB, Vedam SS, Wong JW, Yorke E. The management of respiratory motion in radiation oncology report of AAPM Task Group 76. Med Phys. 2006;33:3874–900.
9. Li R, Jia X, Lewis JH, Gu X, Folkerts M, Men C, Jiang SB. Real-time volumetric image reconstruction and 3D tumor localization based on a single x-ray projection image for lung cancer radiotherapy. Med Phys. 2010;37:2822–6.
10. Li R, Lewis JH, Jia X, Gu X, Folkerts M, Men C, Song WY, Jiang SB. 3D tumor localization through real-time volumetric x-ray imaging for lung cancer radiotherapy. Med Phys. 2011;38:2783–94.
11. Li R, Lewis JH, Jia X, Zhao T, Liu W, Wuenschel S, Lamb J, Yang D, Low DA, Jiang SB. On a PCA-based lung motion model. Phys Med Biol. 2011;56:6009–30.
12. Li R, Lewis JH, Jiang SB. Markerless fluoroscopic gating for lung cancer radiotherapy using generalized linear discriminant analysis. In: IEEE ICMLA. Miami, FL. 2009. p. 468–72.
13. Li R, Sharp G. Robust fluoroscopic tracking of fiducial markers: exploiting the spatial constraints. Phys Med Biol. 2013;58:1789–808.
14. Lin T, Cervino LI, Tang X, Vasconcelos N, Jiang SB. Fluoroscopic tumor tracking for image-guided lung cancer radiotherapy. Phys Med Biol. 2009;54:981–92.
15. Lin T, Li R, Tang X, Dy JG, Jiang SB. Markerless gating for lung cancer radiotherapy based on machine learning techniques. Phys Med Biol. 2009;54:1555–63.
16. Machtay M, Bae K, Movsas B, Paulus R, Gore EM, Komaki R, Albain K, Sause WT, Curran WJ. Higher biologically effective dose of radiotherapy is associated with improved outcomes for locally advanced non-small cell lung carcinoma treated with chemoradiation: an analysis of the Radiation Therapy Oncology Group. Int J Radiat Oncol Biol Phys. 2010;82(1):425–34.
17. Murphy MJ, Martin D, Whyte R, Hai J, Ozhasoglu C, Le QT. The effectiveness of breath-holding to stabilize lung and pancreas tumors during radiosurgery. Int J Radiat Oncol Biol Phys. 2002;53:475–82.

18. Perez CA, Bauer M, Edelstein S, Gillespie BW, Birch R. Impact of tumor control on survival in carcinoma of the lung treated with irradiation. Int J Radiat Oncol Biol Phys. 1986;12:539–47.
19. Perez CA, Stanley K, Rubin P, Kramer S, Brady L, Perez-Tamayo R, Brown GS, Concannon J, Rotman M, Seydel HG. A prospective randomized study of various irradiation doses and fractionation schedules in the treatment of inoperable non-oat-cell carcinoma of the lung. Preliminary report by the Radiation Therapy Oncology Group. Cancer. 1980;45:2744–53.
20. Popescu CC, Olivotto IA, Beckham WA, Ansbacher W, Zavgorodni S, Shaffer R, Wai ES, Otto K. Volumetric modulated arc therapy improves dosimetry and reduces treatment time compared to conventional intensity-modulated radiotherapy for locoregional radiotherapy of left-sided breast cancer and internal mammary nodes. Int J Radiat Oncol Biol Phys. 2010;76:287–95.
21. Seppenwoolde Y, Shirato H, Kitamura K, Shimizu S, van Herk M, Lebesque JV, Miyasaka K. Precise and real-time measurement of 3D tumor motion in lung due to breathing and heartbeat, measured during radiotherapy. Int J Radiat Oncol Biol Phys. 2002;53:822–34.
22. Shah C, Grills IS, Kestin LL, McGrath S, Ye H, Martin SK, Yan D. Intrafraction variation of mean tumor position during image-guided hypofractionated stereotactic body radiotherapy for lung cancer. Int J Radiat Oncol Biol Phys. 2012;82:1636–41.
23. Shirato H, Seppenwoolde Y, Kitamura K, Onimura R, Shimizu S. Intrafractional tumor motion: lung and liver. Semin Radiat Oncol. 2004;14:10–8.
24. Wulf J, Baier K, Mueller G, Flentje MP. Dose-response in stereotactic irradiation of lung tumors. Radiother Oncol. 2005;77:83–7.

Part V
Machine Learning for Quality Assurance

Detection and Prediction of Radiotherapy Errors

13

Issam El Naqa

Abstract

In spite of rigorous regulations for radiotherapy cancer patients' treatment, patient safety may be compromised by relatively rare but deadly errors that can occur during complex treatment planning and delivery of radiotherapy. Therefore, methods for automation of quality assurance (QA) procedures are desired. Here, we present a generic framework for QA using machine learning. Moreover, we demonstrate a new tool for detecting radiotherapy errors using advanced machine learning algorithms. The proposed approach utilizes anomaly detection based on one-class estimation to overcome computational challenges of detecting rare events encountered in currently existing techniques. The proposed anomaly detection approach captures regions in the input space of radiotherapy data where the safe class probability density lives and estimates errors as outliers that reside outside this support region. To model nonlinear support regions, we used a support vector machine (SVM) formalism, in which the QA data is mapped into higher dimensional space using kernel functions to achieve maximal separability and is denoted QA-SVM detector. We demonstrated our method using forty-three treatment plans from patients who received stereotactic body radiation therapy (SBRT) for lung cancer. Our preliminary results indicate a training accuracy of 84 % on cross-validation and testing accuracy of 80 % with 100 % positive predictive value and 80 % negative predictive value.

I. El Naqa
Department of Oncology, McGill University, Montreal, QC, Canada

Department of Radiation Oncology, University of Michigan, Ann Arbor, USA
e-mail: issam.elnaqa@mcgill.ca; ielnaqa@med.umich.edu

13.1 Introduction

Cancer patient treatment outcomes and their safety despite rigorous regulations may be compromised by rare but deadly errors that can occur during complex treatment planning and delivery of radiotherapy, as highlighted by recent editorials in national and international media reports [1]. Quality assurance (QA) in radiotherapy follows recommendations of national and international bodies such as the American Association of Physicists in Medicine (AAPM) and its task group (TG) reports. For instance, TG-40 report and its updated version TG-142 provide a comprehensive QA program for institutional radiation oncology practice accounting for potential risks during planning and delivery of high-energy irradiation and harmonizing the treatment of patients and accommodating advances in technology [2, 3]. The QA process is further complicated for credentialing institutions for multi-institutional clinical trials. A study by the Radiological Physics Center (RPC) anthropomorphic phantom has shown the failure rate could be as high as 28, 14, 9, and 25 % in sites of the head and neck, prostate, liver, and lung, respectively [4]. The Task Group 100 of the AAPM has taken a broad approach to these issues and has been developing a general framework based on failure mode and effect analysis (FMEA) for designing QA protocols [5]. An alternative approach that is better able to account for the complexities in radiotherapy processes and lack of well-defined physics in many cases of these failures could be based on machine learning classification methods. We conjecture that there is a need for such data-driven approaches in radiotherapy QA, which are able to detect such critical errors and mitigate their detrimental impact on patients undergoing radiotherapy. However, the fact that such errors are luckily rare events and do not have well-defined characteristics would make the learning problem challenging and intractable mathematically for classical classification (supervised learning) or clustering (unsupervised learning). Therefore, we propose to pose the radiotherapy error detection/prediction problem as a statistical learning problem from one class (no events) and treat these rare error events as anomalies or outliers. This approach is referred to in machine learning as anomaly detection and would overcome the learning from severely imbalanced classes encountered in classical approaches of classification or clustering.

13.2 Anomaly Detection Using One-Class SVM

The one-class classifier recognizes that there is one class in the data (say, normal performance), while everything else is considered an outlier or anomaly. In this approach, the one-class classifier would give the best functional estimate of dependencies in the radiotherapy system input data X (e.g., beam energy, arrangements, monitor units, dose-volume constraints, etc.) on their class label Y, a category variable that indicates the presence/absence of an anomaly (event error) through a mapping function (linear/nonlinear) $f(X)$. So, instead of solving the extremely hard problem of estimating the multivariate probability density function, which would allow us to solve whatever estimation problem by finding their corresponding marginal distribution, anomaly detection aims to capture domain regions in input space where the probability density lives (its domain support), i.e., a function such that most of the data will live in the

13 Detection and Prediction of Radiotherapy Errors

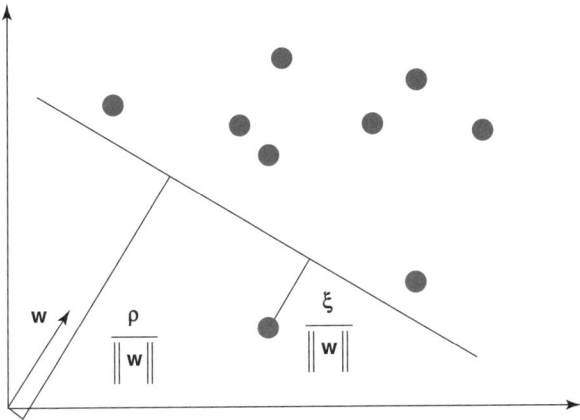

Fig. 13.1 One-class SVM. The problem of anomaly detection becomes separating data from the origin using a hyperplane (adapted with permission)

region where the function is nonzero and outside this support region would be considered an anomaly. Using a support vector machine (SVM) formulation as shown in Fig. 13.1, data are first mapped into a feature space using an appropriate kernel function (e.g., linear, polynomials, radial basis functions, etc.) and then maximally separated from the origin using a proper hyperplane. The hyperplane parameters are determined by solving a quadratic programming problem [6]:

$$\min \frac{1}{2}\|w^2\| + \frac{1}{vl}\sum_{i=1}^{l} \xi_i - \rho \qquad (13.1)$$

subject to:

$$(w \cdot \Phi(x_i)) \geq \rho - \xi_i, \quad i = 1, 2, 3, \ldots, l, \quad \xi_i \geq 0 \qquad (13.2)$$

where w and ρ are hyperplane parameters, Φ is the map from input space to feature space, v is the asymptotic fraction of outliers (errors) allowed, l is the number of training instances, and ξ_i is a slack variable (penalizing misclassifications) as shown in Fig. 13.1. The decision function is given by:

$$f(x) = \text{sgn}(w \cdot \Phi(x) - \rho) \qquad (13.3)$$

The training data could consist of examples of one class (safety class: $f(x) = +1$), and the testing data could contain examples from safety and error classes. An $f(x) = -1$ would indicate an anomaly (error event).

13.3 Application of Anomaly Detection to Radiotherapy QA

The application of machine learning to QA in radiotherapy involves multiple steps: data collection, extraction of relevant features for the safety endpoint of interest, grading of the safety endpoint, selection of appropriate learning algorithm, and defining the training/testing procedure. In Fig. 13.2, we show a corresponding flowchart for the example of the proposed error detection system using one-class SVM

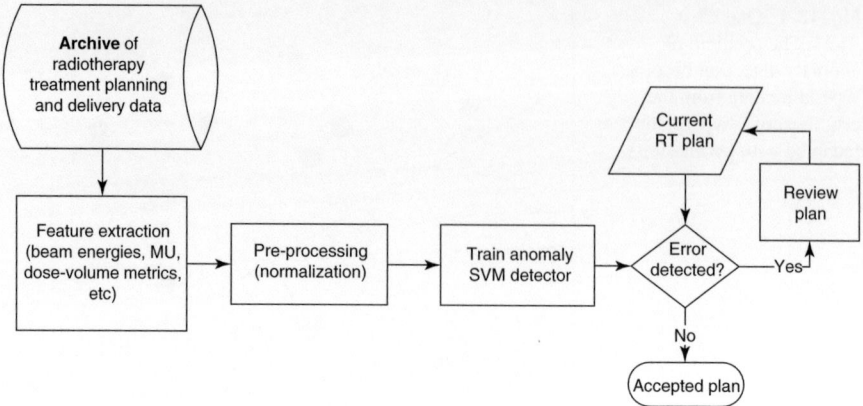

Fig. 13.2 Flowchart for radiotherapy error detection using one-class SVM anomaly detection

for QA of radiotherapy treatment planning. As shown, a one-class SVM anomaly detector is trained using extracted features from existing "safe" data. The extracted features could be treatment technique as well as cancer site dependent. Afterward, the trained anomaly detector is applied to new incoming treatment plans and a flag is turned on by the detector if an error is detected and the plan is sent back for review and this process is repeated till the plan is passed without any errors.

13.3.1 Dataset

To evaluate the proposed QA-SVM system for radiotherapy error detection, we considered a dataset of non-small cell lung cancer (NSCLC) patients who received stereotactic body radiation therapy (SBRT) as part of their treatment. The dataset consisted of 43 successfully treated SBRT patients who received 3 or 5 fractions with 6 or 18 MV beam energies on a linear accelerator (LINAC).

13.3.2 Feature Extraction

Features related to monitor units (MU), beam energies (MV), the number of beams, and the number of fractions, in addition to the percentage lung volume receiving 20 Gy, were extracted from the DICOM files and were used to train the QA-SVM detector. Using principle component analysis (PCA), the top five features with highest explained variance were selected and used subsequently.

13.3.3 Evaluation Results

We used a combination of cases for testing that were considered "safe" and assigned class label "+1" and a set of randomly simulated cases, which were considered

Table 13.1 Evaluation results of QA-SVM using anomaly detection

Sample	Accuracy (%)	PPV (%)	NPV (%)
Training	84	–	–
Testing	80	100	80

"unsafe" by adding different levels of white Gaussian noise to the data. These cases were assigned class label "−1." A radial basis function (RBF) was used for kernel mapping, and the false-positive limit was set at 10 %.

As summarized in Table 13.1, the preliminary results indicate that a training accuracy on cross-validation of 84 % is attained, and a testing accuracy using 5 original trained and 5 unsafe cases as described above was 80 % with 100 % true positive prediction value (PPV) and 80 % true negative prediction value (NPV). The false positives could be attributed to overlap with the positive class and represent border cases that need to be investigated on larger datasets.

> **Conclusions**
> Quality assurance is an important part of safe radiotherapy planning and delivery. The complexity of radiotherapy processes suggests that data-driven learning approaches can provide robust solutions for detection and prediction of errors. In this work, we presented a new approach and tool based on machine learning to overcome the problem of direct modeling of QA errors and rare events in radiotherapy. The tool will be very valuable for automated QA and safety management for patients who undergo radiotherapy treatment.

Acknowledgment The author would like to thank the radiation oncology lung group at Washington University School of Medicine and at McGill University Health Centre for their support. This work was partially supported by NSERC grant RGPIN-397711.

References

1. Bogdanich W. Radiation offers new cures, and ways to do harm. New York: New York Times; 2010.
2. Kutcher GJ, Coia L, Gillin M, Hanson WF, Leibel S, Morton RJ, et al. Comprehensive QA for radiation oncology: report of AAPM Radiation Therapy Committee Task Group 40. Med Phys. 1994;21:581–618.
3. Klein EE, Hanley J, Bayouth J, Yin FF, Simon W, Dresser S, et al. Task Group 142 report: quality assurance of medical accelerators. Med Phys. 2009;36:4197–212.
4. Ibbott GS, Followill DS, Molineu HA, Lowenstein JR, Alvarez PE, Roll JE. Challenges in credentialing institutions and participants in advanced technology multi-institutional clinical trials. Int J Radiat Oncol Biol Phys. 2008;71:S71–5. doi:10.1016/j.ijrobp.2007.08.083.
5. Huq MS, Fraass BA, Dunscombe PB, Gibbons Jr JP, Ibbott GS, Medin PM, et al. A method for evaluating quality assurance needs in radiation therapy. Int J Radiat Oncol Biol Phys. 2008;71:S170–S3.
6. Schölkopf B, Platt J, Shawe-Taylor J, Smola AJ, Williamson RC. Estimating the support of a high-dimensional distribution. Redmond: Microsoft Research; 1999.

Treatment Planning Validation

14

Ruijiang Li and Steve B. Jiang

Abstract

In this chapter, we will discuss the use of machine learning to detect errors in radiotherapy plans and charts. We will cover some general principles and established techniques for detecting errors in radiotherapy. We will discuss the rationale for using machine learning to detect large errors or outliers in radiotherapy treatment plans. As a concrete example, an automated error detection system for radiation treatment plans will be described. The technique was based on unsupervised machine learning, i.e., data clustering, and achieved over 90 % success rates in detecting outliers in over 1,000 treatment plans. Finally, future research directions in the clinical applications of machine learning for treatment planning validation will be briefly discussed.

14.1 Introduction

Adverse events and medical errors could happen in healthcare, resulting in significant patient morbidity and mortality [1]. Such incidents are responsible for 44,000 to 98,000 accidental deaths and over one million excess injuries each year [2, 3]. The severity of medical errors and adverse events in healthcare has made patient safety an

R. Li (✉)
Department of Radiation Oncology, Stanford University,
875 Blake Wilbur Drive, Stanford, CA 94305-5847, USA
e-mail: rli2@stanford.edu

S.B. Jiang
Division of Medical Physics and Engineering, Department of Radiation Oncology,
University of Texas Southwestern Medical Center, 5801 Forest Park Rd., NE3.200,
Dallas, TX 75390-8542, USA
e-mail: Steve.Jiang@UTSouthwestern.edu

urgent and important issue [4]. Significant efforts have been made to improve patient safety in various aspects of healthcare and medical practices [5–8].

Cancer radiotherapy is unique from the point of view of radiation safety, since it is the only application of radiation sources in which very high doses are given deliberately to a part of a human body. When the radiation treatment is planned and delivered with a dose significantly different from the prescribed one, patient outcome can be seriously compromised. For example, if the delivered dose is too low, tumor will not be controlled; if the dose is too high, acute and late complications (even death) can occur.

Incidents that compromise patient safety do happen occasionally, caused by either human error or equipment malfunction [9–13]. In the Panama incident [14], 16 patients were severely overexposed to approximately twice the prescribed dose in the late 2000 and early 2001, resulting in eight treatment-related deaths. Another incident happened in early 2006, involving the death of a young female patient who received a radiation dose of 58 % higher than that intended while undergoing a course of radiotherapy [15]. Therefore, extreme caution is required for the safe and effective use of radiotherapy.

Complete avoidance of human errors is difficult, if not impossible, because radiotherapy is a very complex process, involving a large number of steps from the prescription of the treatment to the delivery of the radiation dose. Many records and communications are involved in those steps, between different professionals and even with the patient. There is a combination of very different activities from manual to sophisticated computer-assisted techniques and high technology equipment.

Conventional approaches widely adopted in the radiotherapy community for reducing medical errors include independent calculation of monitor units (MU) and manual check of the data reported in the treatment chart. These approaches have been proven to be effective in reducing the occurrence of systematic errors before treatment delivery [11, 16, 17]. Other approaches include the use of portal imaging or in vivo dosimetry before or during the treatment to detect errors [12, 18]. Some efforts have been made to develop systematic approaches for collecting, processing, and reporting incidents [12, 13, 19–21].

14.2 Rationale of Using Machine Learning to Detect Errors

Machine learning techniques have recently been used to automatically detect and highlight potential errors in a radiotherapy treatment plan. Because there are hundreds of parameters in a radiotherapy treatment plan and furthermore, the data structure is very complex; it is not feasible to develop an automated tool using rule-based alert logic. Fortunately, most treatment parameters are well correlated due to the fact that in radiotherapy, there are only a limited number of well-established and accepted treatment guidelines presented in task groups' reports. Patterns in the treatment plan parameters can be effectively learned from historical data from all previously treated patients.

In radiotherapy, the treatment information system aggregates all treatment-related patient data into a single, organized, radiotherapy-specific medical chart. It communicates with and controls the treatment machine, i.e., a medical linear accelerator. When a treatment plan is developed, all the treatment parameters are loaded to the information system. At that time, the error detection system will automatically check the plan parameters before the treatment execution. When an outlier treatment parameter is detected, it will be highlighted for human intervention.

Given the relatively large natural variations in the clinical data (between different patients, treatment protocols, institutions), it is not realistic to try to detect small errors at the level of a few percent. We focus on the detection of large errors that may lead to catastrophic consequences, e.g., outliers, in radiotherapy treatment planning.

14.3 Machine Learning Applications for Outlier Detection

Here, we describe some recent applications of machine learning to detect large errors in a radiotherapy treatment plan. The basic idea was to cluster a large number of treatment plans for previously treated patients based on the plan parameters. Then, when checking a new treatment plan, the parameters of the plan will be tested to see whether or not they belong to the established clusters. If not, they will be considered as "outliers" and therefore highlighted to catch the attention of human experts in charge.

In a preliminary study, a simple treatment technique was used to demonstrate the principle [22]. Data for 1,650 prostate cancer patients treated with the "four-field box" technique were used. Both primary and boost treatments were included. A computer code was written to extract all the entries of a patient treatment plan from the IMPAC record and verification system (IMPAC Medical Systems, Inc., Sunnyvale, CA). In this study, to simplify the model, we only considered the most significant eight entries, which are the beam energies and monitor units (MUs) for the four radiation fields for each patient. Those eight entries were the so-called features of a treatment plan in the terminology of computer clustering. They were referred to as E_{AP}, E_{PA}, E_{RL}, E_{LL}, MU_{AP}, MU_{PA}, MU_{RL}, and MU_{LL} for beam energies and MUs for anterior-posterior (AP), posterior-anterior (PA), right lateral (RL), and left lateral (LL) fields, respectively. The beam energies of the linear accelerators used for treating those patients were 6, 10, 18, and 23 MV. The monitor units ranged from about 50 to over 200. To provide equal weight for all the features, we normalized them before clustering. All the features were normalized to have zero mean and unit standard deviation.

14.3.1 Clustering of Treatment Plan Data

Clustering is one of the most important data mining methods applied to uncover patterns and relations in complex medical datasets [23]. The goal of clustering is to separate data into groups, called clusters, such that objects in the same cluster are similar to each other and dissimilar to objects in other clusters. The clustering method we used is K-means clustering [24, 25] (also known as Lloyd's algorithm). K-means strives to minimize the sum-squared-error (SSE) criterion, which is the sum of the squared distance of each data point to its closest cluster center:

$$\text{SSE} = \sum_{j=1}^{k}\sum_{x \in C_j}\left[D(x,\mu_j)\right]^2 \tag{14.1}$$

where C_j is the j-th cluster, μ_j is the center of the j-th cluster, and D stands for the distance between the two points. Each data point, x, is a vector described by the eight features (four beam energy values and four monitor units).

To select the number of clusters, we utilized the Bayesian Information Criterion (BIC) as follows [26]:

$$\text{BIC score} = \ln p(X|\theta) - \frac{M}{2}\ln N \tag{14.2}$$

where $p(X|\theta)$ is the likelihood of the data $X = \{x_1,\ldots,x_N\}$ given the model parameters θ, M is the number of free parameters, N is the number of data points, and ln is the natural logarithm. To compute the likelihood, we assumed the following probabilistic model: $p(x_i) = \sum_{j=1}^{k}\pi_j p(x_i|\theta_j)$, with π_j as the proportion of data points belonging to cluster j and θ_j as the parameters (mean and covariance) for cluster j. We set the means to be the k centroids in K-means. We estimate the covariance matrix (Σ) of each cluster using the sample covariance matrix of the cluster found by K-means.

$$\hat{\Sigma} = \frac{1}{N-1}\sum_{x \in C_j}(x-\mu_j)(x-\mu_j)^T \tag{14.3}$$

where T is the matrix transpose operation. For the mixture weights, π_j, we use the relative size of the cluster (the number of data points in the cluster divided by the total number of data points).

The formula for the log-likelihood is provided below:

$$L(\theta) = \ln p(X|\theta) = \sum_{i=1}^{N}\ln p(x_i) = \sum_{i=1}^{N}\ln\left(\sum_{j=1}^{K}\pi_j p(x_i/\theta_j)\right) \tag{14.4}$$

The point probability of each data point is calculated as:

$$p(x_i|\theta_j) = \frac{1}{(2\pi)^{\frac{d}{2}}|\Sigma_j|^{\frac{1}{2}}}e^{-\frac{1}{2}(x_i-\mu_j)^T\Sigma_j(x_i-\mu_j)} \tag{14.5}$$

where d is the dimension (number of features) of the data. The number of free parameters, M, in our model is $(k-1) + kd + k\frac{d(d+1)}{2}$ where $k-1$ is the number of mixture weights, kd is the number of parameters for the mean vectors, and $k\frac{d(d+1)}{2}$ is the number of parameters for the covariance matrices.

K-means clustering may be trapped into a local optimum minimum. The quality of the final clustering solution depends on the initial selection of the k cluster centers. To avoid local minima, we used a modified version of PCA-Part [27]. PCA-Part is a deterministic initialization method that has been shown to lead K-means to solutions that are close to optimum. PCA-Part is an initialization algorithm that hierarchically splits the data into two in the direction of the largest eigenvector (first principal axis) at each step until k clusters are obtained. K-means clustering aims to minimize the sum-squared-error criterion. The largest eigenvector with the largest eigenvalue is the direction, which contributes to the largest sum-squared-error. Hence, a good candidate direction to project a cluster for splitting is the direction of the cluster's largest eigenvector, which is the basis for PCA-Part initialization. Our modified algorithm integrates the K-means clustering algorithm into the PCA-Part initialization algorithm, in order to refine the results at each iteration. This modification helps the partitions converge to final clusters with smaller *SSE* values.

14.3.2 Outlier Detection

Since there are a large number of patients involved, to avoid statistical bias, we randomly selected 1,000 patients as our training set and used the other 650 patients as testing set. The training set was used to build clusters, while the test set was used to test the model's outlier detection capability.

We assumed that our training set comprises of normal ("correct") treatments. We applied K-means clustering to extract similarity groups from these data. In K-means clustering, we assumed that each cluster came from a Gaussian distribution. Since our training data were examples of normal ("correct") treatment, we tested a new treatment instance as correct or an outlier by testing whether it belongs to any of our Gaussian clusters.

We assigned a rule of classifying a test treatment instance as an outlier if its Euclidean distance from the closest cluster center is greater than a threshold. Because we had a probability distribution model for each cluster, we could set the threshold to assure us of the probability of making a type I error or false positive (i.e., of deciding a point as an outlier when in fact it is normal) is smaller than α. In our experiments, we set the threshold to be 2 sigma (where sigma is the standard deviation), which assures us of the probability of a type I error to be less than 5 %.

For each cluster built based on the training set, we first calculated the mean and standard deviation of all the features (in this study, we have eight features for each treatment plan). To check whether or not a new data point is an outlier, we find a cluster whose center is the closest to the data point using the Euclidean distance and then calculate the difference for each feature between the data point and the cluster center. If the difference is within a preset tolerance (e.g., two standard deviations of that cluster) for all the features, this data point is considered to belong to the cluster. Otherwise, we classified the data point as an outlier.

To measure the quality of the clustering results and how well they could be used to identify outliers, we purposely introduced errors to the test set and used the outlier detection algorithm described above to compute the outlier detection rate. The outlier detection rate is defined as the ratio of the number of data points that are detected as outliers to the total number of data points tested. The outlier detection rates were computed at various error levels for MUs. The error level is defined as the deviation from the original value of an MU feature. For example, 10–20 % error level means that the introduced errors are 10–20 % of the original MU value.

We also introduced errors to the energy features. Unlike the MU features, the possible values of an energy feature are discrete. They can only be one of 6, 10, 18, and 23 MV. Similar to computing the outlier detection rate for MU features, we first randomly selected a patient from the test set and selected a feature out of four energy features (E_{AP}, E_{PA}, E_{RL}, E_{LL}). We then randomly set the value of the energy feature to one of the three values that are different from the original value. The outlier detection procedure was performed to compute the outlier detection rate for each of the four energy features and for all features combined.

We checked the clustering results by visualizing the data. One way to visualize data in dimensions greater than three is to project it to two dimensions and plot the data in that two-dimensional space. We applied principal component analysis (PCA) to reduce the dimensionality. To be able to visualize the clustering results, we projected the data set onto its first three principal components, as shown in Fig. 14.1a–c. By looking at all the three figures, the separation between the clusters becomes quite clear. For example, from Fig. 14.1a, clusters 4 and 7 projected onto principal components 1 and 2 seem to be very close. However, it does not mean that they are actually one cluster. When we looked from Fig. 14.1b, c, where the data were projected onto principal components 1 and 3 and principal components 2 and 3, respectively, clusters 4 and 7 are clearly separated. That means they are two separated clusters. Similar situations exist for clusters 5 and 8 and clusters 1 and 6. Figure 14.1 also shows that indeed our clustering results (the different

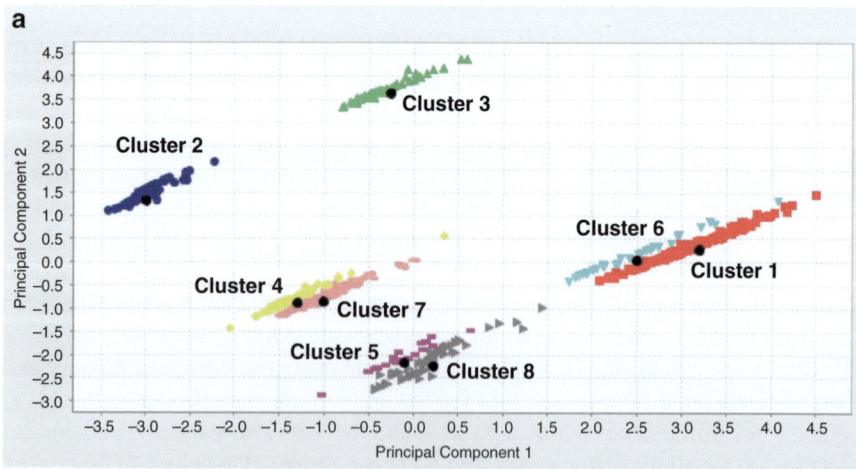

Fig. 14.1 The training data projected onto two-dimensional planes formed by two of its first three principal components: (**a**) principal components 1 and 2; (**b**) principal components 1 and 3; (**c**) principal components 2 and 3 (From Azmandian et al. [22])

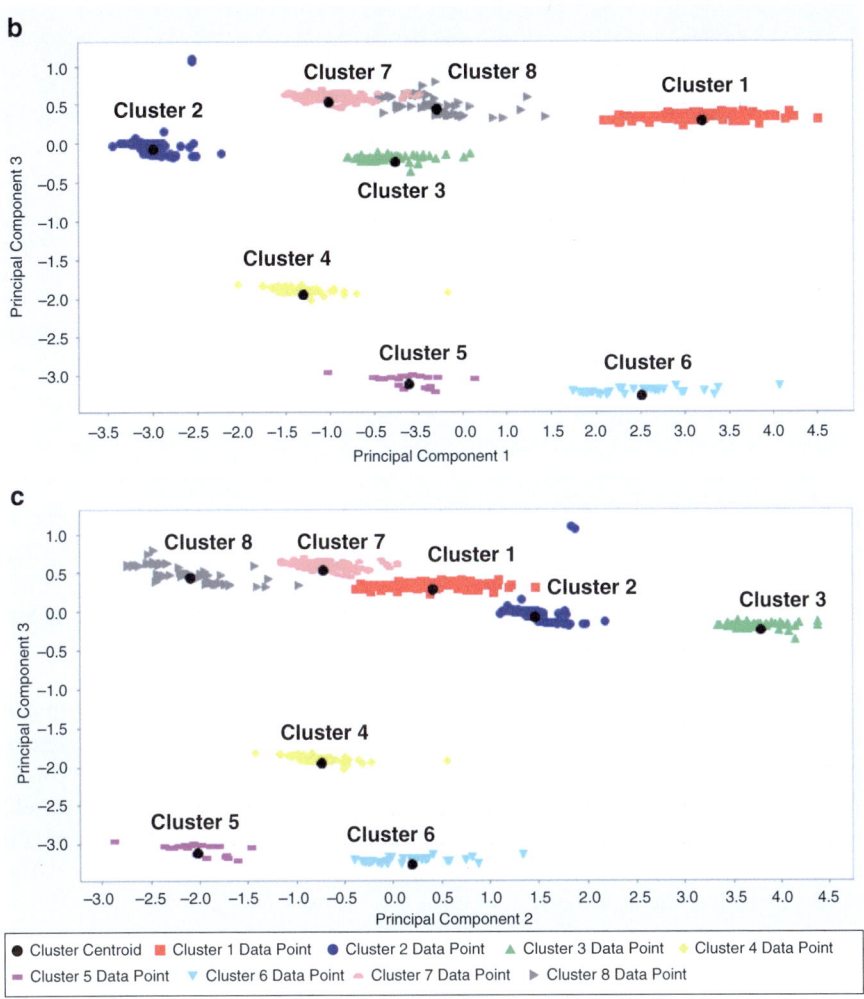

Fig. 14.1 (continued)

colors represent different clusters) make sense, and the cluster means (marked in black dots) are correct.

To find the optimal number of clusters, we ran K-means clustering on the training set for values of k between 2 and 25, scoring each result using the Bayesian Information Criterion (BIC). We found $k = 8$ to produce the best final clustering results. This means, for this group of prostate cancer patients treated using "four-field box" technique, the beam energies and monitor units belong to eight distinct clusters. Again, Fig. 14.2 reveals that indeed there are eight clusters.

The testing results are shown in Fig. 14.2. A few observations can be made from Fig. 14.2: (1) even at the 0 % error level, the detection rate is still about 10 %. This is basically the false-positive rate. (2) The outlier detection rate for the anterior field

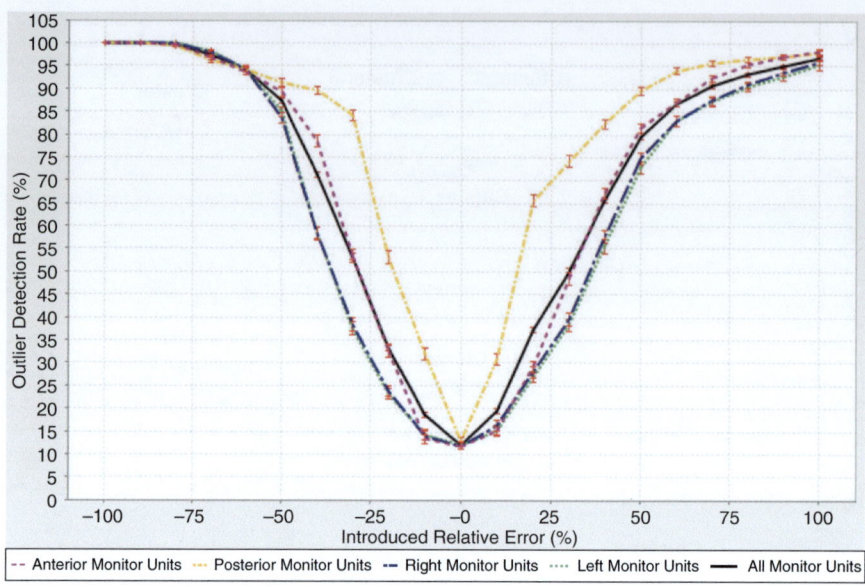

Fig. 14.2 Outlier detection rate as a function of error level for each of MU features and all MU features combined. The error bar shows one standard deviation (From Azmandian et al. [22])

is about the same as that for all fields together. The detection rates for the right and left fields are very similar. The performance of the proposed algorithm decreases from the posterior field, to the anterior field, and to the right and left fields. This is likely due to the fact that the distribution of the monitor units for the posterior field is narrower around peaks than other fields. Also, in all of the clusters, the standard deviation of the posterior MUs is smaller than that of the other MU features. (3) The outlier detection rate as a function of error level is asymmetric, with higher values for negative errors than positive errors. To understand the better performance with negative errors, consider cluster 5 as a simple example. Adding +100 % errors to MUs in cluster 5 (and in essence, doubling their value) will cause the data points to appear to belong to cluster 6 and so they will not be detected as outliers. On the other hand, adding −100 % errors to MUs in clusters 5 (making their values equal to 0) will cause the points to be detected as outliers. As for cluster 6, adding either +100 % or -100 % errors will cause the points to be detected as outliers. Therefore, there is an overall higher outlier detection rate with negative errors.

As shown in Fig. 14.2, the outlier detection rate changes with the error level as one might expect. At ±100 % error level, the detection rate is about 100 %. At ±50 % error level, the detection rate is about 80 %. It seems that the proposed algorithm has a good chance to detect errors at the levels of the Glasgow and Panama incidents [14, 15].

For energy features, the outlier detection rate is 100 % for E_{PA}, E_{RL}, and E_{LL}, while for E_{AP}, it is only 76.9 %, resulting in an overall outlier detection rate of 94.2 % for all four energy features combined. This is because that, for the same cluster, E_{PA}, E_{RL}, and E_{LL} always have the same value, and therefore, changing one will cause the data point to be detected as an outlier, while for E_{AP}, this is not true.

14.4 Summary and Conclusion

We have described some recent applications of machine learning techniques for quality assurance and validation of radiotherapy treatment planning. In particular, we reviewed an automated error detection system for radiation treatment plans based on data clustering. The system was focused on a simplified treatment model, i.e., the "four-field box" technique. The method was successful at detecting large errors in several important parameters in the treatment plans.

Going forward, there are two aspects in the error detection system that will benefit from future development. For many current treatment techniques, especially for intensity-modulated radiation therapy (IMRT) and volumetric-modulated arc therapy (VMAT), there are many more treatment parameters, which makes the data become high dimensional and thus more scattered. For the error detection system to be useful under these more realistic clinical situations, it is expected that substantial algorithm improvements will be needed. Ideally, an error detection system should be self-learning and constantly evolving as more relevant treatment plan data are being collected. The more treatment data available, the more accurate the machine learning is expected to be. Methods that effectively leverage online data and are able to dynamical update itself will need to be developed.

References

1. Malpass A, Helps SC, Runciman WB. An analysis of Australian adverse drug events. J Qual Clin Pract. 1999;19(1):27–30.
2. Kohn LT, et al. To err is human. Washington, DC: National Academy Press; 2000.
3. Weingart SN, et al. Epidemiology of medical error. BMJ. 2000;320(7237):774–7.
4. Corrigan JM, et al. Crossing the quality chasm: a new health system for the 21st century. Washington, DC: National Academy Press; 2001.
5. Baker GR, Norton P. Making patients safer! Reducing error in Canadian healthcare. Healthc Pap. 2001;2(1):10–31.
6. Dennison RD. Creating an organizational culture for medication safety. Nurs Clin North Am. 2005;40(1):1–23.
7. Masotti P, et al. Adverse events in community care: developing a research agenda. Healthc Q. 2007;10(3):63–9.
8. Masotti P, Green M, McColl MA. Adverse events in community care: implications for practice, policy and research. Healthc Q. 2008;12(1):69–76.
9. Calandrino R, et al. Human errors in the calculation of monitor units in clinical radiotherapy practice. Radiother Oncol. 1993;28(1):86–8.
10. Ostrom LT, et al. Lessons learned from investigations of therapy misadministration events. Int J Radiat Oncol Biol Phys. 1996;34(1):227–34.
11. Calandrino R, et al. Detection of systematic errors in external radiotherapy before treatment delivery. Radiother Oncol. 1997;45(3):271–4.
12. Yeung TK, et al. Quality assurance in radiotherapy: evaluation of errors and incidents recorded over a 10 year period. Radiother Oncol. 2005;74(3):283–91.
13. Ekaette EU, et al. Risk analysis in radiation treatment: application of a new taxonomic structure. Radiother Oncol. 2006;80(3):282–7.
14. Akashi M, et al. Investigation of an accidental exposure of radiotherapy patients in Panama. Vienna: International Atomic Energy Agency; 2001.

15. Johnston AM. Unintended overexposure of patient Lisa Norris during radiotherapy treatment at the Beatson Oncology Centre, Glasgow in January 2006, report of an investigation by the inspector appointed by the Scottish Ministers for The Ionising Radiation (Medical Exposures) Regulations 2000. http://www.scotland.gov.uk/Publications/2006/10/27084909/0. 2006.
16. Kutcher GJ, et al. Comprehensive QA for radiation oncology: report of AAPM Radiation Therapy Committee Task Group 40. Med Phys. 1994;21(4):581–618.
17. Fraass B, et al. American Association of Physicists in Medicine Radiation Therapy Committee Task Group 53: quality assurance for clinical radiotherapy treatment planning. Med Phys. 1998;25(10):1773–829.
18. Essers M, Mijnheer BJ. In vivo dosimetry during external photon beam radiotherapy. Int J Radiat Oncol Biol Phys. 1999;43(2):245–59.
19. Dunscombe PB, et al. The Equivalent Uniform Dose as a severity metric for radiation treatment incidents. Radiother Oncol. 2007;84(1):64–6.
20. Yang D, et al. Technical note: electronic chart checks in a paperless radiation therapy clinic. Med Phys. 2012;39(8):4726–32.
21. Moore KL, et al. Vision 20/20: automation and advanced computing in clinical radiation oncology. Med Phys. 2014;41(1):010901.
22. Azmandian F, et al. Towards the development of an error checker for radiotherapy treatment plans: a preliminary study. Phys Med Biol. 2007;52(21):6511–24.
23. Greene D, et al. Ensemble clustering in medical diagnostics. In: CBMS '04: Proceedings of the 17th IEEE symposium on computer-based medical systems (CBMS'04). IEEE, Bethesda, MD, USA; 2004.
24. MacQueen JB. Some methods for classification and analysis of multivariate observations. In: Proceedings of the fifth Berkeley symposium on mathematical statistics and probability, #24: Berkeley, CA, USA, 1. 1967. p. 281–97.
25. Forgy E. Cluster analysis of multivariate data: efficiency vs. interpretability of classifications. Biometrics. 1965;21:768.
26. Schwarz G. Estimating the dimension of a model. Ann Stat. 1978;6(2):461–4.
27. Su T, Dy J. A deterministic method for initializing K-means clustering. In: ICTAI '04: proceedings of the 16th IEEE international conference on tools with artificial intelligence (ICTAI'04), #27: Boca Raton, FL, USA. 2004. p. 784–6.

Treatment Delivery Validation

15

Ruijiang Li

Abstract
This chapter will discuss the application of machine learning techniques for quality assurance and validation of the radiotherapy delivery process. We will discuss issues related to quality assurance and quality control at both the machine-level and the patient-specific measurements. In this chapter, we will demonstrate how machine learning tools can be effectively used for automated treatment verification (both geometrically and dosimetrically) in the context of image-guided and intensity-modulated radiotherapy.

15.1 Introduction

Radiation therapy is a complex process involving multiple steps, including patient simulation and imaging, organ contouring, treatment planning (dose optimization and dose calculation), and treatment delivery. For radiotherapy to be effective, a comprehensive quality assurance (QA) program must be established throughout the radiation therapy process [10]. In particular, to ensure effective treatment, quality must be maintained in all aspects of the delivery process, including performance of the radiation delivery equipment such as linear accelerators, geometric accuracy of target positioning during delivery, and dosimetric agreement between delivered dose and planned dose.

The advent of image-guided intensity-modulated radiotherapy (IMRT) and volumetric-modulated arc therapy (VMAT) has brought their own challenges in terms of comprehensive QA for treatment delivery [5, 6]. These complex treatments

R. Li, PhD
Department of Radiation Oncology, Stanford University,
875 Blake Wilbur Drive, Stanford, CA 94305-5847, USA
e-mail: rli2@stanford.edu

necessitate patient-specific dose measurements in phantoms prior to treating patients, as well as accurate target positioning during dose delivery [9, 20] as a part of quality control (QC) for individual treatment plans. In IMRT and VMAT, patient-specific QC typically quantifies the difference between planned and measured doses at one or more locations in a phantom. This is distinct from QA, which is a system of planned and systematic actions necessary to provide adequate confidence that given requirements for quality are satisfied. These requirements include safe execution of the treatment plan regarding dosage to targets and normal tissues, minimal staff exposure, and treatment monitoring.

Quality control (QC) consists of the measurement of process quality performance, comparison of performance with existing standards, and the actions necessary to keep or regain conformance with the standards; in the context of radiotherapy, this includes setting specifications, measuring performance, comparison with specifications, and, as required, adjusting the process to meet specifications. Within the QA/QC processes, there are random and systematic sources of error. One goal of an optimal QA/QC procedure is to minimize the number and magnitude of systematic errors, by setting action thresholds in an objective and quantitative manner.

The conventional approach to setting action thresholds is to use the mean and standard deviation of a data set obtained by the QA process [4, 19]. Cozzi and Fogliata-Cozzi [4] analyzed breast cancer treatments by in vivo diode measurements. Thresholds on treatment accuracy were set on the allowed difference between the planned dose and the measured dose from diode. Based on the measurements of 421 breast cancer patients, the mean difference and standard deviation, σ of the difference were determined. Action thresholds based on the in vivo dosimetry deviations were set at 1.7 and 3.0 σ. A similar approach was used by Van Esch et al. [19] where 202 cases of breast cancer treatment were used to collate treatment parameters such as medial and lateral gantry angle, field width and length, vertical table position, total MU, etc. The mean and standard deviation of each parameter were calculated, and action thresholds were set at about two to three times the standard deviation depending on the parameter. These conventional approaches will be shown to be ineffective to detect changes in temporal data as acquired in a continual QA/QC process.

In this chapter, we will summarize the recent applications of machine learning tools for validation and verification of the radiation delivery process. Techniques that are developed for both machine-level and patient-specific QA/QC will be discussed. We will demonstrate how machine learning can be used for geometric and dosimetric verification in the context of image-guided and intensity-modulated radiotherapy.

15.2 Validation of Radiation Delivery Parameters

Machine learning tools have been applied to the treatment delivery process to ensure a consistent high-level performance of radiation delivery equipment. Pawlicki et al. [11, 12] reported on the use of an online machine learning technique, e.g., statistical process control [15], for radiotherapy quality assurance of linear accelerators, using both hypothetical and clinical data. They reported results for measurements of output, flatness, and symmetry for a 10 MV photon beam and demonstrated the utility of control charts for detecting changes in the operating point of the beam.

Control charts plot a time series of the data, overlaid with the mean, and upper and lower control limits. The upper and lower control limits correspond to the estimated mean, plus a multiple of the estimated standard deviation of the measurements (see Fig. 15.1). The multiplier of the standard deviation is chosen so that any

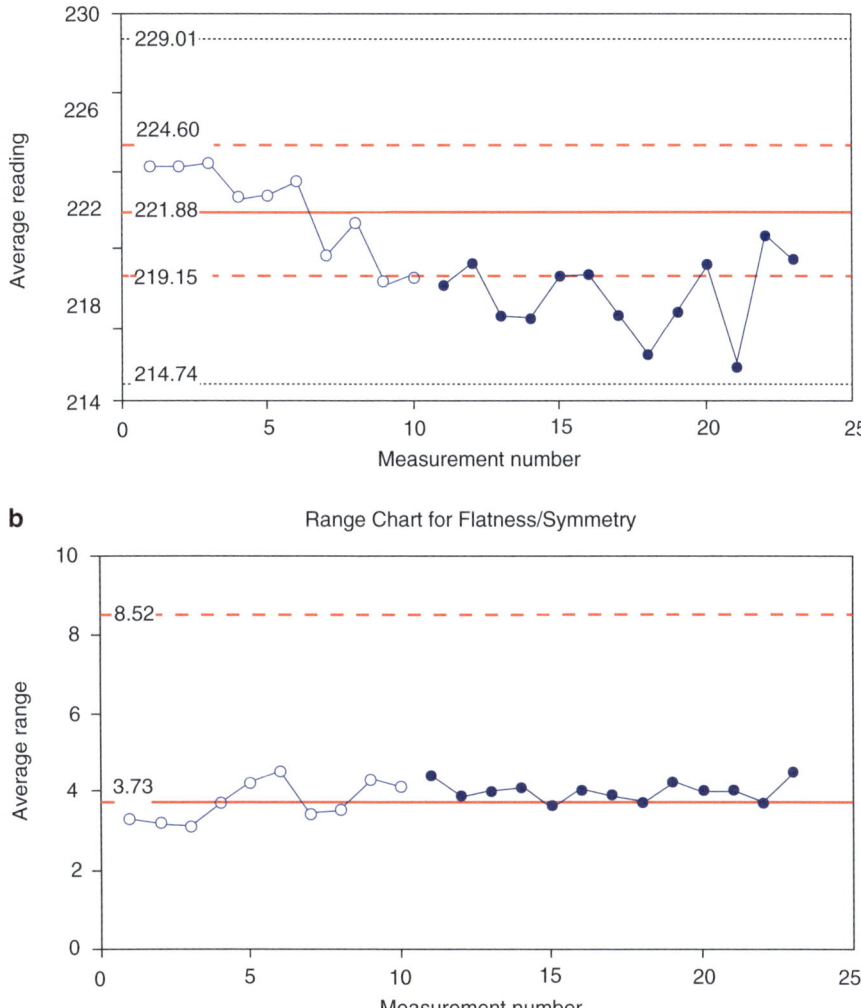

Fig. 15.1 (a) The average chart for the clinical case. Subgroup size of four is used to verify the constancy of the flatness and symmetry. (b) The range chart of the four periphery ion chambers in the RBA-5 device (Reprint from Pawlicki et al. [12])

data falling outside the control line have a strong probability of being attributable to a cause, rather than due to random variation. For normally distributed data, three standard deviations correspond to a 0.00135 probability that the out-of-control variation is due to chance. Roughly speaking, 3 σ corresponds to a 1-in-1,000 probability that the deviation is due to chance. For variations above and below the mean, a 2-in-1,000 probability of incorrectly attributing an out-of-control deviation to chance is a risk that is commonly considered economical to manage.

The reason why a descriptive statistic, such as the standard deviation, is ineffective as a method to set action thresholds for quality assurance is that it tends to obscure the temporal structure of the data. It should be emphasized that clinical requirements will always have precedence over any method of setting action thresholds. If the action thresholds on the process behavior charts are outside the clinical requirements, then the only course of action is to reengineer the process with better equipment and/or procedures.

As dynamic data are continuously generated from a QA process, it is often difficult to identify when a single point displays highly nonrandom behavior and requires further investigation. Process behavior charts separate variation into two sources: variation due to systematic sources for which there is an assignable cause and variation due to random sources for which there is no readily assignable cause. One should search for an assignable cause when a data point exceeds an action threshold on the process behavior chart. After an assignable cause is found, measures can then be incorporated to reduce or eliminate it from the process. When the process is subject to only random variation, the process is predictable, and the limits on the process behavior chart describe the process potential.

15.3 Geometric Verification of Treatment Delivery

Stereotactic body radiotherapy (SBRT) is being increasingly employed as an alternative modality for the treatment of primary and secondary cancers [17, 18]. SBRT has the important advantages of shortened treatment times while delivering higher biologically effective doses. However, normal tissues surrounding the tumors are also exposed to high-dose levels of radiation. Furthermore, cancerous tissue can occasionally move outside the irradiation field, e.g., when the patient has irregular breathing or episodes of coughing. Under these circumstances, malignant tissue will be missed, and more normal tissue than planned will be irradiated. Consequently, the precision requirement of SBRT is high. It is absolutely critical to effectively monitor the target to ensure maximal irradiation of the tumor with minimal irradiation of surrounding normal tissue [8].

Compared with on-board kV imaging, EPID (electronic portal imaging device) acquisition in the cine mode provides a number of advantages for treatment verification purposes: (1) it utilizes the MV treatment beam for imaging and does not involve any additional radiation dose to the patient, and (2) it shows what is actually being irradiated by giving the beam-eye-view of the patient anatomy. Berbeco et al. [1, 2] developed a matching technique for respiratory-gated liver radiotherapy treatment

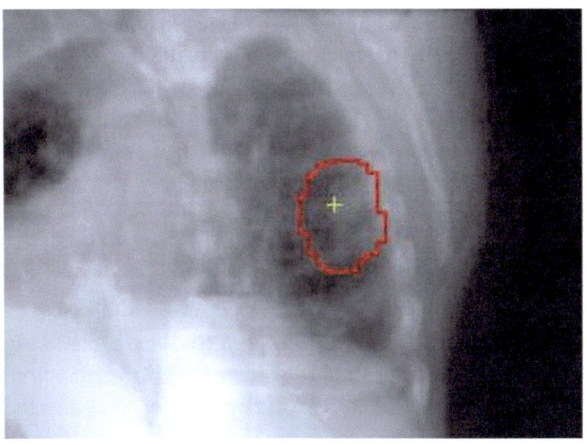 

Fig. 15.2 An original DRR image (*left*) and cine EPID image (*right*) (Reprint from Tang et al. [16])

verification with an EPID in the cine mode. Implanted radiopaque fiducial markers inside or near the target were required for this technique. Markerless techniques were also proposed [13, 14]. However, due to the degraded quality of MV images, it is often very challenging to visualize the tumor target in cine EPID images.

Tang et al. [16] recently proposed a novel approach for SBRT treatment verification using cine EPID images based on a machine learning algorithm. They modeled the treatment verification problem as a two-class classification problem and applied an artificial neural network (ANN) to classify the cine EPID images acquired during the treatment into corresponding classes—with the tumor inside or outside of the beam aperture. Training samples were generated for the ANN using digitally reconstructed radiographs (DRRs) with artificially added shifts in the tumor location—to simulate cine EPID images with different tumor locations (Fig. 15.2). Principal component analysis (PCA) was used to reduce the dimensionality of the training samples and cine EPID images acquired during the treatment. The proposed treatment verification algorithm was tested on five lung SBRT patients in a retrospective fashion. On average, the machine learning algorithm achieved very high classification accuracy, recall rate, and precision rate, all in the high 90 %. For its practical implementation, a comprehensive clinical validation remains to be performed in terms of different tumor volumes, location, and imaging angle. In addition, the algorithm's performance for cine MV images acquired with partial field of view due to modulated treatments such as IMRT or VMAT needs to be evaluated.

15.4 Dosimetric Verification of Treatment Delivery

Patient-specific measurements are typically used to validate the dosimetry of complex treatments such as IMRT and VMAT. To evaluate the dosimetric performance over time of the treatment delivery process, Breen et al. [3] used statistical process control (SPC) concepts to analyze the measurements from 330 head-and-neck (H&N) treatment plans. H&N IMRT cases were planned with the PINNACLE3 treatment planning system (Philips Medical Systems, Madison, WI) and treated on Varian (Palo Alto, CA) or Elekta (Crawley, UK) linacs. As part of regular quality assurance, plans were recalculated on a 20-cm-diameter cylindrical phantom, and ion chamber measurements were made in high-dose volumes (the PTV with highest dose) and in low-dose volumes (spinal cord organ-at-risk, OR). Differences between the planned and measured doses were recorded as a percentage of the planned dose. It was demonstrated that head-and-neck IMRT plans could be delivered with a systematic error of 0.2 % in high-dose volumes and −1.0 % in low-dose volumes.

Statistical process control also provides a means to evaluate adjustments to the process (via beam models) in a robust manner. For IMRT dosimetric verification, measurements in phantom are not the end of the QA process; rather, they are the data upon which a strong foundation of continuous quality improvement can be constructed. Analysis of this large series of H&N IMRT measurements demonstrated that the IMRT dosimetry was stable over time and within accepted tolerances. These data provide useful information for assessing alterations to beam models in the planning system. IMRT is enhanced by the addition of statistical process control to traditional quality control procedures.

Gerard et al. [7] further investigated the use of two complementary machine learning tools—control charts and performance indices—to accurately analyze the dose delivery process in IMRT. Control charts aim at monitoring the process over time using statistical control limits, whereas performance indices aim at quantifying the ability of the process to produce data that are within the clinical specification limits at a precise moment. They showed that three control charts—individual value, moving range, and exponentially weighted moving average (EWMA) control charts, chosen for their capacity of bringing complementary information—allowed an efficient detection of the drifts that occurred in the IMRT dose delivery process for prostate and head-and-neck treatments (see Fig. 15.3). The dose delivery process for prostate treatments was both statistically in control and capable, i.e., its evolution can be predicted within the clinical specification limits. For head-and-neck treatments, the dose delivery process was in control but not statistically capable, when using the current specification limits set at 4 %. This implies that the evolution of the process can be predicted but not within the specification limits. So, as shown by the process performance indices, actions should be undertaken to improve both the process centering and dispersion.

Control charts revealed that some patients' data were outside the specification limits, both for the ionization chamber dose response deviation and for the MLC deviations. Control charts also confirmed that the MLC calibration has a large influence on the dose delivery process, implying that mechanic and dosimetric quality controls of the MLC have to be carefully and regularly performed, in particular after a maintenance operation. Three performance indices, P_p, P_{pk}, and P_{pm}, were shown to identify the reason why data are outside the specification limits and to determine what kind of action should be undertaken.

It should be noted that the application of machine learning to IMRT dose verification is not a replacement for patient-specific dosimetry. Rather, it supports our

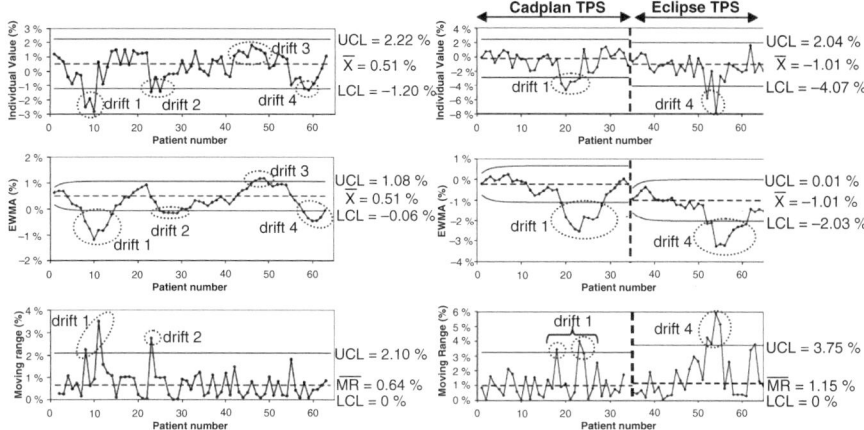

Fig. 15.3 Individual value, EWMA, and moving range control charts for prostate treatments (**a**) and for head-and-neck treatments (**b**)

current practices by allowing the observation of trends in the mean and dispersion of measurements and characterizes, through upper and lower control limits, the envelope in which the process operates. This, in turn, permits the correct interpretation of subsequent measurements, by answering the question, "Is the current measurement under control?" The attribution of out-of-control measurements to specific causes that require attention enhances continuous quality improvement.

15.5 Summary

We have reviewed the recent applications of machine learning techniques for quality assurance and validation of radiotherapy delivery. We discussed both machine-level and patient-specific measurements in the QA/QC process. We showed the promising applications of online machine learning tools to monitor dynamic and sequential data that are acquired during the continual QA/QC process. In addition, we demonstrated how machine learning can be effectively used for automated treatment verification (both geometrically and dosimetrically) in image-guided and intensity-modulated radiotherapy. As the technology of radiotherapy evolves in future, new challenges are expected to arise for QA/QC of radiation delivery, and machine learning will continue to play a key role in this process.

References

1. Berbeco RI, Hacker F, Ionascu D, Mamon HJ. Clinical feasibility of using an EPID in CINE mode for image-guided verification of stereotactic body radiotherapy. Int J Radiat Oncol Biol Phys. 2007;69:258–66.
2. Berbeco RI, Neicu T, Rietzel E, Chen GT, Jiang SB. A technique for respiratory-gated radiotherapy treatment verification with an EPID in cine mode. Phys Med Biol. 2005;50:3669–79.

3. Breen SL, Moseley DJ, Zhang B, Sharpe MB. Statistical process control for IMRT dosimetric verification. Med Phys. 2008;35:4417–25.
4. Cozzi L, Fogliata-Cozzi A. Quality assurance in radiation oncology. A study of feasibility and impact on action levels of an in vivo dosimetry program during breast cancer irradiation. Radiother Oncol: J Eur Soc Therapeut Radiol Oncol. 1998;47:29–36.
5. Ezzell GA, Galvin JM, Low D, Palta JR, Rosen I, Sharpe MB, Xia P, Xiao Y, Xing L, Yu CX. Guidance document on delivery, treatment planning, and clinical implementation of IMRT: report of the IMRT Subcommittee of the AAPM Radiation Therapy Committee. Med Phys. 2003;30:2089–115.
6. Galvin JM, Ezzell G, Eisbrauch A, Yu C, Butler B, Xiao Y, Rosen I, Rosenman J, Sharpe M, Xing L, Xia P, Lomax T, Low DA, Palta J. Implementing IMRT in clinical practice: a joint document of the American Society for Therapeutic Radiology and Oncology and the American Association of Physicists in Medicine. Int J Radiat Oncol Biol Phys. 2004;58:1616–34.
7. Gerard K, Grandhaye JP, Marchesi V, Kafrouni H, Husson F, Aletti P. A comprehensive analysis of the IMRT dose delivery process using statistical process control (SPC). Med Phys. 2009;36:1275–85.
8. Jiang SB. Radiotherapy of mobile tumors. Semin Radiat Oncol. 2006;16:239–48.
9. Keall PJ, Mageras GS, Balter JM, Emery RS, Forster KM, Jiang SB, Kapatoes JM, Low DA, Murphy MJ, Murray BR, Ramsey CR, Van Herk MB, Vedam SS, Wong JW, Yorke E. The management of respiratory motion in radiation oncology report of AAPM Task Group 76. Med Phys. 2006;33:3874–900.
10. Kutcher GJ, Coia L, Gillin M, Hanson WF, Leibel S, Morton RJ, Palta JR, Purdy JA, Reinstein LE, Svensson GK, et al. Comprehensive QA for radiation oncology: report of AAPM Radiation Therapy Committee Task Group 40. Med Phys. 1994;21:581–618.
11. Pawlicki T, Mundt AJ. Quality in radiation oncology. Med Phys. 2007;34:1529–34.
12. Pawlicki T, Whitaker M, Boyer AL. Statistical process control for radiotherapy quality assurance. Med Phys. 2005;32:2777–86.
13. Rottmann J, Aristophanous M, Chen A, Court L, Berbeco R. A multi-region algorithm for markerless beam's-eye view lung tumor tracking. Phys Med Biol. 2010;55:5585–98.
14. Rottmann J, Keall P, Berbeco R. Real-time soft tissue motion estimation for lung tumors during radiotherapy delivery. Med Phys. 2013;40:091713.
15. Shewhart WA. Economic control of quality of manufactured product. New York: D. Van Nostrand Company, Inc.; 1931.
16. Tang X, Lin T, Jiang S. A feasibility study of treatment verification using EPID cine images for hypofractionated lung radiotherapy. Phys Med Biol. 2009;54:S1–8.
17. Timmerman R, Paulus R, Galvin J, Michalski J, Straube W, Bradley J, Fakiris A, Bezjak A, Videtic G, Johnstone D, Fowler J, Gore E, Choy H. Stereotactic body radiation therapy for inoperable early stage lung cancer. JAMA. 2010;303:1070–6.
18. Timmerman RD, Kavanagh BD, Cho LC, Papiez L, Xing L. Stereotactic body radiation therapy in multiple organ sites. J Clin Oncol. 2007;25:947–52.
19. Van Esch A, Bogaerts R, Kutcher GJ, Huyskens D. Quality assurance in radiotherapy by identifying standards and monitoring treatment preparation. Radiother Oncol: J Eur Soc Therapeut Radiol Oncol. 2000;56:109–15.
20. Xing L, Thorndyke B, Schreibmann E, Yang Y, Li T-F, Kim G-Y, Luxton G, Koong A. Overview of image-guided radiation therapy. Med Dosim. 2006;31:91–112.

Part VI
Machine Learning for Outcomes Modeling

Bioinformatics of Treatment Response

16

Issam El Naqa

Abstract

Radiotherapy treatment outcomes are determined by complex interactions among treatment, anatomical, and patient-related variables. A key component of radiation oncology research is to predict at the time of treatment planning, or during the course of fractionated radiation treatment, the probability of tumor eradication and normal tissue risks for the type of treatment being considered for that particular patient. Traditionally, these outcomes are modeled using information about the dose distribution and the fractionation. However, it is recognized that radiation response is multifactorial including clinical prognostic factors and, more recently, inherited genetic variations have been suggested as playing an important role in radiation response. Therefore, recent approaches have utilized increasingly data-driven models incorporating advanced bioinformatics and machine learning tools in which dose-volume metrics are mixed with other patient- or disease-based prognostic factors in order to improve outcomes prediction. Accurate prediction of treatment outcomes would provide clinicians with better tools for informed decision-making about expected benefits versus anticipated risks. In this chapter, we provide an overview of the current status of data-driven outcome modeling techniques for patients who receive radiation treatment with special focus on its big data notion and the emerging role of machine learning approaches to improve outcome modeling and response prediction.

I. El Naqa
Department of Oncology, McGill University, Montreal, QC, Canada

Department of Radiation Oncology, University of Michigan, Ann Arbor, USA
e-mail: issam.elnaqa@mcgill.ca; ielnaqa@med.umich.edu

16.1 Introduction

Recent years have witnessed tremendous technological advances in radiotherapy treatment planning, image guidance, and treatment delivery [1, 2]. Moreover, clinical trials examining treatment intensification in patients with locally advanced cancer have shown incremental improvements in local control and overall survival [3]. However, radiation-induced toxicities remain major dose-limiting factors [4, 5]. Therefore, there is a need for studies directed toward predicting treatment benefit versus risk of failure. Clinically, such predictors would allow for more individualization of radiation treatment plans. In other words, physicians may prescribe a more or less intense radiation regimen for an individual based on model predictions of local control benefit and toxicity risk. Such an individualized regimen would aim toward an optimized radiation treatment response while keeping in mind that a more aggressive treatment with a promised improved tumor control will not translate into improved survival unless severe toxicities are accounted for and limited during treatment planning. Therefore, improved models for predicting both local control and normal tissue toxicity should be considered in the optimal treatment planning design process.

Radiotherapy outcomes are usually characterized by two metrics: the tumor control probability (TCP) and the normal tissue complication probability (NTCP) of surrounding normal tissues [2, 6]. TCP/NTCP models could be used during the consultation period as a guide for ranking treatment options [7, 8]. Alternatively, once a decision has been reached, these models could be included in an objective function, and the optimization problem driving the actual patient's treatment plan can be formulated in terms relevant to maximizing tumor eradication benefit and minimizing complication risk [9–11]. Traditional models of TCP/NTCP models and their variations use information only about the dose distribution and fractionation. However, it is well known that radiotherapy outcomes may also be affected by multiple clinical and biological prognostic factors such as stage, volume, tumor hypoxia, etc. [12, 13] as depicted in Fig. 16.1. Therefore, recent years have witnessed the emergence of data-driven models utilizing informatics techniques, in

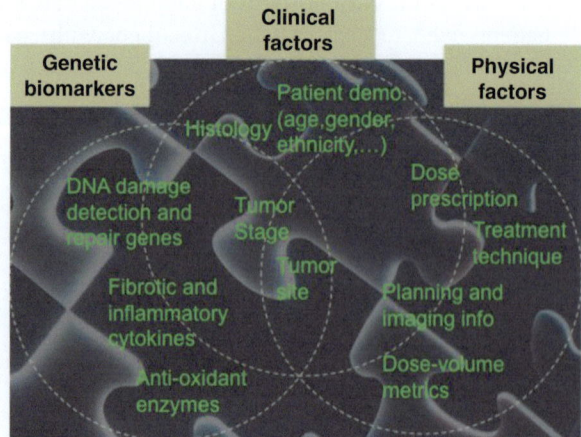

Fig. 16.1 Radiotherapy treatment involves complex interaction of physical, biological, and clinical factors. The successful informatics approach should be able to resolve this interaction "puzzle" in the observed treatment outcome (e.g., local control or toxicity) for each individual patient [21]

which dose-volume metrics are combined with other patient- or disease-based prognostic factors [4, 5, 14–20].

In this chapter, we provide an overview of the current status of data-driven outcome modeling techniques for predicting tumor response and normal tissue toxicities for patients who receive radiation treatment with special focus on the emerging role of machine learning approaches to improve outcome modeling and response prediction. Then, we present examples of radiotherapy data and its big data notion. Finally, we discuss the potentials and challenging obstacles to applying bioinformatics and machine learning strategies to radiotherapy outcome modeling.

16.2 Data-Driven Outcome Modeling

Radiotherapy outcome models could be divided according to the underlying principle into (1) analytical models, which employ biophysical understanding of irradiation effects such as the linear quadratic (LQ) model, and (2) data-driven models, which are phenomenological models and depend on parameters available from the collected clinical and dosimetric data [20]. In the context of data-driven and multivariable modeling of outcomes, the observed treatment outcome (e.g., TCP or NTCP) is considered as the result of mathematical mapping of several dosimetric, clinical, or biological input variables [19]. Mathematically this is expressed as: $f(\mathbf{x}; \mathbf{w}^*): X \to Y$ where R^N (an input variable vector of N dimensions) is composed of the input metrics (dose-volume metrics, patient disease specific prognostic factors, or biological markers). The expression $y_i \in Y$ is the corresponding observed treatment outcome scalar. The variable $\mathbf{w}^*$ includes the optimal parameters of model $f(\cdot)$ obtained by optimizing a certain objective functional. Learning is defined in this context as estimating dependencies from data [22]. The two common types of learning could be applied: supervised and unsupervised. Supervised learning is used when the endpoints of the treatments such as tumor control or toxicity grade are known; these endpoints are provided by experienced oncologists following RTOG or NCI criteria, and it is the most commonly used learning method in outcome modeling. Nevertheless, unsupervised methods such as principal component analysis (PCA) are also used to reduce dimensionality and to aid visualization of multivariate data and selection of learning method parameters [23]. The selection of the functional form of the model $f(\cdot)$ is closely related to the prior knowledge of the problem. In analytical models, the shape of the functional form is selected based on the clinical or biological process at hand; however, in data-driven models, the objective is usually to find a functional form that fits the data [24].

16.3 Radiotherapy as a Big Data Resource

A typical radiotherapy treatment scenario can generate a large pool of "big data" that comprise but are not limited to patient demographics, volumetric dosimetric data about radiation exposure to the tumor and surrounding tissues, and 3D and 4D anatomical and functional disease longitudinal imaging features (radiomics), in addition to genomics and proteomics data derived from peripheral blood and tissue specimens. Accordingly, big data in radiotherapy could be divided based on its nature into four categories: clinical, dosimetric, imaging, and biological. These four categories of radiotherapy big data are described in the following.

16.3.1 Clinical Data

Clinical data in radiotherapy typically refers to cancer diagnostic information (site, histology, stage, grade, etc.) and patient-related characteristics (age, gender, comorbidities, etc.). In some instances, other treatment modalities information (surgery, chemotherapy, hormonal treatment, etc.) would be also classified under this category. The mining of such data could be challenging if the data is unstructured; however, there are good opportunities for natural language processing (NLP) techniques to assist in the organization of data [25].

16.3.2 Dosimetric Data

This type of data is related to the treatment planning process in radiotherapy, which involves radiation dose simulation using computed tomography imaging, specifically dose-volume metrics derived from dose-volume histograms (DVHs) graphs. Dose-volume metrics have been extensively studied in the radiation oncology literature for outcome modeling [14–17, 26, 27]. These metrics are extracted from the DVH such as volume receiving certain dose (Vx); minimum dose to $x\%$ volume (D_x); mean, maximum, and minimum dose; etc. More details are in our review chapter [20]. Moreover, we have developed a dedicated software tool called "Dose response explorer" (DREES) for deriving these metrics and modeling of radiotherapy response [28].

16.3.3 Radiomics (Imaging Features)

kV x-ray computed tomography (kV-CT) has been historically considered the standard modality for treatment planning in radiotherapy because of its ability to provide electron density information for target definition, structures, and heterogeneous dose calculations [2, 29]. However, additional information from other imaging modalities could be used to improve treatment monitoring and prognosis in different cancer types. For example, physiological information (tumor metabolism, proliferation, necrosis, hypoxic regions, etc.) can be collected directly from nuclear imaging modalities such as SPECT and PET or indirectly from MRI [30, 31]. The complementary nature of these different imaging modalities has led to efforts toward combining information to achieve better treatment outcomes. For instance,

PET/CT has been utilized for staging, planning, and assessment of response to radiation therapy [32, 33]. Similarly, MRI has been applied in tumor delineation and assessing toxicities in head and neck cancers [34, 35]. Moreover, quantitative information from hybrid-imaging modalities could be related to biological and clinical endpoints, a new emerging field referred to as "radiomics" [36, 37]. In our previous work, we demonstrated the potential of this new field to monitor and predict response to radiotherapy in head and neck [38], cervix [38, 39], and lung [40] cancers, in turn allowing for adapting and individualizing treatment.

16.3.4 Biological Markers

A biomarker is defined as "a characteristic that is objectively measured and evaluated as an indicator of normal biological processes, pathological processes, or pharmacological responses to a therapeutic intervention" [41]. Biomarkers can be categorized based on the biochemical source of the marker into exogenous or endogenous. Exogenous biomarkers are based on introducing a foreign substance into the patient's body such as those used in molecular imaging as discussed above. Conversely, endogenous biomarkers can further be classified as (1) "expression biomarkers," measuring changes in gene expression or protein levels, or (2) "genetic biomarkers," based on variations, for tumors or normal tissues, in the underlying DNA genetic code. Measurements are typically based on tissue or fluid specimens, which are analyzed using molecular biology laboratory techniques [42]. Expression biomarkers are the result of gene expression changes in tissues or bodily fluids due to the disease or normal tissues' response to treatment [43]. These biomarkers can be further divided into single parameter (e.g., prostate-specific antigen (PSA) levels in blood serum) versus bio-arrays. These can be based on disease pathophysiology or pharmacogenetic studies or they can be extracted from several methods, such as high-throughput gene expression (aka transcriptomics) [44–46], resulting protein expressions (aka proteomics) [47, 48], or metabolites (aka metabolomics) [49, 50]. On the other hand, the inherent genetic variability of the human genome is an emerging resource for studying disposition to cancer and the variability of patient responses to therapeutic agents. These variations in the DNA sequences of humans, in particular single-nucleotide polymorphisms (SNPs), have strong potential to elucidate complex disease onset and response in cancer [51]. Methods based on the candidate gene approach and high throughput (genome-wide associations (GWAS) studies) are currently heavily investigated to analyze the functional effect of SNPs in predicting response to radiotherapy [52–54]. There are several ongoing SNP genotyping initiatives in radiation oncology, including the pan-European GENEPI project [55], the British RAPPER project [56], the Japanese RadGenomics project [57], and the US Gene-PARE project [58]. An international consortium has been also established to coordinate and lead efforts in this area [59]. Examples include the identification of SNPs related to radiation toxicity in prostate cancer treatment [60–62].

16.4 Systems Radiobiology

To integrate heterogeneous big data in radiotherapy, engineering-inspired system approaches would have great potential to achieve this goal. Systems biology has emerged as a new field to apply systematic study of complex interactions to biological systems [63], but its application to radiation oncology, despite this potential, has been unfortunately limited to date [64, 65]. Recently, Eschrich et al. presented systems biology approach for identifying biomarkers related to radiosensitivity in different cancer cell lines using linear regression to correlate gene expression with survival fraction measurements [66]. However, such a linear regression model may lack the ability to account for higher-order interactions among the different genes and neglect the expected hierarchal relationships in signaling transduction of highly complex radiation response. It has been noted in the literature that modeling of molecular interactions could be represented using graphs of network connections as in power line grids. In this case, radiobiological data can be represented as a graph (network) where the nodes represent genes or proteins and the edges may represent similarities or interactions between these nodes. We have utilized such approach based on Bayesian networks for modeling dosimetric radiation pneumonitis relationships [67] and more recently in predicting local control from biological and dosimetric data [68].

In the more general realm of bioinformatics, this systems approach could be represented as a part of a feedback treatment planning system as shown in Fig. 16.2,

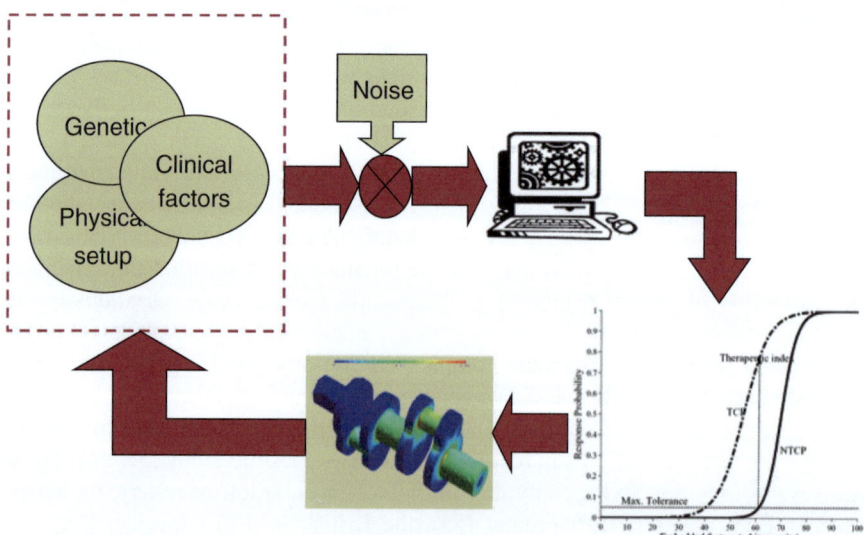

Fig. 16.2 The bioinformatics understanding of heterogeneous variable interactions as a feedback into the treatment planning system to improve patient's outcomes. A heterogeneous list of variables with their noisy characteristics are acquired from retrospective or prospective studies and fed into in a learning algorithm to derive estimates of TCP/NTCP, which is typically corrected based on feedback of newly tested patients or scientific and clinical findings

in which bioinformatics understanding of heterogeneous variables interactions could be used as an adaptive learning process to improve outcome modeling and personalization of radiotherapy regimens.

16.5 Software Tools for Outcome Modeling

Many of the TCP/NTCP outcome modeling methods require dedicated software tools for implementation. Examples of such software tools in the literature are BIOPLAN and DREES. BIOPLAN (BIOlogical evaluation of treatment PLANs) uses several analytical models for evaluation of radiotherapy treatment plans [69], while DREES is an open-source software package developed by our group for dose-response modeling using analytical and data-driven methods [28] presented in Fig. 16.3. It should be mentioned that several commercial treatment planning systems have currently incorporated different TCP/NTCP models, mainly analytical ones that could be used for ranking and biological optimization purposes. A discussion of these models and their quality assurance guidelines is provided in TG-166 [11].

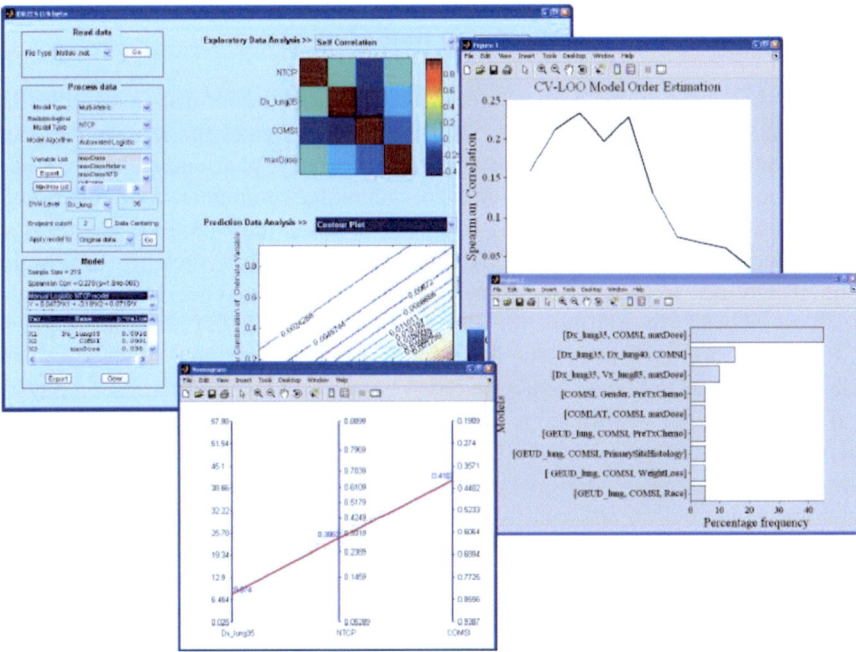

Fig. 16.3 DREES allows for TCP/NTCP analytical and multivariate modeling of outcomes data. The example is for lung injury. The components shown here are Main GUI, model order and parameter selection by resampling methods, and a nomogram of outcome as function of mean dose and location

16.6 Discussion

16.6.1 Data Sharing

Successful outcome modeling requires large datasets to meet statistical requirements, and sharing data is necessary to achieve this purpose. However, data sharing remains an issue for nontechnical issues [70]. Therefore, the Quantitative Analyses of Normal Tissue Effects in the Clinic (QUANTEC) consortium has suggested that cooperative groups adopt a policy of anonymizing clinical trial data and making these data publicly accessible after a reasonable delay. This delay would enable publication of all the investigator-driven, planned studies while encouraging the establishment of key databanks of linked treatment planning, imaging, and outcomes data [71]. An alternative approach is to apply rapid learning as suggested by the Maastro clinic group at Maastricht, in which innovative information technologies are developed that support semantic interoperability and enable distributed learning and data sharing without the need for the data to leave the hospital or the institution [72]. An example of multi-institutional data sharing is developed by the groups of Maastro clinic and the Policlinico Universitario Agostino Gemelli in Rome, Italy (Gemelli) [73].

16.6.2 Lack of Web Resources for Radiobiology

As of today, there are no dedicated web resources for bioinformatics studies in radiation oncology. Nevertheless, radiotherapy biological marker studies can still benefit from existing bioinformatics resources for pharmacogenomic studies that contain databases and tools for genomic, proteomic, and functional analysis as reviewed by Yan [74]. For example, the National Center for Biotechnology Information (NCBI) site hosts databases such as GenBank, dbSNP, Online Mendelian Inheritance in Man (OMIM), and genetic search tools such as BLAST. In addition, the Protein Data Bank (PDB) and the program CPHmodels are useful for protein structure three-dimensional modeling. The Human Genome Variation Database (HGVbase) contains information on physical and functional relationships between sequence variations and neighboring genes. Pattern analysis using PROSITE and Pfam databases can help correlate sequence structures to functional motifs such as phosphorylation [74]. Biological pathway construction and analysis is an emerging field in computational biology that aims to bridge the gap between biomarker findings in clinical studies with underlying biological processes. Several public databases and tools are being established for annotating and storing known pathways such as KEGG and Reactome projects or commercial ones such as the IPA or MetaCore [75]. Statistical tools are used to properly map data from gene/protein differential experiments into the different pathways such as mixed effect models [76] or enrichment analysis [77].

16.6.3 Protecting the Confidentiality and Privacy of Clinical Phenotype Data

QUANTEC offered a solution to radiotherapy digital data (treatment planning, imaging, and outcomes data) accessibility by asking cooperative groups to adopt a policy of anonymizing clinical trial data and making the data publicly accessible after a reasonable delay [71]. With regard to blood or tissue samples, no recommendation was made, however, by extending the same work and making any gene or protein expression assay measurements available under the same umbrella, while raw specimen data could be accessed from the biospecimen resource. For example, in the RTOG biospecimen standard operating procedure (SOP), it is highlighted that biospecimens received by the RTOG Biospecimen Resource are de-identified of all patient health identifiers and are enrolled in an approved RTOG study. Each patient being enrolled by an institution has to qualify and consent to be part of the study before being assigned a case and study ID by the RTOG Statistical Center. No information containing specific patient health identifiers is maintained by the Resource Freezerworks database, which is primarily an inventory and tracking system. In addition, information related to medical identifiers and any code lists could be removed completely from the dataset after a certain period say 10 years or so. Moreover, it has been argued that current measures by the Health Insurance Portability and Accountability Act (HIPPA) of 18 data elements are not sufficient and techniques based on research in privacy-preserving data mining, disclosure risk assessment data de-identification, obfuscation, and protection may need to be adopted to achieve better protection of confidentiality [78].

16.7 Future Research Directions

The ability to maintain high-fidelity large-scale data for radiotherapy studies remains a major challenge despite the high volume of clinical generated data on almost daily basis. As discussed above there have been several ongoing institutional and multi-institutional initiatives such as the RTOG, radiogenomics consortium, and EuroCAT to develop such infrastructure; however, there is plenty of work to be done to overcome issues related to, data sharing hurdles, patient confidentiality issues lack of signaling pathways databases of radiation response, development of cost-effective multicenter communication systems that allows transmission, storage, and query of large datasets such images, dosimetry, and biomarkers information. The use of NLP techniques is a promising approach in organizing unstructured clinical data. Dosimetry and imaging data can benefit from existing infrastructure for Picture Archiving and Communication Systems (PACS) or other medical image databases. Methods based on the new emerging field of systems radiobiology will continue to grow on a rapid pace, but they could also benefit immensely from the development of specialized radiation response signaling pathway databases analogous to the currently existing pharmacogenomics databases. Data sharing among different institutions is a major hurdle, which could be solved through cooperative groups or distributed databases by developing in a cost-effective manner the necessary bioinformatics and communication infrastructure using open-access resources through partnership with industry.

Conclusion

Recent evolution in radiotherapy imaging and biotechnology has generated enormous amount of big data that spans clinical, dosimetric, imaging, and biological markers. This data provided new opportunities for reshaping our understanding of radiotherapy response and outcome modeling. However, the complexity of this data and the variability of tumor and normal tissue responses would render the utilization of advanced bioinformatics and machine learning methods as indispensible tools for better delineation of radiation complex interaction mechanisms and basically a cornerstone to "making data dreams come true" [79]. However, it also posed new challenges for data aggregation, sharing, confidentiality, and analysis. Moreover, radiotherapy data constitutes a unique interface between physics and biology that can benefit from the general advances in biomedical informatics research such as systems biology and available web resources while still requiring the development of its own technologies to address specific issues related to this interface. Successful application and development of advanced data communication and bioinformatics tools for radiation oncology big data so to speak is essential to better predicting radiotherapy response to accompany other aforementioned technologies and usher significant progress toward the goal of personalized treatment planning and improving the quality of life for radiotherapy cancer patients.

References

1. Bortfeld T, Schmidt-Ullrich R, De Neve W, Wazer D, editors. Image-guided IMRT. Berlin: Springer; 2006.
2. Webb S. The physics of three-dimensional radiation therapy: conformal radiotherapy, radiosurgery, and treatment planning. Bristol, UK. Philadelphia: Institute of Physics Pub; 2001.
3. Halperin EC, Perez CA, Brady LW. Perez and Brady's principles and practice of radiation oncology. 5th ed. Philadelphia: Wolters Kluwer Health/Lippincott Williams & Wilkins; 2008.
4. Bentzen SM, Constine LS, Deasy JO, Eisbruch A, Jackson A, Marks LB, et al. Quantitative Analyses of Normal Tissue Effects in the Clinic (QUANTEC): an introduction to the scientific issues. Int J Radiat Oncol Biol Phys. 2010;76:S3–9.
5. Jackson A, Marks LB, Bentzen SM, Eisbruch A, Yorke ED, Ten Haken RK, et al. The lessons of QUANTEC: recommendations for reporting and gathering data on dose-volume dependencies of treatment outcome. Int J Radiat Oncol Biol Phys. 2010;76:S155–60.
6. Steel GG. Basic clinical radiobiology. 3rd ed. London/New York: Arnold/Oxford University Press; 2002.
7. Armstrong K, Weber B, Ubel PA, Peters N, Holmes J, Schwartz JS. Individualized survival curves improve satisfaction with cancer risk management decisions in women with BRCA1/2 mutations. J Clin Oncol. 2005;23:9319–28.

8. Weinstein MC, Toy EL, Sandberg EA, Neumann PJ, Evans JS, Kuntz KM, et al. Modeling for health care and other policy decisions: uses, roles, and validity. Value Health. 2001;4:348–61.
9. Moiseenko V, Kron T, Van Dyk J. Biologically-based treatment plan optimization: a systematic comparison of NTCP models for tomotherapy treatment plans. In: Proceedings of the 14th international conference on the use of computers in radiation therapy, Seoul. 2004.
10. Brahme A. Optimized radiation therapy based on radiobiological objectives. Semin Radiat Oncol. 1999;9:35–47.
11. Allen Li X, Alber M, Deasy JO, Jackson A, Ken Jee KW, Marks LB, et al. The use and QA of biologically related models for treatment planning: short report of the TG-166 of the therapy physics committee of the AAPM. Med Phys. 2012;39:1386–409.
12. Choi N, Baumann M, Flentjie M, Kellokumpu-Lehtinen P, Senan S, Zamboglou N, et al. Predictive factors in radiotherapy for non-small cell lung cancer: present status. Lung Cancer. 2001;31:43–56.
13. Fu XL, Zhu XZ, Shi DR, Xiu LZ, Wang LJ, Zhao S, et al. Study of prognostic predictors for non-small cell lung cancer. Lung Cancer. 1999;23:143–52.
14. Blanco AI, Chao KS, El Naqa I, Franklin GE, Zakarian K, Vicic M, et al. Dose-volume modeling of salivary function in patients with head-and-neck cancer receiving radiotherapy. Int J Radiat Oncol Biol Phys. 2005;62:1055–69.
15. Bradley J, Deasy JO, Bentzen S, El-Naqa I. Dosimetric correlates for acute esophagitis in patients treated with radiotherapy for lung carcinoma. Int J Radiat Oncol Biol Phys. 2004;58:1106–13.
16. Marks LB. Dosimetric predictors of radiation-induced lung injury. Int J Radiat Oncol Biol Phys. 2002;54:313–6.
17. Hope AJ, Lindsay PE, El Naqa I, Bradley JD, Vicic M, Deasy JO. Clinical, dosimetric, and location-related factors to predict local control in non-small cell lung cancer. In: ASTRO 47th annual meeting. Denver. 2005. p. S231.
18. Tucker SL, Cheung R, Dong L, Liu HH, Thames HD, Huang EH, et al. Dose-volume response analyses of late rectal bleeding after radiotherapy for prostate cancer. Int J Radiat Oncol Biol Phys. 2004;59:353–65.
19. El Naqa I, Bradley JD, Lindsay PE, Blanco AI, Vicic M, Hope AJ, et al. Multi-variable modeling of radiotherapy outcomes including dose-volume and clinical factors. Int J Radiat Oncol Biol Phys. 2006;64:1275–86.
20. Deasy JO, El Naqa I. Image-based modeling of normal tissue complication probability for radiation therapy. Cancer Treat Res. 2008;139:215–56.
21. Spencer S, Bonnin DA, Deasy J, Bradley JD, El Naqa I. Bioinformatics methods for learning radiation–induced lung inflammation from heterogeneous retrospective and prospective data. New York: Hindawi Publishing Corporation; 2009.
22. Hastie T, Tibshirani R, Friedman JH. The elements of statistical learning data mining, inference, and prediction: with 200 full-color illustrations. New York: Springer; 2001.
23. Härdle W, Simar L. Applied multivariate statistical analysis. Berlin/New York: Springer; 2003.
24. El Naqa I, Bradley J, Lindsay PE, Hope A, Deasy JO. Predicting radiotherapy outcomes using statistical learning techniques. Phys Med Biol. 2009;54:S9–30.
25. Shivade C, Raghavan P, Fosler-Lussier E, Embi PJ, Elhadad N, Johnson SB, et al. A review of approaches to identifying patient phenotype cohorts using electronic health records. J Am Med Inform Assoc. 2013;21(2):221–30.
26. Hope AJ, Lindsay PE, El Naqa I, Alaly JR, Vicic M, Bradley JD, et al. Modeling radiation pneumonitis risk with clinical, dosimetric, and spatial parameters. Int J Radiat Oncol Biol Phys. 2006;65:112–24.
27. Levegrun S, Jackson A, Zelefsky MJ, Skwarchuk MW, Venkatraman ES, Schlegel W, et al. Fitting tumor control probability models to biopsy outcome after three-dimensional conformal radiation therapy of prostate cancer: pitfalls in deducing radiobiologic parameters for tumors from clinical data. Int J Radiat Oncol Biol Phys. 2001;51:1064–80.

28. El Naqa I, Suneja G, Lindsay PE, Hope AJ, Alaly JR, Vicic M, et al. Dose response explorer: an integrated open-source tool for exploring and modelling radiotherapy dose-volume outcome relationships. Phys Med Biol. 2006;51:5719–35.
29. Khan FM. Treatment planning in radiation oncology. 2nd ed. Philadelphia: Lippincott Williams & Wilkins; 2007.
30. Condeelis J, Weissleder R. In vivo imaging in cancer. Cold Spring Harb Perspect Biol. 2010;2:a003848.
31. Willmann JK, van Bruggen N, Dinkelborg LM, Gambhir SS. Molecular imaging in drug development. Nat Rev Drug Discov. 2008;7:591–607.
32. Bussink J, Kaanders JHAM, van der Graaf WTA, Oyen WJG. PET-CT for radiotherapy treatment planning and response monitoring in solid tumors. Nat Rev Clin Oncol. 2011;8:233–42.
33. Zaidi H, El Naqa I. PET-guided delineation of radiation therapy treatment volumes: a survey of image segmentation techniques. Eur J Nucl Med Mol Imaging. 2010;37:2165–87.
34. Newbold K, Partridge M, Cook G, Sohaib SA, Charles-Edwards E, Rhys-Evans P, et al. Advanced imaging applied to radiotherapy planning in head and neck cancer: a clinical review. Br J Radiol. 2006;79:554–61.
35. Piet D, De Frederik K, Vincent V, Sigrid S, Robert H, Sandra N. Diffusion-weighted magnetic resonance imaging to evaluate major salivary gland function before and after radiotherapy. Int J Radiat Oncol Biol Phys. 2008;71(5):1365–71.
36. Lambin P, Rios-Velazquez E, Leijenaar R, Carvalho S, van Stiphout RG, Granton P, et al. Radiomics: extracting more information from medical images using advanced feature analysis. Eur J Cancer. 2012;48:441–6.
37. Kumar V, Gu Y, Basu S, Berglund A, Eschrich SA, Schabath MB, et al. Radiomics: the process and the challenges. Magn Reson Imaging. 2012;30:1234–48.
38. El Naqa I, Grigsby P, Apte A, Kidd E, Donnelly E, Khullar D, et al. Exploring feature-based approaches in PET images for predicting cancer treatment outcomes. Pattern Recogn. 2009;42:1162–71.
39. Kidd EA, El Naqa I, Siegel BA, Dehdashti F, Grigsby PW. FDG-PET-based prognostic nomograms for locally advanced cervical cancer. Gynecol Oncol. 2012;127:136–40.
40. Vaidya M, Creach KM, Frye J, Dehdashti F, Bradley JD, El Naqa I. Combined PET/CT image characteristics for radiotherapy tumor response in lung cancer. Radiother Oncol. 2012;102:239–45.
41. Group BDW. Biomarkers and surrogate endpoints: preferred definitions and conceptual framework. Clin Pharmacol Ther. 2001;69:89–95.
42. El Naqa I, Craft J, Oh J, Deasy J. Biomarkers for early radiation response for adaptive radiation therapy. In: Li XA, editor. Adaptive radiation therapy. Boca Raton: Taylor & Francis; 2011. p. 53–68.
43. Mayeux R. Biomarkers: potential uses and limitations. NeuroRx. 2004;1:182–8.
44. Nuyten DS, van de Vijver MJ. Using microarray analysis as a prognostic and predictive tool in oncology: focus on breast cancer and normal tissue toxicity. Semin Radiat Oncol. 2008;18:105–14.
45. Ogawa K, Murayama S, Mori M. Predicting the tumor response to radiotherapy using microarray analysis (Review). Oncol Rep. 2007;18:1243–8.
46. Svensson JP, Stalpers LJ, Esveldt-van Lange RE, Franken NA, Haveman J, Klein B, et al. Analysis of gene expression using gene sets discriminates cancer patients with and without late radiation toxicity. PLoS Med. 2006;3, e422.
47. Wouters BG. Proteomics: methodologies and applications in oncology. Semin Radiat Oncol. 2008;18:115–25.
48. Alaiya A, Al-Mohanna M, Linder S. Clinical cancer proteomics: promises and pitfalls. J Proteome Res. 2005;4:1213–22.
49. Tyburski JB, Patterson AD, Krausz KW, Slavik J, Fornace Jr AJ, Gonzalez FJ, et al. Radiation metabolomics. 1. Identification of minimally invasive urine biomarkers for gamma-radiation exposure in mice. Radiat Res. 2008;170:1–14.
50. Spratlin JL, Serkova NJ, Eckhardt SG. Clinical applications of metabolomics in oncology: a review. Clin Cancer Res. 2009;15:431–40.

51. Erichsen HC, Chanock SJ. SNPs in cancer research and treatment. Br J Cancer. 2004;90:747–51.
52. West CML, Elliott RM, Burnet NG. The genomics revolution and radiotherapy. Clin Oncol. 2007;19:470–80.
53. Andreassen CN, Alsner J. Genetic variants and normal tissue toxicity after radiotherapy: a systematic review. Radiother Oncol. 2009;92:299–309.
54. Alsner J, Andreassen CN, Overgaard J. Genetic markers for prediction of normal tissue toxicity after radiotherapy. Semin Radiat Oncol. 2008;18:126–35.
55. Baumann M, Hölscher T, Begg AC. Towards genetic prediction of radiation responses: ESTRO's GENEPI project. Radiother Oncol. 2003;69:121–5.
56. Burnet NG, Elliott RM, Dunning A, West CML. Radiosensitivity, radiogenomics and RAPPER. Clin Oncol. 2006;18:525–8.
57. Iwakawa M, Noda S, Yamada S, Yamamoto N, Miyazawa Y, Yamazaki H, et al. Analysis of non-genetic risk factors for adverse skin reactions to radiotherapy among 284 breast cancer patients. Breast Cancer. 2006;13:300–7.
58. Ho AY, Atencio DP, Peters S, Stock RG, Formenti SC, Cesaretti JA, et al. Genetic predictors of adverse radiotherapy effects: the Gene-PARE project. Int J Radiat Oncol Biol Phys. 2006;65:646–55.
59. West C, Rosenstein BS. Establishment of a Radiogenomics Consortium. Int J Radiat Oncol Biol Phys. 2010;76:1295–6.
60. Kerns SL, Ostrer H, Stock R, Li W, Moore J, Pearlman A, et al. Genome-wide association study to identify single nucleotide polymorphisms (SNPs) associated with the development of erectile dysfunction in African-American men after radiotherapy for prostate cancer. Int J Radiat Oncol Biol Phys. 2010;78:1292–300.
61. Kerns SL, Stock R, Stone N, Buckstein M, Shao Y, Campbell C, et al. A 2-stage genome-wide association study to identify single nucleotide polymorphisms associated with development of erectile dysfunction following radiation therapy for prostate cancer. Int J Radiat Oncol Biol Phys. 2013;85(1):e21–8.
62. Rosenstein BS, West CM, Bentzen SM, Alsner J, Andreassen CN, Azria D, et al. Radiogenomics: radiobiology enters the era of big data and team science. Int J Radiat Oncol Biol Phys. 2014;89:709–13.
63. Alon U. An introduction to systems biology: design principles of biological circuits. Boca Raton: Chapman & Hall/CRC; 2007.
64. Feinendegen L, Hahnfeldt P, Schadt EE, Stumpf M, Voit EO. Systems biology and its potential role in radiobiology. Radiat Environ Biophys. 2008;47:5–23.
65. El Naqa I. Machine learning methods for predicting tumor response in lung cancer. Wiley Interdiscip Rev: Data Min Knowl Discov. 2012;2:173–81.
66. Eschrich S, Zhang H, Zhao H, Boulware D, Lee J-H, Bloom G, et al. Systems biology modeling of the radiation sensitivity network: a biomarker discovery platform. Int J Radiat Oncol Biol Phys. 2009;75:497–505.
67. Oh JH, El Naqa I. Bayesian network learning for detecting reliable interactions of dose-volume related parameters in radiation pneumonitis. In: International conference on machine learning and applications (ICMLA), Miami. 2009.
68. Oh JH, Craft J, Al Lozi R, Vaidya M, Meng Y, Deasy JO, et al. A Bayesian network approach for modeling local failure in lung cancer. Phys Med Biol. 2011;56:1635–51.
69. Sanchez-Nieto B, Nahum AE. Bioplan: software for the biological evaluation of radiotherapy treatment plans. Med Dosim. 2000;25:71–6.
70. Sullivan R, Peppercorn J, Sikora K, Zalcberg J, Meropol NJ, Amir E, et al. Delivering affordable cancer care in high-income countries. Lancet Oncol. 2011;12:933–80.
71. Deasy JO, Bentzen Sr M, Jackson A, Ten Haken RK, Yorke ED, Constine LS, et al. Improving normal tissue complication probability models: the need to adopt a ‚ÄúData-Pooling‚Äù Culture. Int J Radiat Oncol Biol Phys. 2010;76:S151–4.
72. Lambin P, Roelofs E, Reymen B, Velazquez ER, Buijsen J, Zegers CML, et al. 'Rapid learning health care in oncology' – an approach towards decision support systems enabling customised radiotherapy'. Radiother Oncology. 2013;109:159–64.

73. Roelofs E, Dekker A, Meldolesi E, van Stiphout RGPM, Valentini V, Lambin P. International data-sharing for radiotherapy research: an open-source based infrastructure for multicentric clinical data mining. Radiother Oncol. 2014;110(2):370–4.
74. Yan Q. Biomedical informatics methods in pharmacogenomics. Methods Mol Med. 2005;108:459–86.
75. Viswanathan GA, Seto J, Patil S, Nudelman G, Sealfon SC. Getting started in biological pathway construction and analysis. PLoS Comput Biol. 2008;4, e16.
76. Wang L, Zhang B, Wolfinger RD, Chen X. An integrated approach for the analysis of biological pathways using mixed models. PLoS Genet. 2008;4, e1000115.
77. Subramanian A, Tamayo P, Mootha VK, Mukherjee S, Ebert BL, Gillette MA, et al. Gene set enrichment analysis: a knowledge-based approach for interpreting genome-wide expression profiles. Proc Natl Acad Sci U S A. 2005;102:15545–50.
78. Krishna R, Kelleher K, Stahlberg E. Patient confidentiality in the research use of clinical medical databases. Am J Public Health. 2007;97:654–8.
79. Nature Editorial. Making data dreams come true. Nature. 2004;428:239.

Modelling of Normal Tissue Complication Probabilities (NTCP): Review of Application of Machine Learning in Predicting NTCP

17

Sarah Gulliford

Abstract

Predicting normal tissue toxicity following radiotherapy is a multidimensional challenge. The dose received by healthy tissue surrounding the tumour is described using a 3D dose distribution. In addition, patient- and treatment-related factors must also be considered in any predictive model of toxicity. Mixing these complex and disparate data types is a challenge that can be addressed with machine learning. This chapter introduces the concept of normal tissue complication probability (NTCP) and reviews literature related to the use of machine learning in this field.

17.1 NTCP Modelling

The response of normal tissue incidentally and unavoidably irradiated during radiotherapy is the main factor limiting the increase in prescription dose to the tumour. Optimising this trade-off, known as the therapeutic ratio, is the fundamental challenge in radiotherapy (Fig. 17.1). Although complimentary in approach, the complexity of predicting normal tissue response is a higher dimensional problem than predicting local control. The reasons for this are (1) there are usually more than one organ at risk irradiated and protecting all of these structures requires compromise, (2) each structure responds differently to radiotherapy due to the type of cells and the structural and functional organisation of the tissue, and (3) the dose distributions to the surrounding normal tissues are inhomogeneous with gradients

---

S. Gulliford
Joint Department of Physics, Institute of Cancer Research and Royal Marsden National Health Service Foundation Trust, Sutton, UK
e-mail: Sarah.Gulliford@icr.ac.uk

© Springer International Publishing Switzerland 2015
I. El Naqa et al. (eds.), *Machine Learning in Radiation Oncology: Theory and Applications*, DOI 10.1007/978-3-319-18305-3_17

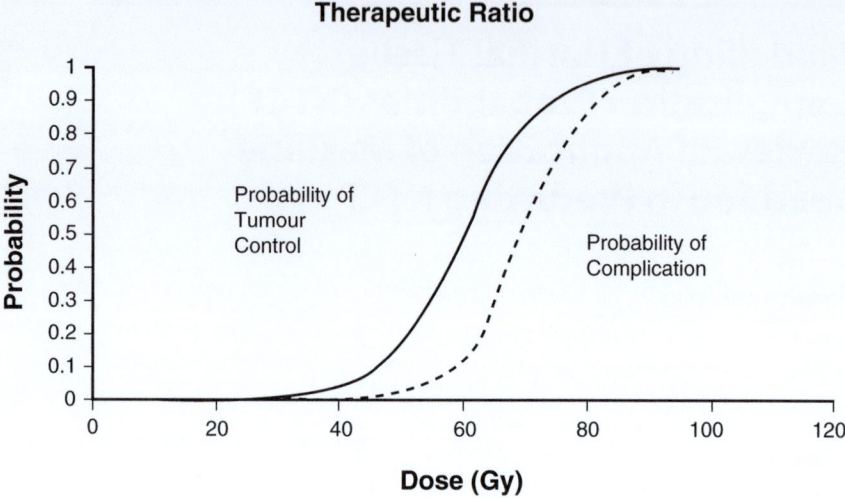

Fig. 17.1 As the dose delivered to a tumour increases, so does the probability of tumour control (TCP). However, the resultant increase in dose to surrounding healthy tissues increases the normal tissue complication probability (NTCP). Balancing TCP against NTCP is known as the therapeutic ratio

Fig. 17.2 An axial slice from a radiotherapy treatment plan of a patient treated for head and neck cancer. The Primary PTV and nodal volume are contoured along with the spinal cord (*red*). The colour wash indicates the dose distribution

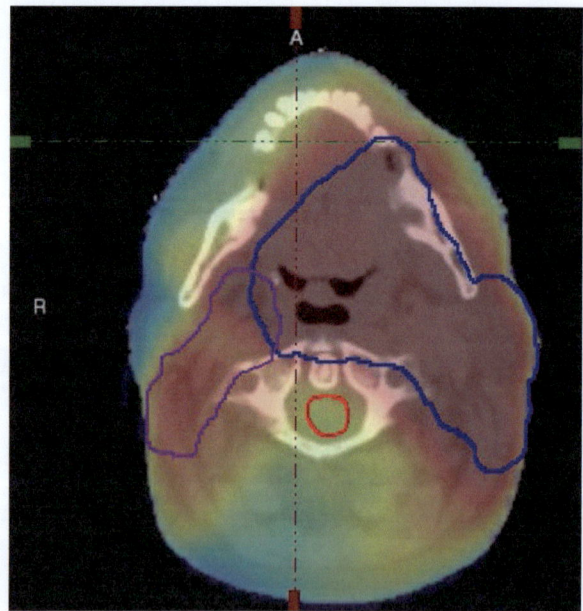

across the tissues commonly related to the proximity of the tumour (Fig. 17.2). This variability results in a large number of potential dose distributions to the structure. Consequently, the dose-volume relationship to toxicity is complex and not well understood.

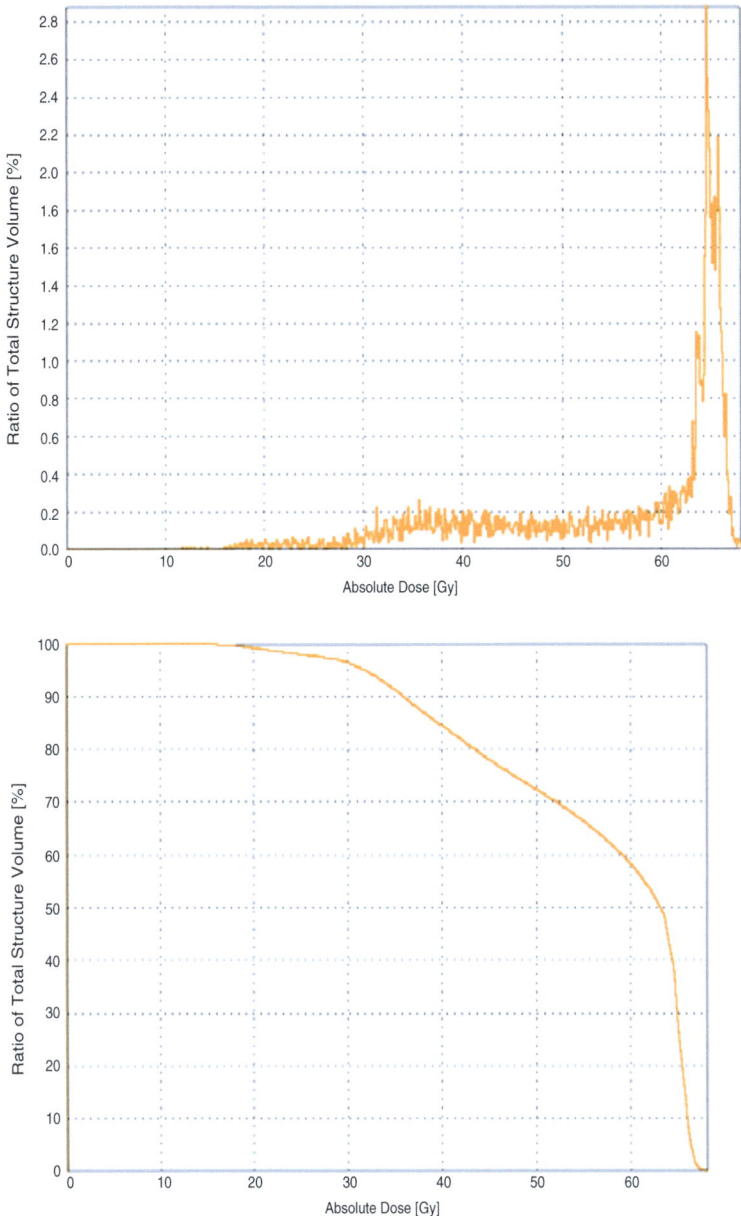

Fig. 17.3 Examples of differential and cumulative dose-volume histograms (DVH) for a normal tissue structure close to the tumour

Typically, the 3D dose distribution to each delineated structure is characterised using a dose-volume histogram (DVH). A differential dose-volume histogram reports the volume (absolute or relative) of a structure which receives a specific dose (Fig. 17.3 top).

Modern treatment planning systems usually calculate histograms with a bin width of ≤0.1 Gy. More commonly, histograms are displayed as cumulative dose-volume histograms where, for each dose level, the volume of the organ or structure receiving at least that dose is reported (Fig. 17.3, bottom). These values are commonly reported as Vx where x is the relevant dose, e.g. V60 is the volume of a structure receiving at least 60 Gy.

Describing the dose distributions in order to model the response of the structure has been explored widely. The QUANTEC report published as a supplement in International Journal of Radiation Oncology, Biology and Physics [38] provided a comprehensive report summarising the published data on the dose-volume response for 16 organs at risk whilst considering the limitations of the data and providing recommendations on how to improve future data collection and analysis. Commonly, the dose measure is quantified as a metric such as maximum or mean dose or volume of the structure receiving a specified dose (V(x)). Once developed and validated, these metrics can be used prospectively as constraints during the treatment planning process. Each treatment plan is assessed prior to treatment in order to ensure safety and to evaluate the likely therapeutic success and risk of complication. In order to assess this risk, the concept of normal tissue complication probability (NTCP) has been developed. It is the probability that a given dose distribution to a defined tissue or structure will result in a quantifiable (unfavourable) response in the patient. The dose-response of tumours to radiation is characterised using a sigmoidal response, and this shape of response is translated as the basis for NTCP models. However, whereas in the case of a tumour where the dose is (ideally) homogeneous, in the case of a normal tissue, the dose distribution is ideally inhomogeneous with as much tissue as possible being spared. The result of this is the challenge of which metric to plot on the abscissa.

17.1.1 NTCP Models

A range of NTCP models have been developed; the most widely known and perhaps the most regularly used is the Lyman-Kutcher-Burman (LKB) model. This model comprises an empirical model of dose-response as a function of irradiated volume [35], the reduction of a dose-volume histogram to a single metric [32] and parameter fits for individual organs at risk [5] based on the tolerance doses summarising clinical knowledge by Emami et al. [18]. Originally, the Lyman model was developed for particle therapy where dose distributions fall off steeply and essentially result in uniform dose D to a percentage of the organ with little dose to the remainder. The tolerance dose parameter $TD_{50}(1)$ or TD5(1) is the 50 or 5 % probability of experiencing toxicity where the whole structure is irradiated. The power law is employed to account for fractional irradiation.

$$\mathrm{NTCP} = \frac{1}{\sqrt{2\Pi}} \int_{-\infty}^{t} e^{-t^2/2} \, dt \qquad (17.1)$$

$$t = \frac{D - TD_{50}(V)}{m * TD_{50}(V)} \quad (17.2)$$

where

$$TD_{50}(v) = TD_{50}(1)/V^n \quad (17.3)$$

$TD_{50}(V)$ is the tolerance dose for a partial volume V. The parameter m is the standard deviation of $TD_{50}(1)$ and n indicates the volume effect of the organ being assessed. $n=0$ indicates a completely 'serial' structure, where the maximum dose dominates outcome and $n=1$ is a 'parallel' structure where the mean dose is related to the outcome.

17.1.2 Dosimetric Data Reduction-Summary Measure

In reality, the dose distribution to an organ at risk is likely to be inhomogeneous. In this case, a reduction is required to translate the inhomogeneous dose distribution to a single metric that results in the same radiation response as a corresponding homogeneous dose distribution. The most commonly used metric is the generalised equivalent uniform dose [47]. Originally developed as the equivalent uniform dose to tumours [46], the concept was extended to include normal tissues. The formula is usually written as

$$\text{gEUD} = \left(\sum V_i D_i^a\right)^{\frac{1}{a}} \quad (17.4)$$

where D_i is the dose in the ith bin of the DVH and V_i is the volume of tissue receiving dose D_i and a is the volume parameter and is equivalent to $1/n$.

Alternative models which consider the functional architecture of the organ/structure have also been employed. The functional subunit (FSU) [63] is a concept which describes either an anatomically defined substructure such as the nephron of a kidney or the largest group of cells which continue to function provided one clonogen survives. In an analogy to electrical circuits, FSU are arranged in either series, parallel or a combination of both (Fig. 17.4). If the architecture of a structure is serial, then lethal damage to just one functional subunit can impair function. An example of this is the spinal cord where damage to a short section of the spine can lead to serious side effects. Consequently, constraining the maximum dose delivered to any part of the structure is used to protect a serial structure from damage. In contrast, organs arranged in parallel have a reserve, whereby a number of functional subunits may be damaged before there is any loss of function. This is true of the liver. In this case, it is the mean dose to the structure that is generally considered. In many cases, the true architecture of an organ is mixed, and the manifestation of the side effects differs.

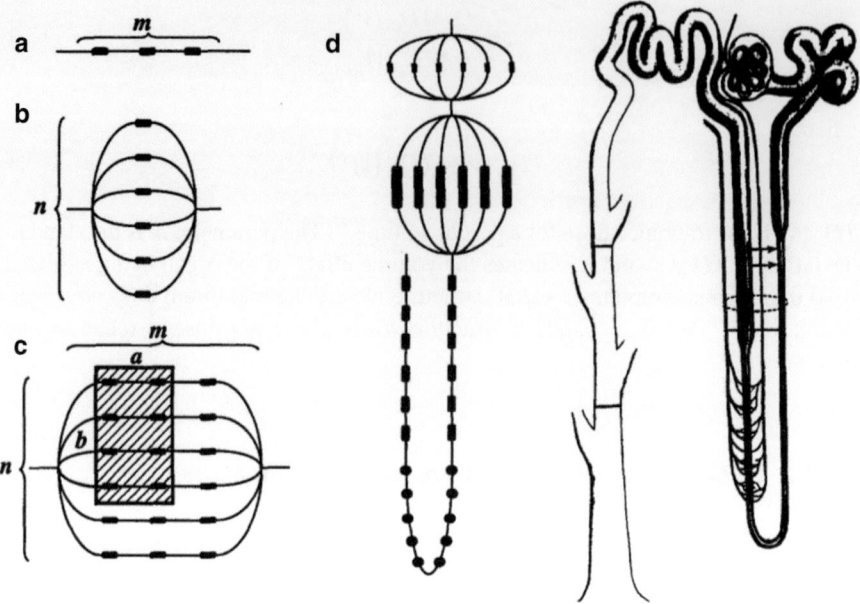

Fig. 17.4 Description of series and parallel functional subunits [30]

The relative seriality model [28] proposed by Kallman considers the dose distribution on a voxel-by-voxel basis and calculates the probability of local damage for each voxel in the treatment plan incorporating the dose-response curve with a tolerance dose D and slope γ before combining these probabilities and weighting according to the parameter S which defines the 'relative seriality' of the organ with a value of 1 indicating a highly serial structure, whilst parallel structures have an s value close to 0.

Niemierko et al. [48] proposed a critical volume model based on FSU. A parallel architecture model proposed by Jackson et al. [25] considers the phenomenological response of functional subunits describing the probability defined in terms of the tolerance dose and slope of the dose-response. Whilst each of these models attempts to model the dose-response relationship for an individual structure, the LKB model is still dominant in the clinic.

17.1.3 Quantification of Toxicity Data

Each organ or normal tissue structure exhibits an individual profile of one or more radiation-induced responses. For example, the rectum is incidentally irradiated (as a normal tissue) in the course of treating a number of pelvic malignancies, including the prostate and endometrium. Rectal toxicity may manifest as loose stools, rectal urgency, pain and frequency in addition to the well-studied endpoint of rectal bleeding [42]. It is thought that the underlying pathophysiology for each of these symptoms may be different. In order to understand the relationship between dose (and other contributing factors) and toxicity, the quality of the toxicity data is vitally important. A number of validated reporting schemes exist. Many of these include questions for specific normal tissues and specific endpoints. However, the fact that there is more than one scoring scheme available suggests that none are perfect and inconsistencies will occur when comparing data and models based on different schemes. In addition, it is important to ensure that the length of follow-up of a patient cohort is sufficiently timed to include all likely events. A cross-sectional analysis at 3 years is likely to yield different results to a cumulative analysis up to 3 years. All of these factors must be taken into account when building models as there is potential for 'garbage in, garbage out'.

17.1.4 Parameter Fitting

Conventionally, models are obtained by fitting a sigmoidal-shaped curve to a measure of dose to predict toxicity. This is achieved using data from retrospective cohorts of patients. Commonly, multivariate logistic regression [24] is performed where the model to predict probability of toxicity is comprised of coefficients describing the contribution of individual explanatory variables to the final model [15]. Maximum likelihood estimation (MLE) [27] is employed to establish the coefficients using optimisation algorithms such as conjugate gradient descent. The outcome predicted by the model is compared to the known outcome, and the error is minimised to find the optimal parameter fit. Logistic regression assumes that the variables in the model are independent and uncorrelated. Since DVH data is neither of these, careful consideration is required on the use of logistic regression. As a result, dosimetric information can be reduced to a summary metric such as mean dose resulting in a compromise of the data included in the model. Statistical techniques of cross validation and bootstrapping are employed to ensure generalisability of the models.

17.1.5 Challenges of NTCP Modelling

Despite many studies on large, high-quality datasets, predicting NTCP remains a challenge. Figure 17.5 presents results from the UK MRC-RT01 study [21]. A retrospective analysis of implementing dose-volume constraints to dose distributions for the rectum following prostate radiotherapy demonstrated that the more constraints a patient failed, the more likely they were to experience toxicity. However, 1/3 of patients who met all the constraints still reported moderate or severe rectal toxicity. There are many potential reasons for this.

1. In addition to the dosimetric response of normal tissues, many other factors contribute to the incidence of toxicity, including patient characteristics, such as comorbidities or previous treatments which may modify the dose-response and

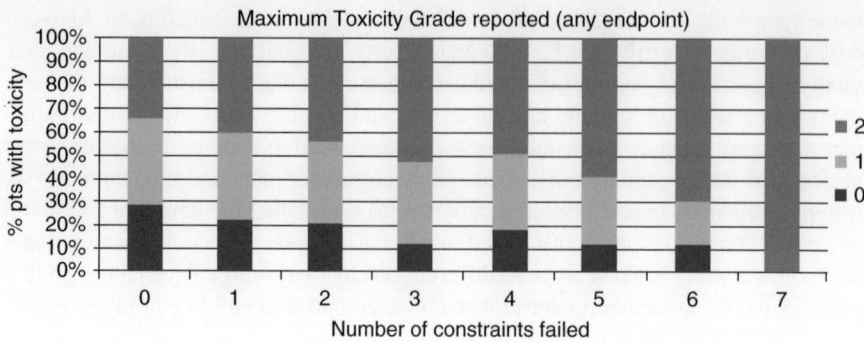

Fig. 17.5 Maximum grade of combined late rectal toxicity, none (0), mild (1) and moderate/severe (2), compared to the number of dose-volume constraints (applied retrospectively) failed [21]

other treatments including chemotherapy which have the potential to cause side effects but may also affect the dose-response of an organ [29].
2. Preliminary data is emerging to indicate that the response of normal tissues is partly determined by genetic susceptibilities. Genome-wide association studies (GWAS) have so far shown inconsistent results when associations between toxicity and single nucleotide polymorphism (SNPs) have been investigated [1].
3. Currently, the 3D dose distribution to an organ is summarised and/or reduced to provide dosimetric information. However, this often results in the loss of spatial information. It is known that many organs contain substructure which is inherent to organ function. A classic example is the kidney [13] where dose to the nephrons is known to be important.
4. Dosimetric data for an organ at risk relies on the contouring of the structure on the treatment planning system. Institutional protocols should be in place to ensure consistency of outlining. However, definitions may vary between institutions, and this is particularly important when applying a model to data from another institution [20].
5. What you see is not what you get (WYSINWYG). In addition to contouring consistency, most NTCP studies use the treatment planning scan to define the organ at risk. Great care is taken at each fraction of radiotherapy to ensure that the treatment plan is reproduced and that the target is irradiated accordingly. However, variation in normal tissues is not necessarily accounted for, so unless an accumulated dose, based on daily imaging, is constructed, there may well be a difference between the dosimetric data reported from the treatment plan and the actual dose to the normal tissue being modelled [26].

Awareness of these challenges and, where possible, incorporating them into the NTCP model will improve the robustness and the generalisability of the resultant models.

17.2 Why Should We Consider Machine Learning Approaches to Dose-Volume Effects?

Machine learning brings a new toolbox to the challenges of predicting NTCP. The concept of allowing a non-linear model to develop without an 'a priori' definition of the relationship between input variables and outcomes removes bias from our limited understanding of the response of normal tissues to radiation and enables us to uncover new information. Many of the considerations for predicting NTCP using machine leaning are common to the different 'flavours' of machine learning. As discussed, the data available includes dosimetric data, patient characteristics, previous health history, other current health conditions (comorbidities), systemic therapy (chemotherapy) and surgery. Little is known about the interaction between these different types of information, and therefore, the flexibility of being able to include variables without understanding higher-order interaction terms is a genuine advantage of machine learning. Many of the publications to date that predict NTCP from dosimetric variables present the data in the form of volume receiving (x) Gy or a reduction of the dose-volume histogram to EUD. The bins of the histograms for an individual patient are known to be highly correlated. Depending on the uniformity of the radiotherapy protocol for the cohort under observation, there is usually an inter-patient correlation to consider. Machine learning approaches are generally well placed to cope with such interactions.

17.2.1 Feature Selection

Feature/variable selection can be regarded as either a preprocessing step or an integral part of model fitting. Where the existence or strength of correlation between individual features and toxicity is unknown, a wide range of possibilities will need to be included in the original input data. It is important to also consider interactions between variables that may contribute to the predictive power of the model.

Advantages of preprocessing feature selection include reduction of model complexity, decrease in computational burden and improved generalisability of unseen data [16].

A wide range of methods for variable selection are available, and a useful summary on this is found in [49]. Within the literature for predicting NTCP using machine learning, undoubtedly one of the most popular is principal component analysis (PCA). Principal components are uncorrelated linear combinations of variables in a given dataset, which account for the variance in the input features in a dataset without reference to the corresponding outcome data, i.e. unsupervised learning. Ideally, data with the same outcome class naturally cluster together, and the clusters are separable from each other. PCA is a particularly attractive feature for DVH-based analysis where variables are known to be highly correlated and has been coupled with conventional statistical models such as logistic regression as well as machine learning methodologies.

A large proportion of the variance in a dataset is often described by the first few principal components. PCA enables reduction to a lower dimension allowing visualisation which can inform researchers on the complexity of the input-output relationship of the data and consequently on the appropriate choice of model. The reduction of dimensionality results in the ability to visualise high-order data. One of the earliest studies using PCA to predict NTCP was published by Dawson et al. [12] who considered PCA for two different organs at risk. PCA was chosen in order to consider all the bins of a DVH without having to reduce to a single metric, such as mean dose, or summary metric such as EUD. The first cohort included 56 head and neck patients where data from the parotid glands was used to predict xerostomia (dryness of the mouth) 12 months after radiotherapy. The dosimetric data was characterised as a cumulative DVH with 1 Gy bins (84 bins in total). The first two principal components explained 94 % of the variance in the DVH. When these were plotted against each other (Fig. 17.6) and labelled according to outcome class, there was a clear separation between the classes indicating that outcome classes were potentially linearly separable. The 1st principal component was shown to correspond to a larger percentage of parotid volume treated to 10–60 Gy. This was approximated as the mean dose, which is commonly used as the constraint to the parotid gland [14]. Logistic regression was applied to the first three principal components in addition to patient sex, age and diagnosis. Only the first principal component was significantly associated with toxicity.

In contrast to these clear-cut results, the other cohort studied was 203 patients who received radiotherapy to either partial or whole liver. Initial PCA analysis of the DVH (again 1 Gy bins of the cumulative DVH) showed separated clusters for patients where the whole liver was irradiated vs. those who received partial liver radiotherapy. Subsequent PCA excluded patients who received >20 Gy to >90 % of the liver volume, reducing the number of patients to 138. The first two principal components were plotted along with the Lyman NTCP model however no separation between clusters was observed. Despite this result the results of logistic regression including the first three principal components and relevant clinical factors demonstrated that only the first principal component was significantly associated with toxicity.

Following on from the work by Dawson, Bauer et al. [2] explored the use of PCA to quantify rectal bleeding in a cohort of prostate cancer patients treated with radiotherapy. As with the previous study, the intention was to reduce the degrees of freedom in the rectal dose-volume histograms to characterise those with or without toxicity. The paper gives a very helpful explanation of the background to PCA.

However, unlike other studies on this subject, the authors state that direct implementation of PCA forfeits ease of interpretation as the individual principal components do not represent unique dose-volume combinations that are associated with outcome, although they acknowledge that some insight into relevant features of the DVH may be ascertained. Consequently, the authors propose the use of a varimax rotation, an orthogonal rotation applied to the subset of principal components that account for most of the variance in the dataset. The varimax rotation maximises sparseness of the subset, and only small regions of each mode (component) remain

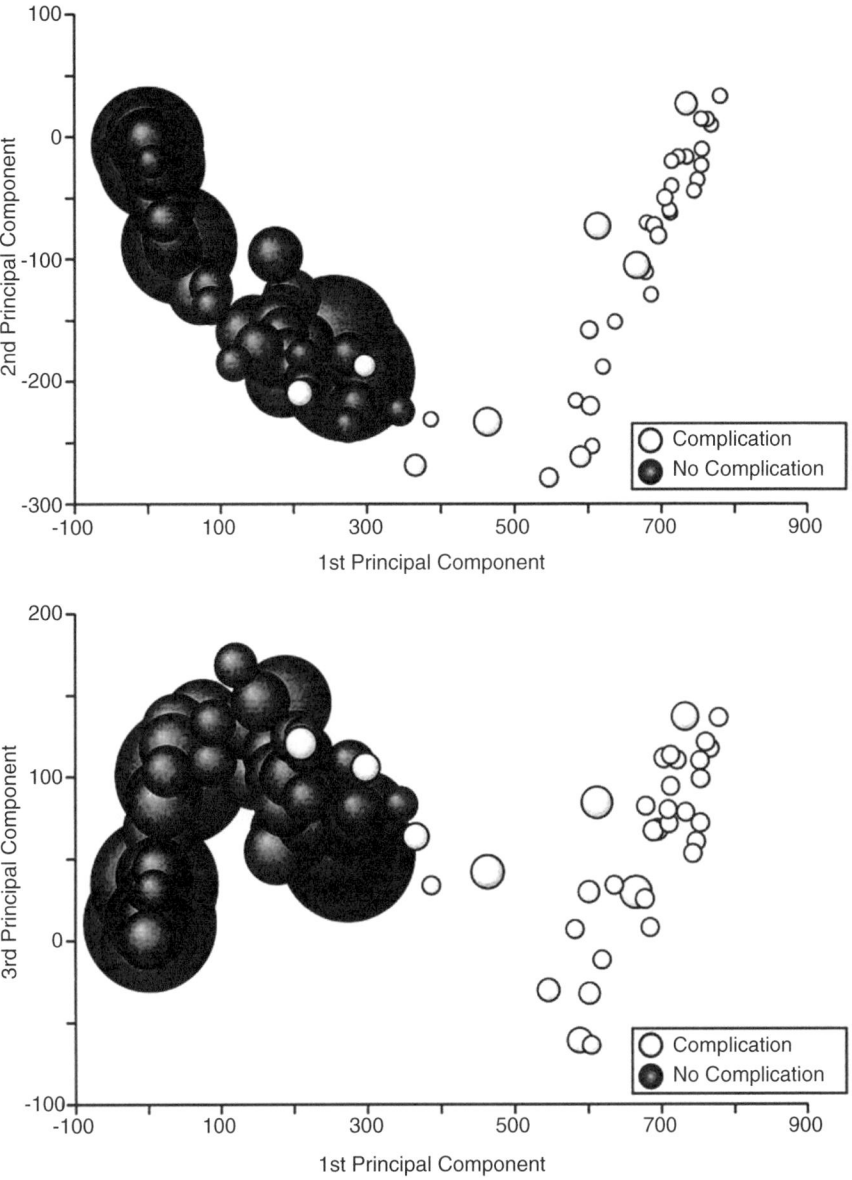

Fig. 17.6 Demonstrating linear separability of data describing xerostomia based on parotid gland dose distributions (Taken from Dawson et al. [12])

large allowing identification of specific regions of the DVH. However, the process reintroduces correlation which must be accounted for. A subsequent study by Sohn et al. [56] applied PCA to a cohort of 262 prostate cancer patients who were treated with a different treatment planning technique. Here, the conventional four-field 'box' beam arrangement was used. However, an adaptive approach based on imaging over the first week of treatment was employed. Fifty patients reported late rectal bleeding CTCAE v. $3 \geq G2$. As with the previous study, the bins of the cumulative DVH provided the input features; however, in this case, the bin width was 0.1 Gy resulting in 850 variables. 93.5 % of the variation was accounted for by the first two principal components. This increased to 96.1 % when the 3rd principal component was also included. The 1st principal component was correlated with much of the DVH, whilst the 2nd principal component was considered to be related to the volume of the rectum in the high-dose region where all of the treatment beams overlapped. The 3rd principal component was correlated with 2 distinct regions 40–45 and 70 Gy. Again, this was attributed to the treatment technique. Although the first three principal components accounted for most of the variation and were interpretable, when plotted no obvious clusters were observed. Univariate logistic regression analysis indicated that only the 2nd principal component was significantly associated with rectal bleeding. Multivariate models including the first two and the first three principal components were both shown to be statistically significant. The first principal component was shown to correlate both with mean dose and independently with V60, whilst the 3rd principal component correlated with the maximum dose.

The use of PCA to predict both rectal and bladder toxicity following prostate radiotherapy was reported by Skala et al. [55]. In this study, responses from 437 patients to a postal questionnaire (using RTOG grading) sent out following radiotherapy were analysed. The DVH data were characterised in 1 Gy bins and were analysed using both absolute (volume in cc) and relative (% of volume) descriptors. PCA results were tested for correlation with toxicity $\geq G2$ using the Mann-Whitney test, but none of the principal components was statistically significant. Standard descriptors of dose Dmax, V50, V60 and V70 were also tested, and again none were found to be statistically significant. The incidence of rectal toxicity $\geq G2$ reported in the study was very low (~3 %), and therefore, the lack of statistical significance is unsurprising. Bladder toxicity was slightly higher (~10 %); however, historically, correlating dosimetry with toxicity of the bladder has been much more challenging with variable results [62]. It is important to emphasise that the lack of correlations is most likely related to the data itself and that the use of a more sophisticated technique will not necessarily improve the results.

Another study by Vesprini et al. [61] describes using the same methodology as Skala on a cohort of 102 prostate cancer patients who received hypo-fractionated radiotherapy (3 Gy per fraction) to predict the incidence of both acute and late bladder and rectal toxicity. Association between dosimetric descriptors, both conventional and principal components, and toxicity was assessed using Pearson's correlation coefficient. None of the dosimetric predictors for the rectum were correlated with acute rectal toxicity. However, the bladder V40, V50 and the 3rd principal component were correlated to acute genitourinary (GI) toxicity. In contrast, all of the conventional descriptors and the 1st principal component were statistically significant for late rectal toxicity, and none of the bladder variables were related to late genitourinary (GU) toxicity. The interpretation of principal component 1 was not presented, but the results were shown to overlap with those provided by the conventional dosimetric variables. It was suggested that principal component results did not necessarily add extra information on the relationship between the rectal DVH and rectal toxicity.

A more recent publication on the use of PCA in radiotherapy incorporates spatial information into the relationship between dosimetry and toxicity. Liang et al. [33] used PCA to identify patterns of irradiation of the bone marrow in the pelvic region which were likely to increase acute haematologic toxicity. White blood cell count nadir was used as an indicator for acute haematological toxicity in a cohort of 37 patients treated with chemo-radiotherapy for cervical cancer. The dose distribution for each patient was standardised by mapping each treatment planning CT, via deformable registration, onto a pelvic bone template. The corresponding dose distributions were interpolated and mapped onto the template. The dose to each voxel in the standard image was calculated and considered as a predictor variable. The template ensured the same number of voxels for each patient, and these voxels were sampled systematically, left-right, anterior-posterior and superior-inferior, to form a row vector for each patient containing 44,146 elements. For each patient, the same element referred to the same voxel. Clearly, this dataset would benefit from dimensionality reduction. As with some of the previous studies, since all of the variables were measured using the same scale (Gy), PCA was performed with the covariance matrix. Of the 36 non-zero eigenvalues with corresponding eigenvectors, 5 were statistically correlated with acute haematologic toxicity using univariate logistic regression. Although the first PC accounted for over 20 % of the variation, the principal components shown to be correlated to toxicity were the 12th, 23rd, 24th, 25th and 31st principal components, and combined together, they accounted for just 4.2 % of the variation in the dataset. The results of the regression were used to test if the resultant dose space was related to toxicity. Acute haematological toxicity was defined by dichotomising the white blood cell nadir as $<2,000/\mu ml$ for no toxicity ($n=23$) vs. $\geq 2,000/\mu ml$ for toxicity ($n=14$). Difference maps of the dose distribution were projected onto the pelvic bone template for those with/without the defined toxicity and compared with the voxels which were shown to be statistically significant in the regression model. There was good agreement between the two assessments (Fig. 17.7). This mapping approach allowed the visualisation of important anatomical regions of active bone marrow which could be avoided using intensity modulated radiotherapy (IMRT).

17.2.2 General Considerations

The use of machine learning is often favoured where the underlying relationship between the data is unknown and there is a need for future prospective evaluation of data. This is exactly the case for normal tissue complication probability. Generally, the dose-response of organs at risk is not well quantified, particularly for specific endpoints. This needs to be improved in order to optimise the use of available technology and to further increase the rate of successful cancer treatments. In the

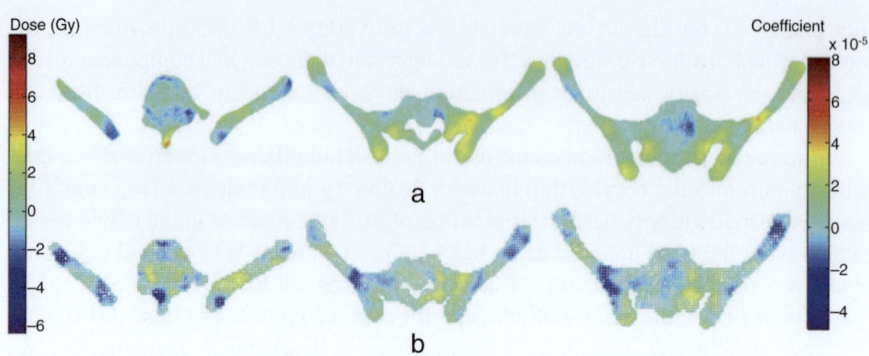

Fig. 17.7 The top row indicates areas of pelvic bone marrow correlated to acute haematologic toxicity dichotomised as white blood cell nadir <or > /2,000 μml. The bottom row represents the regression coefficients produced after PCA (Taken from Liang et al. [33])

meantime, we prospectively evaluate every treatment plan going through the clinic, and the development of knowledge-based tools to facilitate this process is highly desirable. Therefore, the ability of a trained model to generalise unseen data is imperative. Techniques to ensure this include cross validation and bootstrapping which reduce the dependency of a final model on a specific training dataset. The use of an independent (relevant) test set, to measure model performance, once the model has been finalised, should also be regarded as standard practice. It is important to appreciate the extent to which the model can generalise. If a model is trained on data from a centre, then a well-built model should be able to reflect the toxicity experience of that centre. However, it may not be able to predict toxicity for a similar cohort of patients from a neighbouring centre where subtle changes in treatment technique, toxicity reporting or patient demographic may render the model irrelevant.

Since the intention of radiotherapy is to keep the incidence of toxicity to a minimum, the balance of toxicity/no toxicity in the dataset may be very unbalanced with only a small number of patients reporting toxicity. Whilst this is generally good news for the patient, it is a challenge to model building. A number of approaches exist to try to account for this. Firstly, the ratio of toxicity/nontoxicity cases should be standardised across training groups, for example, stratified cross validation, and in the independent test set. It is also possible to promote the number of cases within the dataset for the underrepresented class [30].

17.2.3 Assessing Model Accuracy

The performance of NTCP models is often quantified using the receiver-operator curve (ROC) analysis which quantifies the ability of a continuous variable to predict for a dichotomised outcome by considering every possible cut-point in the continuous variable and calculating the resultant sensitivity and specificity [57]. Sensitivity (true positive rate (TPR)) and specificity (true negative rate (TNR)) are calculated from the confusion matrix (contingency table) of predicted vs. known outcome classes for a given dataset and cut-point. The resultant plot of sensitivity against 1-specifity for all possible cut-points is known as the ROC curve. The area under the curve (AUC) indicates the probability that the model would rank a randomly selected positive case higher than a randomly selected negative case. Alternatively, Matthews correlation coefficient [37], also calculated from the confusion matrix of a binary classification problem, is an alternative approach to quantifying the predictive power of the model. It is regarded as being particularly useful in situations where the classes are of different sizes.

It is defined as

$$\text{MCC} = \frac{(\text{TP} \times \text{TN} - \text{FN} \times \text{FP})}{\sqrt{((\text{TN}+\text{FN})(\text{TP}+\text{FP})(\text{TN}+\text{FP})(\text{TP}+\text{FN}))}} \qquad (17.5)$$

where TP is the number of true positives, TN true negatives, FN false negatives and FP false positives.

An MCC value of 1 indicates a perfect classification, 0 a random classification and −1 a wholly inverted classification.

Once the model has been finalised, it is useful to evaluate the importance of each input feature in making the prediction. Some model types, for example, decision trees, lend themselves to interpretation, whilst others such as artificial neural networks are regarded as impenetrable black boxes. Even in this case, it is possible to investigate the role of each input by using techniques such as leave one out (LOO) where data for each input feature is removed and the effect of the predictive power of the model reassessed.

17.3 Classic Machine Learning Approaches

There are many flavours of machine learning; however, most of the literature related to predicting NTCP is from the more established techniques. These can be broadly separated into supervised and unsupervised learning approaches including. Conventionally, a model relates a number of variables to an outcome or classification; this is supervised learning. In contrast, unsupervised learning finds patterns and groupings among the input variables only; these groupings should then naturally reflect the classification of the data. The following sections consider the use of supervised learning approaches, artificial neural networks and support vector machines, and unsupervised learning techniques for prediction of NTCP.

17.3.1 Artificial Neural Networks

Artificial neural networks (ANNs) are one of the classic machine learning approaches dating back to the seminal work of McCulloch and Pitts [39]. With the analogy of the way the human brain works, it is tempting to think that the knowledge of an experienced clinician or medical physicist can be easily transferred. It has been a popular choice for applications relating to predicting the response of normal tissues to radiotherapy. One of the earliest papers was published by Munley et al. [44] who trained a feedforward, back-propagation, neural network to predict symptomatic lung injury following radiotherapy. Ninety-seven patients were included in the neural network of which 25 had a clinician assessed symptomatic lung injury. Patients with a number of tumour sites were included. Although 2/3 of the patients were treated for lung tumours, the inclusion of other tumour sites increased the diversity of the dose distributions and confounding factors in the training cohort. The neural network had 29 inputs corresponding to pretreatment features which described a range of variable types including patient characteristics (age, race, sex, smoking status); disease characteristics (tumour site and central lung tumour); baseline assessment (heterogeneity of SPECT scan adjacent to and away from the tumour, diffusion capacity of carbon monoxide (DLCO), forced expiration volume in 1 s (FEV1), haemoglobin, chronic obstructive pulmonary disease (COPD)); chemotherapy and dosimetry which included dose-volume histogram reduction using both the Lyman [36] and Kutcher method [32]; volume of lung receiving 10 Gy (V10), V20, V30, V40, V50, V60, V70 and V80; and the full and effective dose to lungs and the lung volume. Each input was scaled 0–1. The architecture included two to five hidden nodes and a single output node each with a sigmoidal activation function. Training was performed using the leave-one-out approach where each patient case was taken out and the neural network retrained. Training was terminated when the ROC analysis was maximised. The final result was an AUC of 0.833 +/−0.04. This result was compared with multivariate logistic regression which resulted in an AUC of 0.813 +/−0.064 and the dose-volume histogram reduction method of Kutcher which yielded an AUC of 0.521 +/−0.08. The influence of each input variable was assessed by retraining the neural network with the leave-one-out approach applied to each variable and ranked by assessing the deterioration in AUC after a fixed number of iterations. The top five variables were found to be heterogeneous SPECT (apart from the tumour), haemoglobin, histogram reduction (Kutcher), COPD and age, the first three of these were also the top three ranked variables using multivariate logistic regression. It is clear from these results that combining dosimetric and clinical information enabled the most accurate prediction of toxicity. The use of a leave-one-case-out approach to train the neural network is likely to result in overfitting, but using a leave-one-input-out approach to investigate the contribution of individual features allowed useful insight into the prediction of toxicity. Following on from the early work by Munley, Su et al. [58] used data from 142 non-small-cell lung cancer patients from the same institution (Duke University Medical Centre) to predict radiation pneumonitis $\geq$ grade 2 also using ANN. Thirty-one of these patients were included in the previous study. This study compared 3 different approaches to segmenting the training and testing data and only considered 8 dosimetric input features describing the volume of lung receiving 10 Gy stepping up in increments of 10 Gy up to 80 Gy. As previously, a leave-one-out approach was employed to train ANN_1 on all but one case and testing on the omitted case. The predictive success was characterised by AUC which was reported to be 0.85. Two further approaches were tested. ANN_2 used 2/3 of the available data for training and 1/3 for testing. The allocation of data was essentially random as patients were ordered in alphabetical order in terms of their last name. Finally, ANN_3 was intended to improve the quality of the training data by ensuring the maximum variation in input parameters for the cases reporting toxicity where once again 2/3 of the cases were used for training. The respective AUC for ANN_2 and ANN_3 were 0.68 and 0.81 demonstrating that careful consideration of the cases provided for training can have a statistically significant improvement in predictive accuracy. The ability to generalise the unseen cases should also be improved compared to the leave-one-out method. A comparison with standard predictive models of V20 and mean lung dose and LKB model (TD5/5 23 Gy, m 0.17 and n 0.86) [5] demonstrated that each of these models yielded an AUC of around 0.5, no better than chance, although the authors acknowledged that a fairer comparison would have been to derive the parameters for their own data using maximum likelihood estimation.

In 2007, Chen et al. [8] reported results for a larger cohort of lung cancer patients from the same institution, Duke University Medical Centre, North Carolina. Radiation-induced pneumonitis ($\geq$ grade 2) was reported in 34 out of 235 patients, all of whom were treated using 3D conformal radiotherapy. ANNs were constructed using an algorithm that successively pruned and grew the input features and hidden nodes, using a training-validation cohort to assess improvement (or otherwise) of each successive iteration. To avoid local minima, weights and bias were trained from five randomised initial sets and the lowest error used overall. Weights were constrained to ensure reasonable responses between input variables and outcome. For example, weights connecting dosimetric variables were constrained to have a positive value only. The authors acknowledged that this approach prohibits a complimentary subtractive effect between variables but suggest that this will safeguard against detrimental overfitting. 93 potential input variables were available. Dosimetric information included V6 to V60 in 2 Gy increments and gEUD varying from 0.4 to 4 in increments of 0.1. The mean dose to the heart was also included. Since many of the dosimetric variables are highly correlated, the training rules ensured that once a variable had been incorporated into the model, no other highly correlated variables (>0.95) were eligible for inclusion in the model. The inclusion of non-dosimetric variables was justified by citing previous analysis of normal tissue response which was shown to be modified by interaction with chemotherapy [40] and age [34]. A wide range of non-dosimetric variables, similar to the previous publications, were included covering patient demographics, treatment information and pre-radiotherapy assessment of lung function. A tenfold cross-validation approach was used to ensure that the results were generalisable, whilst a 2nd approach using all patient data for training was developed for prospective testing. Leave-one-out analysis was used on this 2nd architecture to assess the influence of individual-chosen variables. Comparison of models was performed using ROC analysis. For the ANN trained using cross validation, the optimised architecture containing only dosimetric variables resulted in an ROC of 0.67 for the independent test when non-dosimetric variables were added to the model construction; this improved to 0.76. Each of the ANN developed using cross validation contained different variables; however, the authors highlight that often highly correlated variables were represented in each model. The model trained for prospective testing included 6 variables, V16, gEUD $a=3.5$, gEUD $a=1$, forced expiration volume in 1 s (FEV1), carbon monoxide diffusion capacity of the lung (DLCO%) (both of which were assessed prior to radiotherapy) and induction chemotherapy. All input features except FEV1 and induction chemo were shown to be individually statistically significant. It is clear from these results that different parts of the dose distribution were included in the final model despite dosimetric correlation being constrained. This result suggests that different parts of the dose distribution are important in predicting toxicity. We will consider this again with later publications.

To date, we have considered neural networks where features from the dose distribution have been based on the cumulative dose distribution. The disadvantage of using dose-volume histograms is that all spatial information is discarded. It is known that each organ at risk has an internal structure and function and that this is important for both damage and repair. For example, it has been shown that sparing the superficial gland of both parotids reduces the incidence of xerostomia compared to sparring one parotid completely and irradiating the other. This is thought to be because the majority of parotid stem cells are located in the superficial lobe whilst the deep lobe is predominantly the ductal structure [41]. It is also considered that the nephrons are the most sensitive structure within the kidney [13]. Very little work has been done to incorporate spatial information into prediction of normal tissue toxicity. One example is the paper by Büttner et al. [4] where a dose-surface map of the rectum was used to provide the input features to an ensemble of neural networks which predicted rectal bleeding following prostate radiotherapy. A dose-surface map is generated by unfolding the cylindrical structure of the rectum outlined in the treatment planning system. A number of unfolding methodologies have been suggested. In this study, a slicewise method was chosen whereby the rectal contour outlined on each slice of the treatment planning CT was virtually unfolded by cutting at the most posterior point. The maps were normalised on a slice-by-slice basis to produce maps as shown in Fig. 17.8. Since the dose in the adjacent pixels is correlated, four locally connected neural network architectures were constructed. The first connected a row of 3 neighbouring pixels to each node in the hidden layer with an overlap of 1 pixel. The second connected a 3×3 group of pixels to the 1st hidden layer where a group of 4×4 nodes was connected to the 2nd hidden layer. In the 3rd architecture, a group of 3×3 pixels was connected to the 1st hidden layer. These nodes were connected to the 2nd hidden layer row by row with no overlap. Finally, in the fourth architecture, the input nodes were connected in the same way as the 2nd architecture, i.e. 3×3 group of pixels linked to the hidden nodes. The weights between each group were shared making the presumption that a global dose-response could be modelled. In comparison, a fully connected ANN using the dose-surface histogram values, i.e. the area of the DSM receiving x Gy, was constructed with 35 inputs characterising the dose between 5 and 73 Gy.

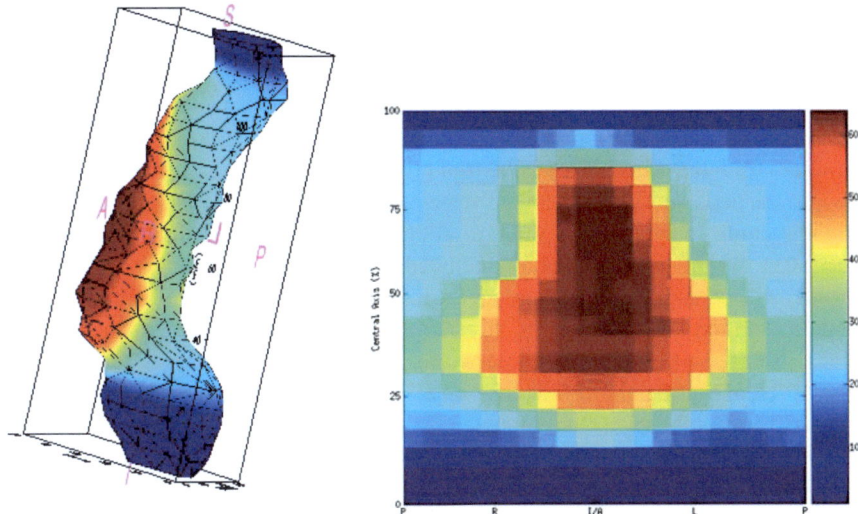

Fig. 17.8 Example dose distribution to the rectum shown as a mesh based on the contours delineated on the treatment planning CT and as a slicewise-unfolded, normalised dose-surface map (DSM)

An ensemble approach [22] was employed to train the ANN-based classifier. An ensemble is a group of independently trained ANN, each of which contributes to the output prediction. Ensembles should be less susceptible to overfitting and 'choosing an unrepresentative' local minima. In this study, an ensemble of 250 ANN was constructed. Each ANN was trained using a different sample of cases from the training data with independent initialisation of the weights in each ANN. Since the incidence of rectal bleeding was relatively low (53/329 patients), 20 % of the patients who did not report rectal bleeding and 75 % of the patients who did report rectal bleeding were randomly chosen for each ANN. Expert ensemble was developed by sequentially adding ANN and evaluated using the area under the ROC curve for predictions on a subset of patients from the training set. If the AUC improved when predictions from the newest ANN were added, then the ANN was added to the ensemble. This process was repeated three times, and ANN that was included in all three ensembles was incorporated into the expert ensemble. This whole process was repeated for each fold of the tenfold cross validation.

Architecture 2 was shown to produce the best predictive results with an AUC of 0.61 for all ANNs and 0.64 for the expert ensemble; this was compared to AUC of 0.59 for the dose-surface histogram-based ANN. In order to assess the influence of the data partition resulting from cross validation, the cross-validation partitioning was repeated 100 times and the most promising locally connected architecture (2) retrained. The mean AUC was 0.65 +/−0.018.

Compared to other studies, the AUC is relatively low. However, the improvement in the AUC when spatial information was incorporated suggests that using spatial information improves the input information and that overall shortcomings may well be due to a lack of non-dosimetric data or the fact that the radiotherapy dose distribution (either DVH or DSM) from the treatment planning scan is not representative of the actual dose distribution received by the patient over the course of the fractionated treatment. The study by Tomastis et al. [59] combined dosimetric and clinical/treatment variables to predict late rectal bleeding for a large cohort of patients ($n=718$) from the AIROPROS 0102 trial [19]. The dosimetric information included dose to the pelvic nodes and seminal vesicles; ICRU dose; mean and maximum dose to the rectum; rectal V50Gy, V60Gy, V70Gy and V75Gy; and finally rectal EUD ($n=0.03$) [53]. Clinical variables included diabetes, hormonal therapy, haemorrhoids, use of anticoagulants/antiaggregants, previous abdominal surgery, pelvic node irradiation and seminal vesicle irradiation. A genetic algorithm was used for feature selection [43]. Five variables were chosen: EUD, previous abdominal surgery, presence of haemorrhoids, use of anticoagulants and androgen deprivation. Fourfold cross validation was employed. In each split, half of the patients were used for training, a quarter to validate training and a quarter as an independent test set. Stratification was employed to ensure that the number of cases who reported toxicity ($n=52$) was balanced in each group. The number of hidden nodes in the architecture varied between 1 and 10, and a leave-one-out approach was used in training. Assessment of the ANN was performed using area under the ROC curve. The leave-one-out training method resulted in an AUC of 0.730 which reduced to 0.704 when tested on the validation cohort. The cross-validation AUC resulted in an AUC of 0.714; this is in comparison with an AUC of 0.636 for a logistic regression model using the same variables and fitted in the same way. The importance of each variable was tested by replacing the variable with the average value in each case. It was found that EUD was the most important variable followed by previous abdominal surgery, haemorrhoids, anticoagulants and finally androgen deprivation. This study demonstrates that adding clinical factors is likely to improve the predictive abilities for rectal bleeding. However, caution is required when one class of outcome is underrepresented as this can skew the AUC results.

17.3.2 Support Vector Machines (SVM)

Support vector machines are a class of machine learning that attempt to find a boundary plane that separates two classification outcomes in feature space. When the cases are linearly separable, this is relatively straightforward; however, more often than not, when considering prediction of normal tissue toxicity, the cases are not linearly separable. In this situation, the variables can be transformed into a higher dimensional feature space where the cases may be separated by a hyperplane. This is achieved using a non-linear kernel such as a polynomial or radial basis function. Each data point represents a vector of the variables included in the model. The dual optimisation of separating the cases whilst improving fitting accuracy results in a balanced trade-off. This is computationally intensive to solve; however, it is possible to characterise the prediction function using only a subset of training data. The cases used to define the boundary between classes are known as support vectors. Unlike other approaches to machine learning, SVM maximises the distance between the two classes rather than minimising the mean square error, and it is permissible for a defined number of cases to be on the 'wrong side' of the boundary. The framework of a SVM implicitly includes higher-order interactions between variables without having to predefine what they are.

In a publication complimentary to their work using neural networks (discussed in the previous section), Chen et al. describe using support vector machines to predict pneumonitis [6] on the same dataset reported for ANN [8]. A radial basis kernel function was chosen for the SVM in preference to a sigmoid or polynomial kernel as the increase in free parameters might result in overfitting. SVM were constructed using only dosimetric variables and separately with all available variables. Parameter values C and σ were predetermined using a grid search. Variable selection was performed using a similar approach to the ANN study whereby variables were added and substituted iteratively employing a tenfold cross validation. Although each of the ten results was independent, there was a large crossover between the input variables selected. For the SVM trained using only dosimetric variables, EUD with $a = 1.1$ (1), 1.3 (8) and 1.4 (1) were chosen along with V48Gy (3) and V50Gy (10). The overall AUC was 0.71. Similarly, for the SVM trained with dosimetric and non-dosimetric variables, EUD $a = 1.2$, (1) 1.3(7) and 1.4(2) were chosen along with induction chemotherapy (chosen in all tenfolds), tumour location (9), gender (8) and two histological variables: adenocarcinoma vs. not (2) and small cell vs. not (1). As with the previous ANN study, highly correlated inputs were not permitted in the same model. Therefore, each fold of the SVM using dosimetric variables chose an EUD close to mean lung dose ($a = 1$) and a higher dose constraint. Induction chemotherapy was also featured in every fold. Tumour location and gender were also strongly represented with the histological variables to a lesser extent. This level of consistency between folds is reassuring for generalisability. The AUC for the SVM including dosimetric and non-dosimetric variables was 0.76. A LOO approach was employed to investigate the importance of individual variables in the SVM_{all} model. The AUC was reduced by 0.19 with the exclusion of EUD and by 0.09 for induction chemotherapy. The importance of these two variables was consistent with the results from the previous ANN study. However, the contribution of other variables demonstrates the risk of overfitting if techniques such as cross validation are not employed.

El Naqa et al. describe the use of non-linear kernel-based approaches for predicting normal tissue toxicities [16] highlighting the challenges of mixed models built from different data types including dosimetric metrics, patient characteristics and disease-/treatment-based prognostic factors. They recommend the use of kernel-based methods, specifically support vector machines, citing the following advantages over other machine learning approaches: ability to adapt to artificial intelligence, ability to avoid excessive overfitting and ability to maintain computational efficiency of classical statistical methods, and in summary, they state that SVM overcome the stigma of a black box due to rigorous mathematical foundations. Preprocessing of the data is achieved using PCA which also allows visualisation of the higher dimensional data. Examples from two clinical datasets were presented. The first was a small cohort of 55 head and neck cancer patients where a model is developed to predict xerostomia which results from a lack of salivary production following radiotherapy. Clinical variables included patient age, gender, ethnicity, treatment, Karnofsky performance, chemotherapy, stage and histology. In addition, a previously developed dosimetric model which predicts salivary function using the dose to the parotid gland with a factor of 0.054/Gy [3] is incorporated. It was observed that the groups of patients with and without xerostomia were reasonably separated, and it was subsequently demonstrated that a linear kernel produced a model which was not bettered by either radial basis function or polynomial Kernel. The authors comment that this is 'not the norm in radiotherapy' as exemplified by the 2nd dataset presented. Data of 219 patients treated with radiotherapy for non-small-cell lung cancer (NSCLC) were used to predict radiation pneumonitis (RTOG grade3). Dosimetric characterisation of the dose to the lung was achieved using volume receiving x Gy (Vx). Vx with increments of 10 Gy from 10 to 80 Gy was included. Using these variables, it was demonstrated that the classes could not be separated using PCA. Using SVM, it was demonstrated that an improvement in model performance was observed with increasing order of polynomial. A separate model was developed which included non-dosimetric variables including patient, disease and treatment variables. In addition, the dosimetric descriptors were expanded to include Dx (the volume of lung receiving a minimum dose x). In total, 58 variables were included. The top 30 variables were selected using recursive feature elimination SVM. Variable pruning was used to account for multicolinearity of correlated variables. The model resulted in an MCC of 0.22 and contained 6 variables. A further SVM was developed using three variables discerned from a previous study using model order selection with resampling logistic regression. The resultant SVM with a radial basis function kernel had an MCC of 0.34. The improvement in this value is attributed to the ability of SVM to account for interactions between model variables.

In a subsequent, more comprehensive publication, El Naqa et al.[17] expand on the data presented. Often in radiotherapy, the incidence of complications can be quite low. Conventionally, an SVM cost function treats the two potential classes equally; however, to account for the imbalance between classes, different weights can be assigned to the samples in the two different classes with a higher penalty weight assigned to the underrepresented class.

As such, the penalty term is expanded to

$$C\sum_{i=1}^{i=n} \xi_i = C^+ \sum_{i=z^+} \xi_i + C^- \sum_{i=z^-} \xi_i \qquad (17.6)$$

In addition to the datasets studied in the previous publication, data predicting acute oesophagitis in a cohort of 166 NSCLC patients was also presented. Finally, data from a multi-institutional RTOG study (9311) was used as an independent validation set to predict radiation pneumonitis. As previously reported, the best model to predict xerostomia was a linear classifier which yielded an MCC value of 0.64. The model to predict oesophagitis included concurrent chemotherapy and dosimetric information in the form of Vx. No pre-model variable selection was performed. Optimal performance was achieved using a radial basis function with $\sigma=2$ and $C=100$ and yielded an MCC of 0.43.

It was previously demonstrated that the highest value of MCC for radiation pneumonitis (0.21) was achieved using a radial basis function kernel with $\sigma=5$ and $C=100$ and that an MCC of 0.34 was obtained by using parameters from a previous multimetric approach in SVM. Data from the RTOG study (excluding the data used for training from Washington State St Louis (WUSTL)) was used to test the generalisability of the model to the independent data resulting in a reduced MCC of 0.15. A subsequent model using only the mean lung dose and centre of mass of the tumour (superior-inferior direction) (COM-SI) resulted in an MCC of 0.28 when tested on the unseen RTOG data. The advantages of using an ensemble of support vector machines are explored by Schiller et al. [54]. Using the radiation pneumonitis data from WUSTL, the differences in AUC for differing sizes of ensembles of SVM were compared using Student's t-test. The results indicated that the AUC was statistically significantly improved for larger ensembles.

17.3.3 Self Organising Maps

Self-organising maps are an unsupervised form of machine learning. Unsupervised learning clusters similar data together based on the input features with no reference to corresponding output data. Similar to PCA, self-organising maps reduce the dimensionality of the data. Proposed by Kohonen [31], self-organising maps are regularised grids of neurons which are trained by adapting weights. Each neuron contains information on the physical location and the weights which can be considered as typical values of the input features for that neuron. Neighbouring neurons will be more similar than distant nodes. Once trained, subsequent cases are mapped onto the SOM by finding the neuron with the most similar weights. The weights can be initialised randomly; however, the process may be speeded up by performing PCA and using the first two principal components to initialise the weights. Unlike PCA, the use of self-organising maps to predict normal tissue complication probability is very sparse. The most prominent example is the study by Chen et al. [7] which is complementary to their studies using ANN and SVM. Using the same dataset of 219 lung cancer patients of whom 34 reported radiation pneumonitis, a self-organising map was trained. As with previous studies, two models were developed SOM_{dose} which included dosimetric variables describing the mean dose to the lung and heart, volume of lung receiving x Gy and EUD with varying values of a and SOM_{all} which also incorporated the non-dosimetric variables such as chemotherapy status, tumour information and baseline lung function. Once the weights in an SOM are initialised, each case is presented to the map. Two parameters which steer the learning of the SOM are neighbourhood distance and learning rate. The neighbourhood distance defines the acceptable difference between the weights of an input and the weights associated with each neuron in order to decide if the patient case belongs to a particular node. In this study, similarity was assessed using the Euclidean distance. The other parameter is the learning rate which in the context of SOM defines how much information from the input vector (i.e. how many of the input variables) are used in training. Once a case has been assigned to a neuron, the associated weights are updated and the process is repeated iteratively. In this study, a cross-validation approach was used. One fold of data was removed for independent testing and one group was used to test the efficacy of the SOM trained on the other eight groups. A map of 4×3 neurons was found to be optimal, and input variables were included using trial and substitution. Each model was evaluated using the ninth group of data, and a variable was accepted if the AUC increased. Training was terminated when no new input variables were added to the model. Once trained, the outcome information was introduced, and the probability of radiation pneumonitis $\geq$grade 2 (P) on each neuron was calculated as follows:

$$P = N_p / N_n + N_p \qquad (17.7)$$

where N_p is the number of patients assigned to the neuron who experienced toxicity and N_n is the number of patients assigned to the node who did not experience toxicity.

The map can then be used prospectively by finding the appropriate neuron for each new patient and then using the probability assigned to that node. Evaluation of each model was performed for the 10th group of data using AUC and repeated for each fold of cross validation. The resultant AUC was 0.67 for SOM_{dose} and 0.73 for SOM_{all}. The difference between the two AUCs was shown to be statistically significant ($p<0.05$). The influence of the cross-validation groups was tested by repeating the splitting of the data 200 times and retraining the SOM_{all} model. Remarkably, the AUC was 0.724 (SD=0.017) suggesting a very consistent outcome. The variables included in at least one fold of the cross validation of SOM_{dose} were EUD $a=0.7$, 0.8, 0.9 and 1 and V40, V42 and V44. For SOM_{all}, the features selected were EUD $a=0.9$, 1 and 1.1, chemotherapy, histology and tumour location. As mentioned previously, EUD with $a=1$ is the mean dose which has been previously considered as being predictive of radiation pneumonitis. When this variable was removed from the model, the decrease in AUC was shown to be statistically significant. The only other variable shown to produce a statistically significant decrease on exclusion was chemotherapy. These results are consistent with the two other publications by the same group.

17.3.4 Bayesian Networks

Bayesian networks have become a popular statistical approach to challenging non-linear problems. Bayesian networks are presented using directed acyclic graphs which summarise the joint probability distribution between a set of variables. The network is optimised by finding the conditional probabilities on each node which best represents the dataset. Oh et al. [50] describe using a Bayesian network to detect interaction of dose-volume-related parameters to predict radiation pneumonitis. The dataset comprised information of a cohort of 209 patients treated with radiotherapy for non-small-cell lung cancer. Forty-eight of the patients were subsequently diagnosed with radiation pneumonitis. Input features included clinical features and dosimetric features characterised as Vx and Dx (minimum dose to the hottest $x\%$ volume). In all, 160 features were available, and the first step was to reduce the number of variables in the model. Information gain-based approach was employed for feature selection. Subsequently, the number of input features was reduced from 43. The Bayesian classifier assigns each case to the class with the highest posterior probability, determined by Bayes' theorem. We have discussed previously that dose-volume data is highly correlated; however, a naïve Bayesian classifier presumes that all features are mutually independent. Therefore, Oh et al. also implemented a tree augmented naïve Bayes classifier which allows connections between features, to overcome this challenge. Given the potential number of networks that may exist for a given dataset, it is not feasible to find an exact solution, and approximate solutions are usually employed. In this case, both hill climbing and the K2 algorithms with random ordering were implemented with the maximum number of parents allowed on each node equal to three. The Bayesian networks were evaluated using the BDe score metric [23]. Tenfold cross validation was employed, and each network was assessed after 30 iterations. The performance of each network was assessed using Matthews correlation coefficient. There was reasonable consistency between the different models with MCC between 0.25 and 0.3. Unexpectedly, the tree augmented naïve Bayes classifier was reported to be inferior in predictive power to the naïve Bayesian classifier. One of the advantages of a Bayesian classifier approach is that it is inherently visual, and therefore, relationship between variables can be observed. In this study, the dosimetric features relating to the heart and lung were shown to be clustered separately. Demonstrating that not only is there a relationship between the heart and lung but also between the variables for each organ.

17.3.5 Decision Trees

Decision trees are constructed using recursive partitioning analysis which optimises successive dichotomisation of input variables resulting in a tree-like structure used for classification. Each tree is 'grown' by starting at the root and splitting the training cases into two, maximally separated, classes. This branching continues until a terminal node (leaf) is reached. Each leaf has an associated probability of being assigned to a specific class. In the case of NTCP, this is the probability of experiencing a defined toxicity. Once trained, prospective cases can be tested, by following the appropriate path along branches eventually ending at a leaf.

Das et al. [11] describe using decision trees to augment prediction of the classic Lyman NTCP [35] by producing a combined prediction. Using the same dataset as described previously by Chen et al. [6–8], decision trees with potential dosimetric and non-dosimetric factors were built using tenfold cross validation with a balanced representation of cases experiencing radiation pneumonitis in each fold. The model was constructed using the AdaBoost algorithm which sequentially increases the number of weighted predictive units in the model. The first predictive unit contained only the Lyman model; the subsequent predictive units contained both the Lyman model and a decision tree. The predictive error ε for each predictive unit was calculated as the sum of individual patient errors (deviation from binary outcome) multiplied by patient weights. The weight of the predictive unit and patient weights were updated and propagated to the next iteration. The success of the split was assessed using the Gini index split threshold criterion [22] which was expressed in this study as

$$Ns\left(1 - p^2_{inj,s} - p^2_{uninj,s}\right) - N_L\left(1 - p^2_{inj,L} - p^2_{uninj,L}\right) - N_R\left(1 - p^2_{inj,R} - p^2_{uninj,R}\right) \quad (17.8)$$

where S refers to the node being split, L and R refer to the left and right branches, N is the number of cases and p is the proportion of patients. The subscript inj refers to patients who reported radiation pneumonitis and uninj refers to patients who did not. The variables were ranked best to worst based on the Gini index. Only those variables with a Gini index >80 % of the maximal Gini reduction were included in the model. Only three nodes were allowed on the decision tree in each predictive unit to avoid overfitting. Direction rules were implemented for a subset of variables to ensure that splits were logical, for example, dose variables and disease stage were forced in a positive direction, i.e. higher value associated with increased risk of injury. AUC was used to assess the predictive accuracy of the model as successive predictive units were added. It was demonstrated that there was no further increase in AUC after 11 units. This model resulted in an AUC of 0.72 compared to predictions made solely using the Lyman NTCP model which yielded an AUC of 0.63.

A simplified model was constructed (Fig. 17.9) where the Lyman NTCP value was combined with the value on the appropriate terminal node to provide an overall predictive value. This simplified model was shown to have an AUC of 0.75 and included the use of induction chemotherapy, histology (squamous vs. others), gender and number of fraction per day. More recently, Palma et al. [51] used recursive partitioning analysis to predict radiation pneumonitis on a cohort of patients identified from an international meta-analysis. Data from 836 patients who underwent concurrent chemo-radiation therapy for non-small-cell lung cancer (NSCLC) from 12 different institutions in Europe, North America and Asia were collected. Patients were randomly assigned to either training or validation groups (2/3 vs. 1/3). Initially, univariate logistic regression was used to identify input features that were predictive of radiation pneumonitis. These features were independently assessed using multivariate stepwise logistic regression and recursive partitioning analysis. The incidence of radiation pneumonitis was reported as 29.8 % which was scored using a number of different scoring schemes where in each case grade 2 or greater was counted as a radiation pneumonitis event.

Chemotherapy regimen, age >65 years, V20 and mean lung dose were the variables used in the recursive partitioning model which defined the three risk groups. A statistically significant difference between the risk of pneumonitis between the risk groups was observed for both the training and validation cohorts. The results of this study are strengthened by the inhomogeneity of the dataset, although no quantification is made of predictive accuracy for comparison with other model-based studies.

17 Modelling of Normal Tissue Complication Probabilities (NTCP)

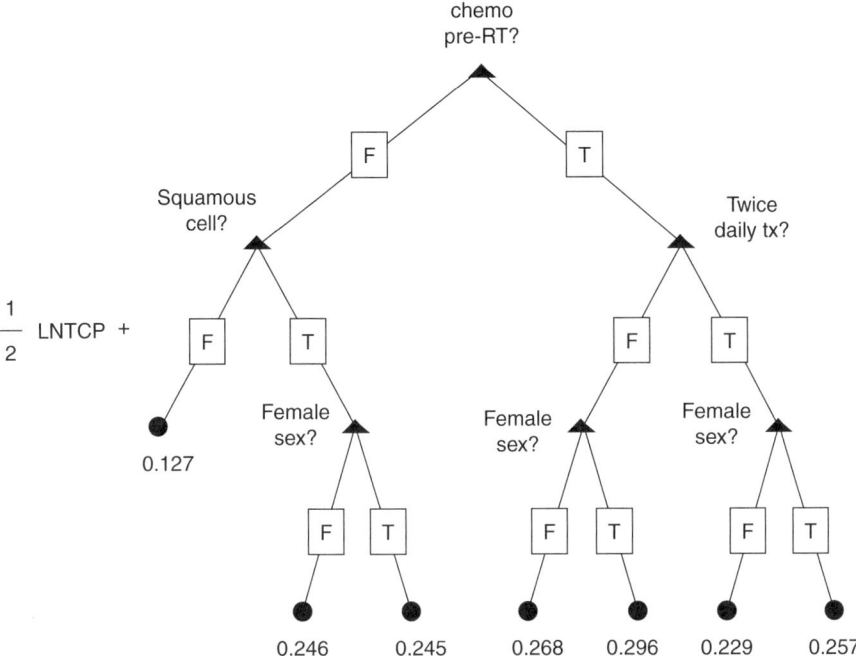

Fig. 17.9 Predictive model combining Lyman NTCP with a decision tree (Taken from Das et al. [11])

17.3.6 Hybrid Models and Comparative Studies

Each of the models here has shown strengths and weaknesses. None has been shown to be the perfect predictor. The question is whether an improvement can be made by combining predictions from different models to give 'the best of both worlds'. A useful illustration of this is the paper by Das et al. [9] who suggest that fusion of predictions from disparate models obtains a more realistic and robust estimate of the ground truth and that, where consensus exists between models, this reinforces the predictions. The results of the four previous studies discussed earlier in this chapter are combined to give a consensus prediction of the risk of radiation-induced pneumonitis using predictions from independently trained decision trees, neural network, support vector machines and self-organising maps. Each model incorporated dosimetric and non-dosimetric features from the same pool of available input variables; individual reports [6–8] demonstrated that no two models chose the same set of variables. In this study, the prediction of each model was averaged to generate an analogue prediction value. One hundred random divisions of the data in to tenfold cross validation were used to make predictions from each of the model types. These outcomes were converted to a binary value of 0 (no toxicity) and 1 (toxicity) prior to averaging to account for differences in scaling between the outputs of each type of classifier. These results were combined to produce an analogue prediction which was averaged over the four models. The resultant model was shown to have an AUC which converged at 0.79 when 10 randomly selected predictions were chosen for each model; this was an improvement of the results of each of the individual classifiers. The Spearman correlation between any two of the predictions for each model was shown to be high (≥ 0.9) for all models except SVM, whilst correlations between models were much lower. This emphasises the benefit of repeated cross validation and the combination of different classifiers. The importance of individual input features was tested using reverse rank method whereby the patient predictions were ranked highest to lowest risk of pneumonitis based on the consensus prediction. The values of one input variable were then reversed so that the value for the top-ranked patient was substituted with the bottom-ranked patient and vice versa. The predictions were recalculated and the ranking recalculated. The Spearman correlation coefficient was used to compare the pre- and post-switch rankings (which were resampled 10^5 times). A large negative coefficient would indicate a large impact on the predictions from the variable in question. As with previous publications, highly correlated variables (Pearson's coefficient >0.9) were excluded from being added to a model where another correlated feature was already present. Therefore, groups of dosimetric variables were grouped together. The largest negative coefficient was observed when two groups of dosimetric variables and induction chemotherapy were reversed. Female gender and squamous cell histology were also shown to be important. The dosimetric groups represented I EUD (a 0.5–1.2) and vol >20–30 Gy and II EUD (a 1.2–3). Subsequently, the consensus variables were fitted to a logistic regression probability function. This translation of the consensus of machine learning into an easily interpretable model enables the transfer of learned knowledge into the clinical context.

A more recent study by Nalbantov et al. [45] combined predictions from ten different models to predict radiation-induced acute dysphagia (swallowing difficulties). Each model was assigned equal voting rights and tested on a prospective cohort of patients. The results were compared to predictions made by physicians. All were given the same 'input' information which included age, gender, WHO performance status, mean and maximum dose to the oesophagus, overall treatment time and concurrent/sequential chemotherapy. Predictions of acute dysphagia $\geq$G3 (CTCAE) [60] were made using naïve Bayes, bagging, Bayesian networks, boosting, penalised logistic regression, radial basis function network, random forest, linear support vector machine and LASSO and for a combined model with equal voting rights. The combined model resulted in a higher AUC (0.77) for the independent prospective validation cohort than for any of the individual models. The corresponding AUC for the physicians was 0.53.

Other studies have chosen not to create hybrid models but have made a direct comparison between machine learning approaches. Pella et al. [52] presented a comparison between models based on ANN and SVM to predict acute toxicity for a cohort of 321 patients who received prostate radiotherapy. Both techniques were chosen for the flexibility that allows both dosimetric and clinical variables to be considered in the same model. The input features were selected by the authors based on clinical knowledge and appear to be limited compared to those in other studies we have considered. The dose distribution to the rectum was quantified by the dose received (30 and 60 %) by the rectum (D30 and D60, respectively) and the absolute volume (cc) of the rectum on the planning scan. The dose distribution to the bladder was described using only the dose received by 50 % of the bladder and the absolute bladder volume (cc) from the treatment planning scan. Unusually, a single outcome of either GI or GU toxicity ≥ grade 2 was used; this choice was justified by the perceived low incidence of both GI (37 %) and GU (11.5 %) toxicity in the cohort. The artificial neural network architecture was optimised using a genetic algorithm. The optimised ANN was reported to have two hidden layers with 47 neurons in the first hidden layer with a sigmoid activation function and 44 neurons also with a sigmoid activation function. A linear activation function was used in the output layer. The ROC for the optimised ANN was 0.697. In comparison, the optimal SVM was found to have used a polynomial kernel of the 9th order which resulted in an AUC of 0.717. Both of these values related to a subset of 30 patients withheld from training. It should be noted that the optimisation of both ANN and SVM chose parameters that could lead to overfitting. An ANN with 13 inputs but nearly 100 hidden nodes is likely to be overfitted as is an SVM using a 9th-order polynomial. Since no cross validation was employed, it is impossible to infer how well these models would generalise. No statistical comparison was made between the AUC for the two techniques; this may be again due to the singular nature of the result. Another study by Oh et al. [49] directly compares machine learning methods for outcome prediction of radiation pneumonitis. Comparison is made between both feature selection techniques and classification methods. The feature selection methods were SVM-recursive feature elimination, correlation-based feature selection, chi-square feature selection and information gain. Classifiers included SVM, decision tree, random forest and naive Bayesian. Matthews correlation coefficient was employed to assess performance. Each method was tested on a cohort of 209 NSCLC patients from Washington University School of Medicine of whom 48 reported radiation pneumonitis (which was also reported in the study of Bayesian networks from the same group). Data included clinical variables such as demographics and disease stage and dosimetric variables quantified as *Vx* volume receiving *x* Gy and D_x dose received by *x*% of volume. Some input features were ranked highly by more than one feature selection approach, but generally, there was significant variability between feature selection methods. The feature selection was combined with each of the classification methods starting with the highest rank variable models and subsequently increasing the number of variables. It was observed that SVM with a radial basis function or polynomial kernel function consistently resulted in the highest Matthews correlation coefficient values. Whilst caution is needed when comparing models since results may be data specific, it is useful to consider the relative success of different approaches. Of note is the variability in the results of the feature selection. It is not stated if any adjustment was made for correlated inputs which may have affected the results.

17.4 Summary

This chapter has reviewed many studies which have implemented machine learning to further knowledge in NTCP. Considering the total number of publications on NTCP, machine learning has had a limited impact on the field. Here we consider why this is the case and how that might be addressed. Machine learning, particularly artificial neural networks, is traditionally regarded as being mystical black boxes where it is impossible to interpret the underlying model. Although it is challenging to interpret the weights of a black box, it is not impossible, whilst other machine learning techniques, for example, decision trees, are considerably more transparent. There are a wide variety of machine learning techniques, and deciding which one is appropriate can be daunting. The suite of publications from Duke University [6–8, 11] and comparative papers by Oh [49] and Pella [52] are insightful. It is not wise to necessarily take the AUC measure as the comparative standard between models as this may well be data specific. However, it is useful to consider the congruence of the features selected by the final model. In some cases, combining different models improves predictive accuracy particularly where input features are potentially highly correlated. In this case, an ensemble may facilitate similar information being used in slightly different forms. Alternatively, a hybrid approach can result in the best of all worlds. The flip side is that these models are inherently complex and may suffer from a lack of generalisability if not carefully trained. In addition, it may be more challenging to interpret the role of individual input features when many are distributed throughout the model. Many of the studies presented in this chapter have indicated that the results from machine learning were superior to standard techniques. This may be in part due to the flexible approach to combining different data types that are available. However, only in rare cases does the AUC exceed 0.8. Although this is considered to be a very good result for both classic statistical and machine learning approaches in the medical arena, ideally, every patient would have a valid prediction. The reasons why we reach this glass ceiling are complex but essentially result from a failure to fully reflect the patient experience. No model can predict an outcome from data that is not provided as an input. The amount of data available for each patient is exploding as genetic information is incorporated into the studies. In addition, the dose distribution to organs at risk is insufficiently characterised by DVH, and steps to improve this by including spatial information will further increase the number of input features. Machine learning is a knowledge transfer tool allowing clinicians to present all the data that they regard as relevant to a specific prediction situation. Clearly, medical understanding evolves daily, and therefore, predictive models will need to continuously be updated to include this increased knowledge. Machine learning approaches are well equipped to deal with big data, and it is hoped that in the future, the understanding of the response of normal tissues following cancer treatment including radiotherapy will be well understood and reliable knowledge-based models will be used as standard in the clinic.

References

1. Barnett GC, Coles CE, Elliott RM, Baynes C, Luccarini C, Conroy D, Wilkinson JS, Tyrer J, Misra V, Platte R, Gulliford SL, Sydes MR, Hall E, Bentzen SM, Dearnaley DP, Burnet NG, Pharoah PDP, Dunning AM, West CM. Independent validation of genes and polymorphisms reported to be associated with radiation toxicity: a prospective analysis study. Lancet Oncol. 2012;13:65–77. doi:10.1016/S1470-2045(11)70302-3.

2. Bauer JD, Jackson A, Skwarchuk M, Zelefsky M. Principal component, Varimax rotation and cost analysis of volume effects in rectal bleeding in patients treated with 3D-CRT for prostate cancer. Phys Med Biol. 2006;51:5105–23. doi:10.1088/0031-9155/51/20/003.
3. Blanco AI, Chao KS, El Naqa I, Franklin GE, Zakarian K, Vicic M, Deasy JO. Dose-volume modeling of salivary function in patients with head-and-neck cancer receiving radiotherapy. Int J Radiat Oncol Biol Phys. 2005;62:1055–69. doi:10.1016/j.ijrobp.2004.12.076.
4. Buettner F, Gulliford SL, Webb S, Partridge M. Using dose-surface maps to predict radiation-induced rectal bleeding: a neural network approach. Phys Med Biol. 2009;54:5139–53. doi:10.1088/0031-9155/54/17/005.
5. Burman C, Kutcher GJ, Emami B, Goitein M. Fitting of normal tissue tolerance data to an analytic function. Int J Radiat Oncol Biol Phys. 1991;21:123–35.
6. Chen S, Zhou S, Yin FF, Marks LB, Das SK. Investigation of the support vector machine algorithm to predict lung radiation-induced pneumonitis. Med Phys. 2007;34:3808–14. doi:10.1118/1.2776669.
7. Chen SF, Zhou SM, Yin FF, Marks LB, Das SK. Using patient data similarities to predict radiation pneumonitis via a self-organizing map. Phys Med Biol. 2008;53:203–16. doi:10.1088/0031-9155/53/1/014.
8. Chen SF, Zhou SM, Zhang JN, Yin FF, Marks LB, Das SK. A neural network model to predict lung radiation-induced pneumonitis. Med Phys. 2007;34:3420–7. doi:10.1118/1.2759601.
9. Das SK, Chen SF, Deasy JO, Zhou SM, Yin FF, Marks LB. Combining multiple models to generate consensus: application to radiation-induced pneumonitis prediction. Med Phys. 2008;35:5098–109. doi:10.1118/1.2996012.
10. Das SK, Chen SF, Deasy JO, Zhou SM, Yin FF, Marks LB. Decision fusion of machine learning models to predict radiotherapy-induced lung pneumonitis. In: Seventh international conference on machine learning and applications, proceedings. IEEE Computer Society, Los Alamitos, CA. 2008b. p. 545–50. doi:10.1109/Icmla.2008.122.
11. Das SK, Zhou S, Zhang J, Yin FF, Dewhirst MW, Marks LB. Predicting lung radiotherapy-induced pneumonitis using a model combining parametric Lyman probit with nonparametric decision trees. Int J Radiat Oncol Biol Phys. 2007;68:1212–21. doi:10.1016/j.ijrobp.2007.03.064.
12. Dawson LA, Biersack M, Lockwood G, Eisbruch A, Lawrence TS, Ten Haken RK. Use of principal component analysis to evaluate the partial organ tolerance of normal tissues to radiation. Int J Radiat Oncol Biol Phys. 2005;62:829–37. doi:10.1016/j.ijrobp.2004.11.013.
13. Dawson LA, Kavanagh BD, Paulino AC, Das SK, Miften M, Li XA, Pan C, Ten Haken RK, Schultheiss TE. Radiation-associated kidney injury. Int J Radiat Oncol Biol Phys. 2010;76:S108–15. doi:10.1016/j.ijrobp.2009.02.089.
14. Deasy JO, Moiseenko V, Marks L, Chao KS, Nam J, Eisbruch A. Radiotherapy dose-volume effects on salivary gland function. Int J Radiat Oncol Biol Phys. 2010;76:S58–63. doi:10.1016/j.ijrobp.2009.06.090.
15. El Naqa I, Bradley J, Blanco AI, Lindsay PE, Vicic M, Hope A, Deasy JO. Multivariable modeling of radiotherapy outcomes, including dose-volume and clinical factors. Int J Radiat Oncol Biol Phys. 2006;64:1275–86. doi:10.1016/j.ijrobp.2005.11.022.
16. El Naqa I, Bradley JD, Deasy J. Nonlinear Kernel-based approaches for predicting normal tissue toxicities. In: Seventh international conference on machine learning and applications, Proceedings. IEEE Computer Society, Los Alamitos, CA. 2008. p. 539–44. doi:10.1109/Icmla.2008.126.
17. El Naqa I, Bradley JD, Lindsay PE, Hope AJ, Deasy JO. Predicting radiotherapy outcomes using statistical learning techniques. Phys Med Biol. 2009;54:S9–30. doi:10.1088/0031-9155/54/18/S02.
18. Emami B, Lyman J, Brown A, Coia L, Goitein M, Munzenrider JE, Shank B, Solin LJ, Wesson M. Tolerance of normal tissue to therapeutic irradiation. Int J Radiat Oncol Biol Phys. 1991;21:109–22.
19. Fellin G, Rancati T, Fiorino C, Vavassori V, Antognoni P, Baccolini M, Bianchi C, Cagna E, Borca VC, Girelli G, Iacopino B, Maliverni G, Mauro FA, Menegotti L, Monti AF, Romani F, Stasi M, Valdagni R. Long term rectal function after high-dose prostate cancer radiotherapy: results from a prospective cohort study. Radiother Oncol. 2014;110:272–7. doi:10.1016/j.radonc.2013.09.028.

20. Groom N, Wilson E, Lyn E, Faivre-Finn C. Is pre-trial quality assurance necessary? Experiences of the CONVERT Phase III randomized trial for good performance status patients with limited-stage small-cell lung cancer. Br J Radiol. 2014;87:20130653. doi:10.1259/bjr.20130653.
21. Gulliford SL, Foo K, Morgan RC, Aird EG, Bidmead AM, Critchley H, Evans PM, Gianolini S, Mayles WP, Moore AR, Sanchez-Nieto B, Partridge M, Sydes MR, Webb S, Dearnaley DP. Dose-volume constraints to reduce rectal side effects from prostate radiotherapy: evidence from MRC RT01 Trial ISRCTN 47772397. Int J Radiat Oncol Biol Phys. 2010;76:747–54. doi:10.1016/j.ijrobp.2009.02.025.
22. Hastie TT, Tibshirani R, Friedman J. The elements of statistical learning: data mining, inference and prediction. New York: Springer; 2002.
23. Heckerman D, Geiger D, Chickering DM. Learning Bayesian Networks – the combination of knowledge and statistical-data. Machine Learning. 1995;20:197–243. doi:10.1007/Bf00994016.
24. Hosmer Jr DW, Lemeshow S, Sturdivant RX. Applied logistic regression. New York: Wiley; 2013.
25. Jackson A, Ten Haken RK, Robertson JM, Kessler ML, Kutcher GJ, Lawrence TS. Analysis of clinical complication data for radiation hepatitis using a parallel architecture model. Int J Radiat Oncol Biol Phys. 1995;31:883–91. doi:10.1016/0360-3016(94)00471-4.
26. Jaffray DA, Lindsay PE, Brock KK, Deasy JO, Tome WA. Accurate accumulation of dose for improved understanding of radiation effects in normal tissue. Int J Radiat Oncol Biol Phys. 2010;76:S135–9. doi:10.1016/j.ijrobp.2009.06.093.
27. James G, Witten D, Hastie T, Tibshirani R. An introduction to statistical learning. New York: Springer; 2013.
28. Kallman P, Agren A, Brahme A. Tumor and normal tissue responses to fractionated nonuniform dose delivery. Int J Radiat Biol. 1992;62:249–62. doi:10.1080/09553009214552071.
29. Kasibhatla M, Kirkpatrick JP, Brizel DM. How much radiation is the chemotherapy worth in advanced head and neck cancer? Int J Radiat Oncol Biol Phys. 2007;68:1491–5. doi:10.1016/j.ijrobp.2007.03.025.
30. Klement RJ, Allgauer M, Appold S, Dieckmann K, Ernst I, Ganswindt U, Holy R, Nestle U, Nevinny-Stickel M, Semrau S, Sterzing F, Wittig A, Andratschke N, Guckenberger M. Support vector machine-based prediction of local tumor control after stereotactic body radiation therapy for early-stage non-small cell lung cancer. Int J Radiat Oncol Biol Phys. 2014;88:732–8. doi:10.1016/j.ijrobp.2013.11.216.
31. Kohonen T. Essentials of the self-organizing map. Neural Netw. 2013;37:52–65. doi:10.1016/j.neunet.2012.09.018.
32. Kutcher GJ, Burman C, Brewster L, Goitein M, Mohan R. Histogram reduction method for calculating complication probabilities for three-dimensional treatment planning evaluations. Int J Radiat Oncol Biol Phys. 1991;21:137–46.
33. Liang Y, Messer K, Rose BS, Lewis JH, Jiang SB, Yashar CM, Mundt AJ, Mell LK. Impact of bone marrow radiation dose on acute hematologic toxicity in cervical cancer: principal component analysis on high dimensional data. Int J Radiat Oncol Biol Phys. 2010;78:912–9. doi:10.1016/j.ijrobp.2009.11.062.
34. Lind PA, Wennberg B, Gagliardi G, Rosfors S, Blom-Goldman U, Lidestahl A, Svane G. ROC curves and evaluation of radiation-induced pulmonary toxicity in breast cancer. Int J Radiat Oncol Biol Phys. 2006;64:765–70. doi:10.1016/j.ijrobp.2005.08.011.
35. Lyman JT. Complication probability as assessed from dose-volume histograms. Radiat Res Suppl. 1985;8:S13–9.
36. Lyman JT, Wolbarst AB. Optimization of radiation therapy, III: a method of assessing complication probabilities from dose-volume histograms. Int J Radiat Oncol Biol Phys. 1987;13:103–9.
37. Matthews BW. Comparison of the predicted and observed secondary structure of T4 phage lysozyme. Biochim Biophys Acta. 1975;405:442–51.
38. Marks LB, Ten Haken RK, Martel MK. Guest editor's introduction to QUANTEC: a users guide. Int J Radiat Oncol Biol Phys. 2010;76(3 Suppl):S1–S2.
39. McCullough WS, Pitts W. A logical calculus of the ideas imminent in nervous activity. Bull Math Biol. 1943;52:99–115.

40. Mcdonald S, Rubin P, Phillips TL, Marks LB. Injury to the lung from cancer-therapy – clinical syndromes, measurable end-points, and potential scoring systems. Int J Radiat Oncol Biol Phys. 1995;31:1187–203. doi:10.1016/0360-3016(94)00429-O.
41. Miah AB, Schick U, Bhide SA, Guerrero-Urbano MT, Clark CH, Bidmead AM, Bodla S, Del Rosario L, Thway K, Wilson P, Newbold KL, Harrington KJ, Nutting CM. A phase II trial of induction chemotherapy and chemo-IMRT for head and neck squamous cell cancers at risk of bilateral nodal spread: the application of a bilateral superficial lobe parotid-sparing IMRT technique and treatment outcomes. Br J Cancer. 2015;112:32–8. doi:10.1038/bjc.2014.553.
42. Michalski JM, Gay H, Jackson A, Tucker SL, Deasy JO. Radiation dose-volume effects in radiation-induced rectal injury. Int J Radiat Oncol Biol Phys. 2010;76:S123–9. doi:10.1016/j.ijrobp.2009.03.078.
43. Mitchell M. An introduction to genetic algorithms. Cambridge, MA: MIT; 1998.
44. Munley MT, Lo JY, Sibley GS, Bentel GC, Anscher MS, Marks LB. A neural network to predict symptomatic lung injury. Phys Med Biol. 1999;44:2241–9.
45. Nalbantov G, Oberije C, Lambin P, De Ruysscher D, Dekker A. Combining the predictions for radiation-induced dysphagia in lung cancer patients from multiple models improves the prognostic accuracy of each individual model. J Thorac Oncol. 2011;6:S549.
46. Niemierko A. Reporting and analyzing dose distributions: a concept of equivalent uniform dose. Med Phys. 1997;24:103–10. doi:10.1118/1.598063.
47. Niemierko A. A generalized concept of equivalent uniform dose (EUD). Med Phys. 1999;26:1100.
48. Niemierko A, Goitein M. Modeling of normal tissue-response to radiation - the critical volume model. Int J Radiat Oncol Biol Phys. 1993;25:135–45.
49. Oh JH, Al-Lozi R, El Naqa I. Application of machine learning techniques for prediction of radiation pneumonitis in lung cancer patients. In: Eighth international conference on machine learning and applications, proceedings. IEEE Computer Society, Los Alamitos, CA. 2009. p. 478–83. doi:10.1109/Icmla.2009.118.
50. Oh JH, El Naqa I. Bayesian network learning for detecting reliable interactions of dose-volume related parameters in radiation pneumonitis. In: Eighth International Conference on Machine Learning and Applications, Proceedings. IEEE Computer Society, Los Alamitos, CA. 2009. p. 484–8.
51. Palma DA, Senan S, Tsujino K, Barriger RB, Rengan R, Moreno M, Bradley JD, Kim TH, Ramella S, Marks LB, De Petris L, Stitt L, Rodrigues G. Predicting radiation pneumonitis after chemoradiation therapy for lung cancer: an international individual patient data meta-analysis. Int J Radiat Oncol Biol Phys. 2013;85:444–50. doi:10.1016/j.ijrobp.2012.04.043.
52. Pella A, Cambria R, Riboldi M, Jereczek-Fossa BA, Fodor C, Zerini D, Torshabi AE, Cattani F, Garibaldi C, Pedroli G, Baroni G, Orecchia R. Use of machine learning methods for prediction of acute toxicity in organs at risk following prostate radiotherapy. Med Phys. 2011;38:2859–67.
53. Rancati T, Fiorino C, Fellin G, Vavassori V, Cagna E, Casanova Borca V, Girelli G, Menegotti L, Monti AF, Tortoreto F, Delle Canne S, Valdagni R. Inclusion of clinical risk factors into NTCP modelling of late rectal toxicity after high dose radiotherapy for prostate cancer. Radiother Oncol. 2011;100:124–30. doi:10.1016/j.radonc.2011.06.032.
54. Schiller TW, Chen YX, El Naqa I, Deasy JO. Improving clinical relevance in ensemble support vector machine models of radiation pneumonitis risk. Eighth international conference on machine learning and applications, proceedings. IEEE Computer Society, Los Alamitos, CA. 2009. p. 498–503. doi:10.1109/Icmla.2009.74.
55. Skala M, Rosewall T, Dawson L, Divanbeigi L, Lockwood G, Thomas C, Crook J, Chung P, Warde P, Catton C. Patient-assessed late toxicity rates and principal component analysis after image-guided radiation therapy for prostate cancer. Int J Radiat Oncol Biol Phys. 2007;68:690–8. doi:10.1016/j.ijrobp.2006.12.064.
56. Sohn M, Alber M, Yan D. Principal component analysis-based pattern analysis of dose-volume histograms and influence on rectal toxicity. Int J Radiat Oncol Biol Phys. 2007;69:230–9. doi:10.1016/j.ijrobp.2007.04.066.
57. Streiner DL, Cairney J. What's under the ROC? An introduction to receiver operating characteristics curves. Can J Psychiatry. 2007;52:121–8.

58. Su M, Miften M, Whiddon C, Sun X, Light K, Marks L. An artificial neural network for predicting the incidence of radiation pneumonitis. Med Phys. 2005;32:318–25.
59. Tomatis S, Rancati T, Fiorino C, Vavassori V, Fellin G, Cagna E, Mauro FA, Girelli G, Monti A, Baccolini M, Naldi G, Bianchi C, Menegotti L, Pasquino M, Stasi M, Valdagni R. Late rectal bleeding after 3D-CRT for prostate cancer: development of a neural-network-based predictive model. Phys Med Biol. 2012;57:1399–412. doi:10.1088/0031-9155/57/5/1399.
60. Trotti A, Colevas AD, Setser A, Rusch V, Jaques D, Budach V, Langer C, Murphy B, Cumberlin R, Coleman CN, Rubin P. CTCAE v3.0: development of a comprehensive grading system for the adverse effects of cancer treatment. Semin Radiat Oncol. 2003;13:176–81. doi:10.1016/S1053-4296(03)00031-6.
61. Vesprini D, Sia M, Lockwood G, Moseley D, Rosewall T, Bayley A, Bristow R, Chung P, Menard C, Milosevic M, Warde P, Catton C. Role of principal component analysis in predicting toxicity in prostate cancer patients treated with hypofractionated intensity-modulated radiation therapy. Int J Radiat Oncol Biol Phys. 2011;81:e415–21. doi:10.1016/j.ijrobp.2011.01.024.
62. Viswanathan AN, Yorke ED, Marks LB, Eifel PJ, Shipley WU. Radiation dose-volume effects of the urinary bladder. Int J Radiat Oncol Biol Phys. 2010;76:S116–22. doi:10.1016/j.ijrobp.2009.02.090.
63. Withers HR, Taylor JM, Maciejewski B. Treatment volume and tissue tolerance. Int J Radiat Oncol Biol Phys. 1988;14(4):751–759.

Modeling of Tumor Control Probability (TCP)

18

Issam El Naqa

Abstract

Modeling of tumor control probability is an important task for predicting response in radiotherapy. Most early methods have focused on using biophysical analysis based on understanding irradiation effects from in vitro cell culture. However, it has been recognized that clinical tumor response is multifactorial and involves a complex interaction of physical, biological, and clinical surrogates that data-driven approaches such as machine-learning algorithms would play a prominent role. In this chapter, we present using different examples the process of applying machine learning to modeling TCP and demonstrate its efficacy compared to existing methods and its potential to improving our understanding of tumor response.

18.1 Introduction

Recent years have witnessed tremendous technological advances in radiotherapy treatment planning, image guidance, and treatment delivery [1, 2]. Moreover, clinical trials examining treatment intensification in patients with locally advanced cancer have shown incremental improvements in local control and overall survival [3]. Radiotherapy outcomes are traditionally modeled using information about the dose distribution and the fractionation [4]. However, it is well known that radiotherapy outcomes is multifactorial and may also be affected by multiple clinical and biological prognostic factors such as stage, volume, tumor hypoxia, etc. [5, 6]. Therefore, recent years have witnessed the emergence of data-driven models utilizing

I. El Naqa
Department of Oncology, McGill University, Montreal, QC, Canada

Department of Radiation Oncology, University of Michigan, Ann Arbor, USA
e-mail: issam.elnaqa@mcgill.ca; ielnaqa@med.umich.edu

informatics techniques, in which dose–volume metrics are combined with other patient- or disease-based prognostic factors [7–15]. These approaches have utilized data-driven models incorporating advanced bioinformatics tools in which dose–volume metrics are mixed with other patient- or disease-based prognostic factors in order to improve outcomes prediction [16]. The accurate prediction of tumor response would provide patients and their treating clinicians with better tools for informed decision-making about expected benefits versus anticipated risks and higher likelihood of improved outcomes, in which machine-learning methods are expected to play a prominent role.

18.2 Tumor Control Probability

Tumor control is strictly defined by the probability of the extinction of clonogenic tumor cells at the end of treatment [17]. Several radiobiological models have been proposed in the literature to model TCP. The linear-quadratic model (LQ) is the most frequently used model for including the effects of repair between treatment fractions. The LQ model is based on clonogenic cell survival curves and is parameterized by the radiosensitivity ratio (α/β). It is thought that it quantifies the effects of both unrepairable damage and repairable damage susceptible to misrepair after tumor sterilization by radiation [18, 19]:

$$\mathrm{SF} = \exp\left(-\left((\alpha + \beta * d) * D + In2 * t / T_{pot}\right)\right) \quad (18.1)$$

where d is the fraction size, D is the total delivered dose, t is the difference between the total treatment time (T) and the lag period before accelerated clonogen repopulation begins (T_K), and T_{pot} is the potential doubling time of the cells. The ratio $\ln 2/T_{pot}$ is referred to as the repopulation parameter. Several variations of this model have been proposed including a Poisson-based [20] and a birth–death model [21]. Among the most commonly used LQ-based TCP models [22] is:

$$TCP = \exp\left(-N \exp\left(-\left((\alpha + \beta * d) * D + \ln 2 * t / T_{pot}\right)\right)\right) \quad (18.2)$$

A detailed review of analytical methods for TCP in radiation treatment has been recently published [23].

18.3 Machine Learning for TCP Modeling

Machine learning allows for exploiting nonlinear patterns in the data that may not be directly tractable from using analytical or phenomenological models. There are several steps into development of a TCP model using machine learning as shown in the examples below using dosimetric, clinical, imaging, and biological data in lung cancer.

18.4 Example 1: Dosimetric and Clinical Variables

18.4.1 Data Set

A set of 56 patients diagnosed with non-small cell lung cancer (NSCLC) and who have discrete primary lesions, complete dosimetric archives, and follow-up information for the endpoint of local control (22 locally failed cases) is used. The patients were treated with three-dimensional conformal radiation therapy (3D-CRT) with a median prescription dose of 70 Gy (60–84 Gy). The dose distributions were corrected for heterogeneity using Monte Carlo simulations [24]. The clinical data included age, gender, performance status, weight loss, smoking, histology, neoadjuvant and concurrent chemotherapy, stage, number of fractions, tumor elapsed time, tumor volume, and prescription dose. Treatment planning data were de-archived and potential dose–volume histogram (DVH) prognostic metrics were extracted using CERR [25]. These metrics included Vx (percentage volume receiving at least x Gy), where x was varied from 60 to 80 Gy in steps of 5 Gy, mean dose, minimum and maximum doses, and center of mass location in the craniocaudal (COMSI) and lateral (COMLAT) directions. This resulted in a set of 23 candidate variables to model TCP. The modeling process using nonlinear statistical learning starts by applying dimensionality reduction technique such as principal component analysis (PCA) to visualize the data in two-dimensional space and assess the separability of low-risk from high-risk patients. Separable cases could be modeled by linear kernels while non-separable cases are modeled by nonlinear kernels that allow for separability of the data but at the expense of increased dimensionality. This step could be preceded by a variable selection process and the generalizability of the model is evaluated using resampling techniques as discussed below [26].

18.4.2 Data Exploration

In Fig. 18.1a, we show a correlation matrix representation of the selected candidate variables with clinical TCP and cross-correlations among themselves using Spearman's rank correlation coefficient (rs). Note that many DVH-based dosimetric variables are highly cross-correlated, which complicate the analysis of such data. In Fig. 18.2b, we summarize the PCA analysis of this data by projecting it into two-dimensional space for visualization purposes. The plots show that two principal components are able to explain 70 % of the data and reflect a relatively high overlap between patients with and without local control, indicating potential benefit from using nonlinear kernel methods.

18.4.3 Logistic Regression Modeling Example

The multimetric model building using logistic regression is performed using a two-step procedure to estimate model order and parameters. In each step, a sequential

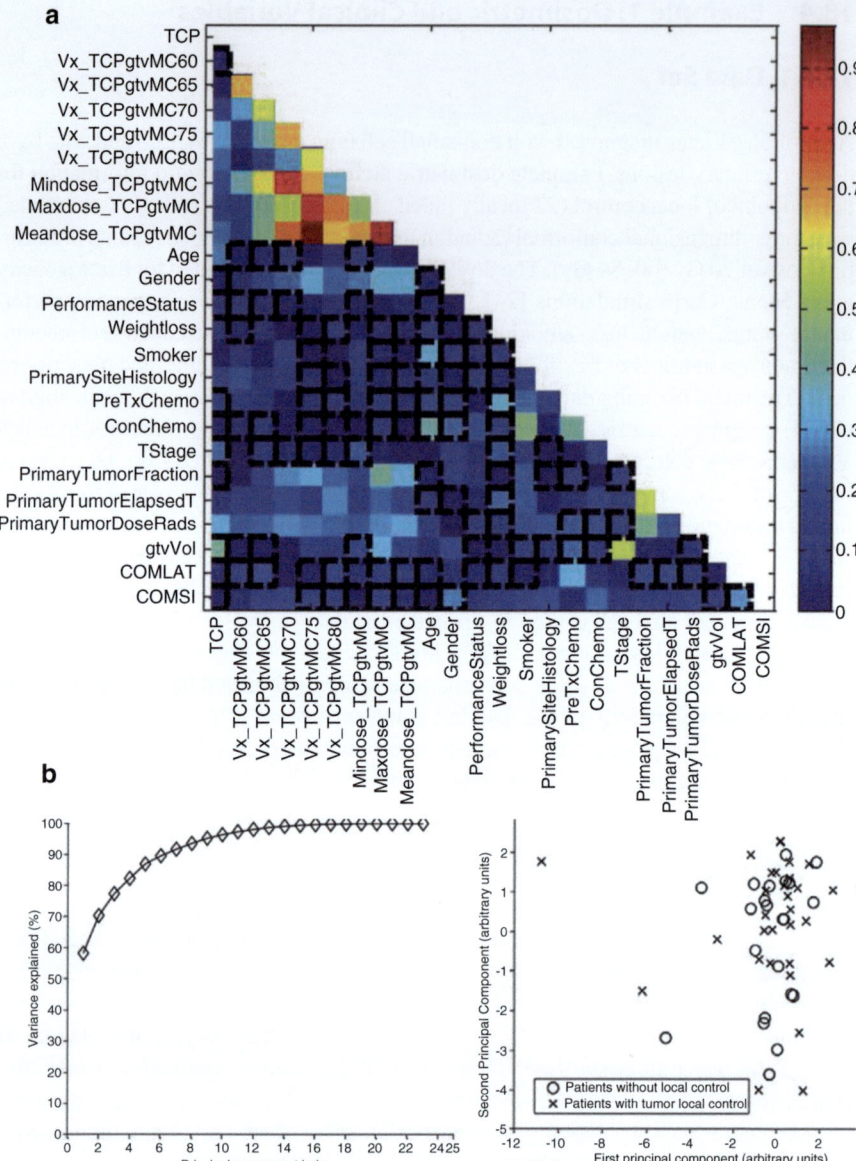

Fig. 18.1 (**a**) Correlation matrix showing the candidate variable correlations with TCP and among the other candidate variables. (**b**) Visualization of higher dimensional data by principal component analysis (PCA). *Left* The variation explanation versus principal component (PC) index. *Right* The data projection into the first two principal component space. Note the cases overlap

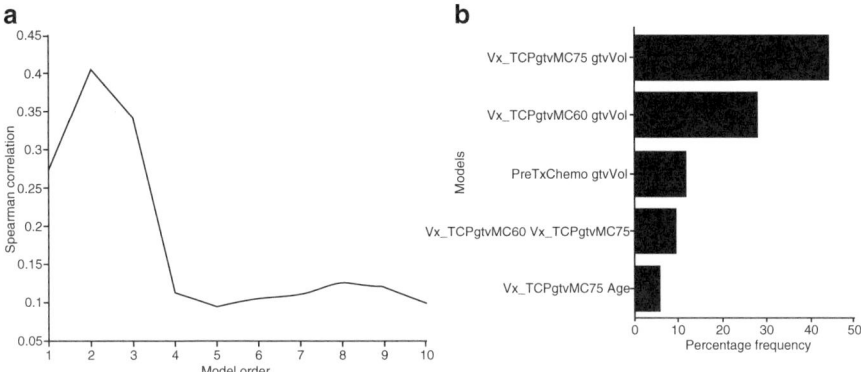

Fig. 18.2 TCP model building using logistic regression. (**a**) Model order selection using LOO-CV. (**b**) Model parameters estimation by frequency selection on bootstrap samples

forward selection strategy is used to build the model by selecting the next candidate variable from the available pool (23 variables in our case) based on increased significance using Wald's statistics [12]. In Fig. 18.2a, we show the model order selection using the LOO-CV procedure. It is noticed that a model order of two parameters provides the best predictive power with Spearman rank correction coefficient (rs = 0.4). In Fig. 18.2b, we show the optimal model parameters' selection frequency on bootstrap resampling (280 samples were generated in this case). A model consisting of GTV volume ($\beta = -0.029, p = 0.006$) and GTV V75 ($\beta = +2.24, p = 0.016$) had the highest selection frequency (45 % of the time). The model suggests that increase in tumor volume would lead to failure, as one would expect due to increase in the number of clonogens in larger tumor volumes. The V75 metric is related to dose coverage of the tumor, where it is noticed that patients who had less than 20 % of their tumor covered by 75 Gy were at higher risk of failure. However, a drawback of this logistic regression approach is that it does not automatically account for possible interactions between these metrics nor does it account for higher-order nonlinearities.

18.4.4 Kernel-Based Modeling Example

To account for potential nonlinear interactions as revealed by the PCA, we will apply kernel-based methods using support vector machines (SVM). Moreover, we will use the same variables selected by the logistic regression approach. We have demonstrated recently that such selection is more robust than other competitive techniques such as the recursive feature elimination (RFE) method used in microarray analysis. In this case, a vector of explored variables is generated by concatenation. The variables are normalized using the z-scoring approach to have a zero mean and unity variance [27]. We experimented with different kernel forms; best results are shown for the radial basis function (RBF) in Fig. 18.3a. The figure shows that the optimal kernel parameters are obtained with an RBF width $\sigma=2$ and regularization parameter $C=10{,}000$. This resulted in a predictive power on LOO-CV rs$=0.68$, which represents 70 % improvement over the logistic regression analysis results. This improvement could be further explained by examining Fig. 18.3b, which shows how the RBF kernel tessellated the variable space nonlinearly into different regions of high and low risks of local failure. Four regions are shown in the figure representing high/low risks of local failure with high/low confidence levels, respectively. Note that cases falling within the classification margin have low confidence prediction power and represent intermediate-risk patients, i.e., patients with "border-like" characteristics that could belong to either risk group [26].

18.4.5 Comparison with Other Known Models

For comparison purposes with mechanistic TCP models, we chose the Poisson-based TCP model and the cell kill equivalent uniform dose (cEUD) model. The Poisson-based TCP parameters for NSCLC were selected according to Willner et al. work [28], in which the sensitivity to dose per fraction ($\alpha/\beta=10$ Gy), dose for 50 % control rate (D50$=74.5$ Gy), and the slope of the sigmoid-shaped dose–response at D50 ($\gamma_{50}=3.4$). The resulting correlation of this model was rs$=0.33$. Using D50$=84.5$ and $\gamma_{50}=1.5$ [29, 30] yielded an rs$=0.33$ also. For the cEUD model, we selected the survival fraction at 2 Gy (SF2$=0.56$) according to Brodin et al. [31]. The resulting correlation in this case was rs$=0.17$. A summary plot of the different methods predictions as a function of binned patients into equal groups is shown in Fig. 18.4. It is observed that the best performance was achieved by the nonlinear (SVM-RBF). This is particularly observed for predicting patients who are at high risk of local failure.

18.5 Use of Imaging Features

Pretreatment or posttreatment information from anatomical or functional/molecular imaging could be used to monitor and predict treatment outcomes in radiotherapy. For instance, changes in tumor volume on computed tomography (CT) have been used to predict radiotherapy response in NSCLC patients [32, 33]. On the other hand, functional/molecular imaging, in particular positron emission tomography (PET) with fluorodeoxyglucose (FDG), has received special attention as a potential prognostic factor for predicting radiotherapy efficacy [34–37]. For instance, high FDG-PET intensity has been shown to correlate with poor local control in lung cancer [38–41]. In our previous work, new features based on image morphology, intensity, and texture/roughness can provide a more complete characterization of uptake heterogeneity [37]. Recently, we have shown that in addition to PET features, CT-derived features (from the gross target volume) may also improve prediction of local tumor response as shown in Fig. 18.5 [42].

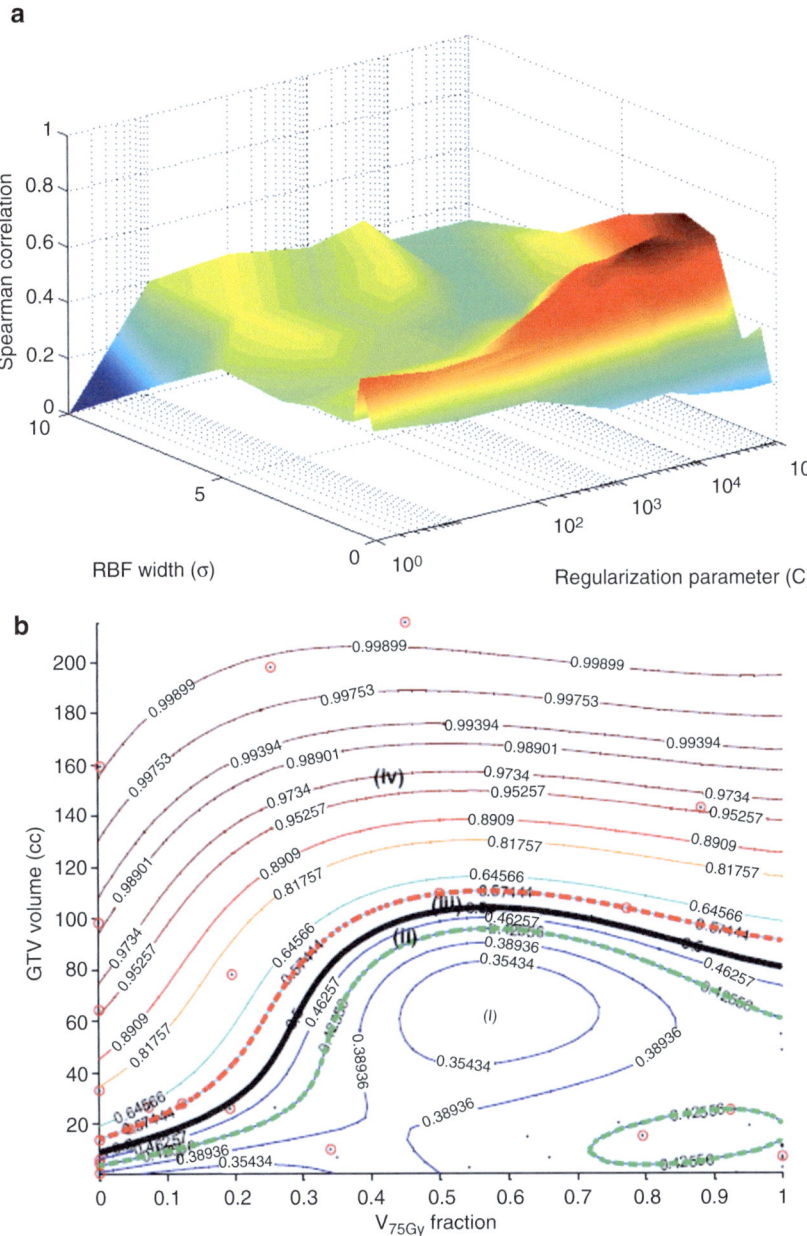

Fig. 18.3 Kernel-based modeling of TCP in lung cancer using the GTV volume and V75 with support vector machine (SVM) and a radial basis function (RBF) kernel. Scatter plot of patient data (*black dots*) being superimposed with failure cases represented with red circles. (**a**) Kernel parameter selection on LOO-CV with peak predictive power attained at $\sigma=2$ and $C=10,000$. (**b**) Plot of the kernel-based local failure (1-TCP) nonlinear prediction model with four different risk regions: (i) area of low-risk patients with high confidence prediction level, (ii) area of low-risk patients with lower confidence prediction level, (iii) area of high-risk patients with lower confidence prediction level, and (iv) area of high-risk patients with high confidence prediction level. Note that patients within the "margin" (cases ii and iii) represent intermediate-risk patients, which have border characteristics that could belong to either risk group

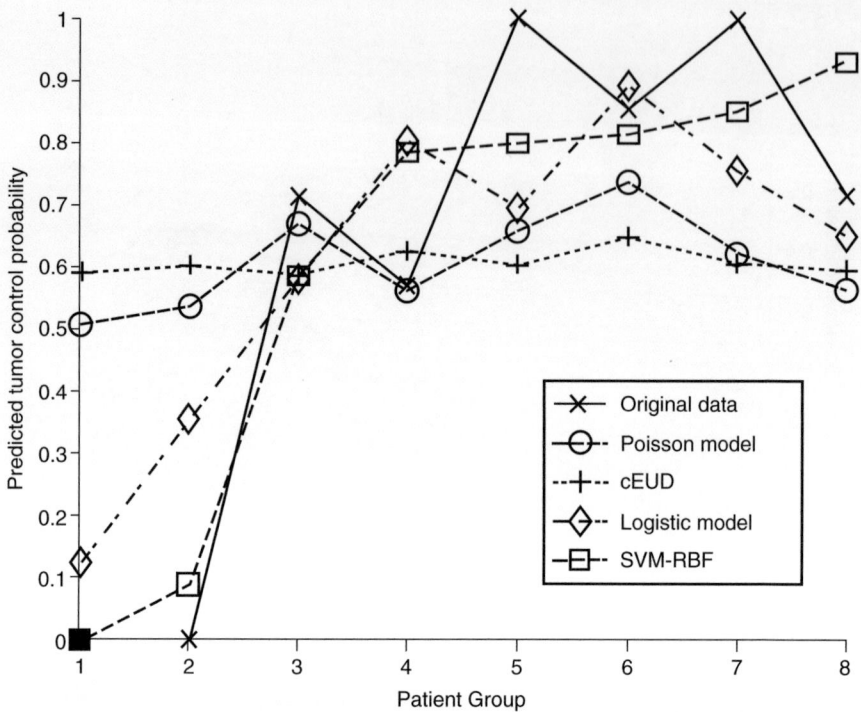

Fig. 18.4 A TCP comparison plot of different models as a function of patients being binned into equal groups using the model with highest predictive power (SVM-RBF). The SVM-RBF is compared to Poisson-based TCP, cEUD, and best two-parameter logistic model. It is noted that prediction of low-risk (high-control) patients is quite similar; however, the SVM-RBF provides a significant superior performance in predicting high-risk (low-control) patients

18.6 Use of Biological Markers

A biomarker is defined as a characteristic that is objectively measured and evaluated as an indicator of normal biological processes, pathological processes, or pharmacological responses to a therapeutic intervention [43]. Biomarkers can be imaging biomarkers as discussed in section 19.5 or measurements of gene expression or protein levels from tissue or fluid specimens. For instance, blood-based protein expression of hypoxia [44] and inflammation [45] were shown to be predictive of tumor response to radiotherapy. Therefore, we conducted a comparison study of physical factors, biological factors extracted from blood sera, and a combined

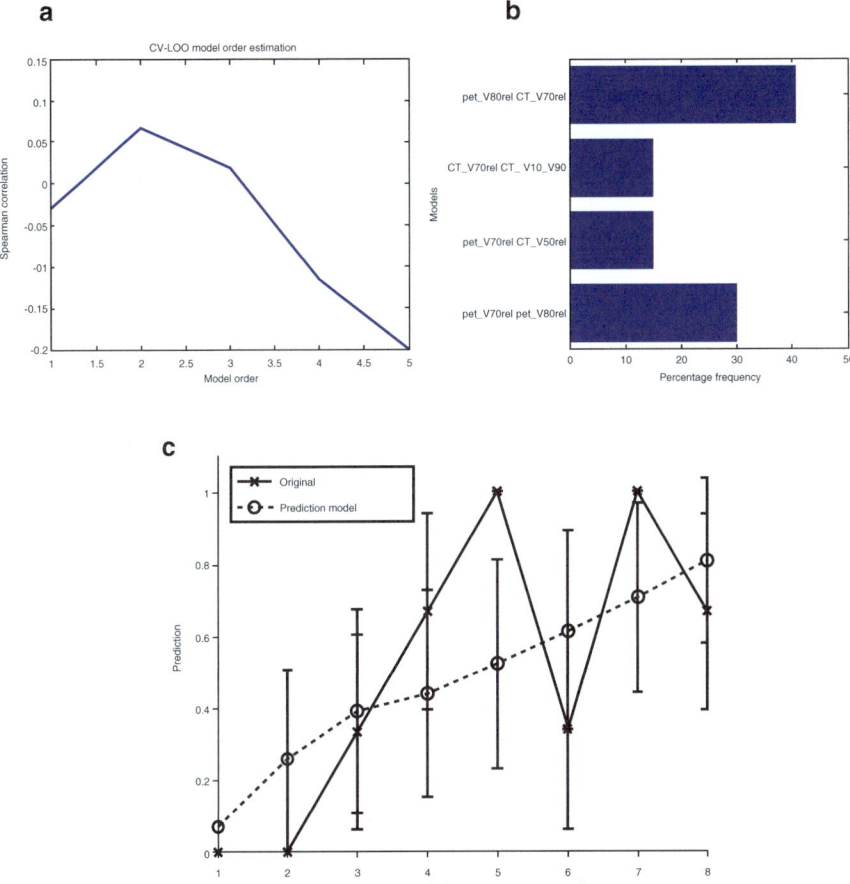

Fig. 18.5 Multimetric modeling of locoregional failure from PET/CT features. (**a**) Model order selection using leave-one-out cross-validation. (**b**) Most frequent model selection using bootstrap analysis. (**c**) Plot of locoregional failure probability as a function of patients binned into equal-size groups showing the model prediction and the original data

model of local control in NSCLC patients. In order to account for the hierarchal relationship between the different variables, we utilized a graphical Bayesian network (BN) framework. A BN is a probabilistic graphical model of outcomes in which the variables (dosimetric, clinical, and biological) are presented as nodes in the graph and their conditional dependencies are represented by directed acyclic graph as shown in Fig. 18.6 [46].

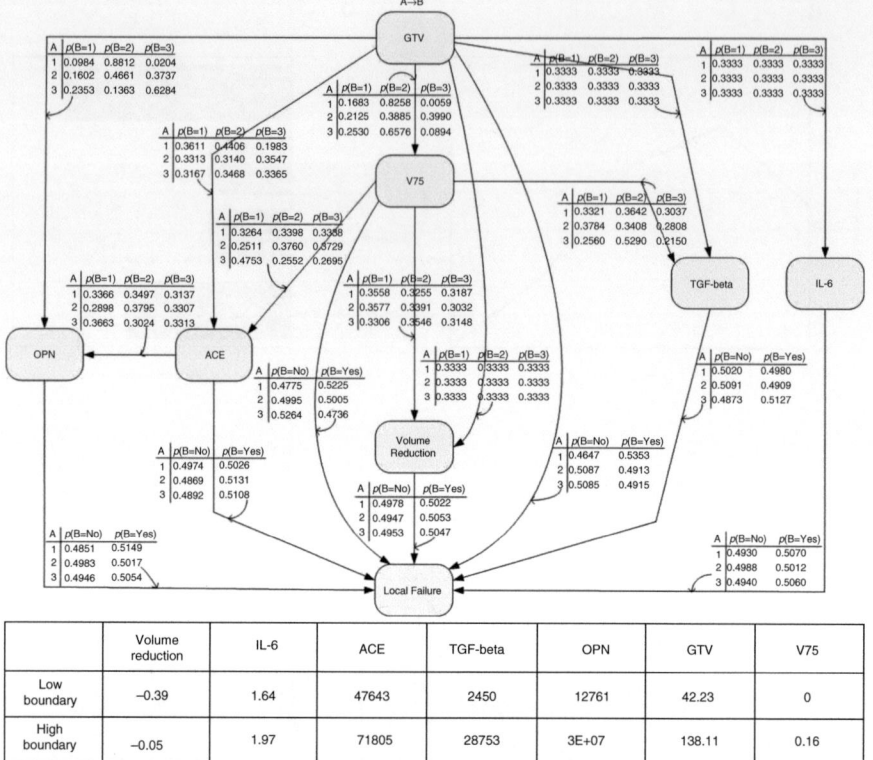

Fig. 18.6 *Top* Bayesian network with probability tables for combined biomarker proteins and physical variables for modeling local tumor control in NSCLC. *Bottom* The binning boundaries for each variable

Conclusions

Recent evolution in imaging and biotechnology has provided new opportunities for reshaping our understanding of radiotherapy response. However, the complexity of radiation-induced effects and the variability of tumor and normal tissue responses would render the utilization of machine-learning algorithms as indispensible tools for better delineation of these complex interaction mechanisms such as the case in modeling TCP. Machine-learning algorithms based on PCA allow for analyzing the complexity of such interaction and reduce the dimensionality of the problem. The use of kernel-based methods such as SVM demonstrated superior ability to predicting local control in NSCLC compared to the state of the art. Moreover, methods based on Bayesian networks allowed for combining physical and biological variables while accounting for the hierarchal relationships between the different variables yielding improved models.

References

1. Bortfeld T, Schmidt-Ullrich R, De Neve W, Wazer D, editors. Image-guided IMRT. Berlin: Springer; 2006.
2. Webb S. The physics of three-dimensional radiation therapy: conformal radiotherapy, radiosurgery, and treatment planning. Bristol/Philadelphia: Institute of Physics Pub; 2001.
3. Halperin EC, Perez CA, Brady LW. Perez and Brady's principles and practice of radiation oncology. 5th ed. Philadelphia: Wolters Kluwer Health/Lippincott Williams & Wilkins; 2008.
4. Moissenko V, Deasy JO, Van Dyk J. Radiobiological modeling for treatment planning. In: Van Dyk J, editor. The modern technology of radiation oncology: a compendium for medical physicists and radiation oncologists. Madison: Medical Physics Publishing; 2005. p. 185–220.
5. Choi N, Baumann M, Flentjie M, Kellokumpu-Lehtinen P, Senan S, Zamboglou N, et al. Predictive factors in radiotherapy for non-small cell lung cancer: present status. Lung Cancer. 2001;31:43–56.
6. Fu XL, Zhu XZ, Shi DR, Xiu LZ, Wang LJ, Zhao S, et al. Study of prognostic predictors for non-small cell lung cancer. Lung Cancer. 1999;23:143–52.
7. Blanco AI, Chao KS, El Naqa I, Franklin GE, Zakarian K, Vicic M, et al. Dose-volume modeling of salivary function in patients with head-and-neck cancer receiving radiotherapy. Int J Radiat Oncol Biol Phys. 2005;62:1055–69.
8. Bradley J, Deasy JO, Bentzen S, El-Naqa I. Dosimetric correlates for acute esophagitis in patients treated with radiotherapy for lung carcinoma. Int J Radiat Oncol Biol Phys. 2004;58:1106–13.
9. Marks LB. Dosimetric predictors of radiation-induced lung injury. Int J Radiat Oncol Biol Phys. 2002;54:313–6.
10. Hope AJ, Lindsay PE, El Naqa I, Bradley JD, Vicic M, Deasy JO. Clinical, dosimetric, and location-related factors to predict local control in non-small cell lung cancer. ASTRO 47th annual meeting. Denver; 2005. p. S231.
11. Tucker SL, Cheung R, Dong L, Liu HH, Thames HD, Huang EH, et al. Dose-volume response analyses of late rectal bleeding after radiotherapy for prostate cancer. Int J Radiat Oncol Biol Phys. 2004;59:353–65.
12. El Naqa I, Bradley JD, Lindsay PE, Blanco AI, Vicic M, Hope AJ, et al. Multi-variable modeling of radiotherapy outcomes including dose-volume and clinical factors. Int J Radiat Oncol Biol Phys. 2006;64:1275–86.
13. Deasy JO, El Naqa I. Image-based modeling of normal tissue complication probability for radiation therapy. Cancer Treat Res. 2008;139:215–56.
14. Bentzen SM, Constine LS, Deasy JO, Eisbruch A, Jackson A, Marks LB, et al. Quantitative Analyses of Normal Tissue Effects in the Clinic (QUANTEC): an introduction to the scientific issues. Int J Radiat Oncol Biol Phys. 2010;76:S3–9. doi:10.1016/j.ijrobp.2009.09.040. S0360-3016(09)03300-8 [pii].
15. Jackson A, Marks LB, Bentzen SM, Eisbruch A, Yorke ED, Ten Haken RK, et al. The lessons of QUANTEC: recommendations for reporting and gathering data on dose-volume dependencies of treatment outcome. Int J Radiat Oncol Biol Phys. 2010;76:S155–60. doi:10.1016/j.ijrobp.2009.08.074. S0360-3016(09)03299-4 [pii].
16. El Naqa I. Machine learning methods for predicting tumor response in lung cancer. Wiley Interdisciplinary Reviews: Data Mining and Knowledge Discovery. 2012;2:173–81. doi:10.1002/widm.1047.
17. Munro TR, Gilbert CW. The relation between tumour lethal doses and the radiosensitivity of tumour cells. Br J Radiol. 1961;34:246–51. doi:10.1259/0007-1285-34-400-246.
18. Hall EJ, Giaccia AJ. Radiobiology for the radiologist. 6th ed. Philadelphia: Lippincott Williams & Wilkins; 2006.
19. Joiner M, Kogel A. Basic clinical radiobiology. 4th ed. London: Hodder Arnold; 2009.
20. Goitein M. Tumor control probability for an inhomogeneously irradiated target volume. In: Zink S, editor. Evaluation of treatment planning for particle beam radiotherapy. Bethesda: National Cancer Institute; 1987.

21. Zaider M, Minerbo GN. Tumour control probability: a formulation applicable to any temporal protocol of dose delivery. Phys Med Biol. 2000;45:279–93.
22. Hall EJ. Radiobiology for the radiologist. 4th ed. Philadelphia: J.B. Lippincott; 1994.
23. Zaider M, Hanin L. Tumor control probability in radiation treatment. Med Phys. 2011;38:574–83.
24. Lindsay PE, El Naqa I, Hope AJ, Vicic M, Cui J, Bradley JD, et al. Retrospective monte carlo dose calculations with limited beam weight information. Med Phys. 2007;34:334–46.
25. Deasy JO, Blanco AI, Clark VH. CERR: a computational environment for radiotherapy research. Med Phys. 2003;30:979–85.
26. El Naqa I, Deasy J, Mu Y, Huang E, Hope A, Lindsay P, et al. Datamining approaches for modeling tumor control probability. Acta Oncol. 2010;49(8):1363–73.
27. Kennedy R, Lee Y, Van Roy B, Reed CD, Lippman RP. Solving data mining problems through pattern recognition. Upper Saddle River, NJ, London: Prentice Hall; 1998.
28. Willner J, Baier K, Caragiani E, Tschammler A, Flentje M. Dose, volume, and tumor control prediction in primary radiotherapy of non-small-cell lung cancer. Int J Radiat Oncol Biol Phys. 2002;52:382–9.
29. Martel MK, Ten Haken RK, Hazuka MB, Kessler ML, Strawderman M, Turrisi AT, et al. Estimation of tumor control probability model parameters from 3-D dose distributions of non-small cell lung cancer patients. Lung Cancer. 1999;24:31–7.
30. Mehta M, Scrimger R, Mackie R, Paliwal B, Chappell R, Fowler J. A new approach to dose escalation in non-small-cell lung cancer. Int J Radiat Oncol Biol Phys. 2001;49:23–33. doi:S0360-3016(00)01374-2 [pii].
31. Brodin O, Lennartsson L, Nilsson S. Single-dose and fractionated irradiation of four human lung cancer cell lines in vitro. Acta Oncol. 1991;30:967–74.
32. Seibert RM, Ramsey CR, Hines JW, Kupelian PA, Langen KM, Meeks SL, et al. A model for predicting lung cancer response to therapy. Int J Radiat Oncol Biol Phys. 2007;67:601–9.
33. Ramsey CR, Langen KM, Kupelian PA, Scaperoth DD, Meeks SL, Mahan SL, et al. A technique for adaptive image-guided helical tomotherapy for lung cancer. Int J Radiat Oncol Biol Phys. 2006;64:1237–44.
34. Borst GR, Belderbos JS, Boellaard R, Comans EF, De Jaeger K, Lammertsma AA, et al. Standardised FDG uptake: a prognostic factor for inoperable non-small cell lung cancer. Eur J Cancer. 2005;41:1533–41.
35. Levine EA, Farmer MR, Clark P, Mishra G, Ho C, Geisinger KR, et al. Predictive value of 18-fluoro-deoxy-glucose-positron emission tomography (18F-FDG-PET) in the identification of responders to chemoradiation therapy for the treatment of locally advanced esophageal cancer. Ann Surg. 2006;243:472–8.
36. Ben-Haim S, Ell P. 18F-FDG PET and PET/CT in the evaluation of cancer treatment response. J Nucl Med. 2009;50:88–99. doi:10.2967/jnumed.108.054205.
37. El Naqa I, Grigsby PW, Apte A, Kidd E, Donnelly E, Khullar D, et al. Exploring feature-based approaches in PET images for predicting cancer treatment outcomes. Pattern Recognit. 2009;42:1162–71.
38. Mac Manus MP, Hicks RJ, Matthews JP, Wirth A, Rischin D, Ball DL. Metabolic (FDG-PET) response after radical radiotherapy/chemoradiotherapy for non-small cell lung cancer correlates with patterns of failure. Lung Cancer. 2005;49:95–108. doi:10.1016/j.lungcan.2004.11.024. S0169-5002(04)00625-7 [pii].
39. Yamamoto Y, Nishiyama Y, Monden T, Sasakawa Y, Ohkawa M, Gotoh M, et al. Correlation of FDG-PET findings with histopathology in the assessment of response to induction chemoradiotherapy in non-small cell lung cancer. Eur J Nucl Med Mol Imaging. 2006;33:140–7.
40. Pieterman RM, van Putten JW, Meuzelaar JJ, Mooyaart EL, Vaalburg W, Koeter GH, et al. Preoperative staging of non-small-cell lung cancer with positron-emission tomography. N Engl J Med. 2000;343:254–61.
41. Wong CY, Schmidt J, Bong JS, Chundru S, Kestin L, Yan D, et al. Correlating metabolic and anatomic responses of primary lung cancers to radiotherapy by combined F-18 FDG PET-CT imaging. Radiat Oncol. 2007;2:18. doi:10.1186/1748-717X-2-18. 1748-717X-2-18 [pii].

42. Vaidya M, Creach KM, Frye J, Dehdashti F, Bradley JD, El Naqa I. Combined PET/CT image characteristics for radiotherapy tumor response in lung cancer. Radiother Oncol. 2012;102: 239–45. doi:10.1016/j.radonc.2011.10.014. S0167-8140(11)00626-8 [pii].
43. Group BDW. Biomarkers and surrogate endpoints: preferred definitions and conceptual framework. Clin Pharmacol Ther. 2001;69:89–95. doi:10.1067/mcp.2001.113989. S0009-9236(01) 63448-9 [pii].
44. Le Q-T, Chen E, Salim A, Cao H, Kong CS, Whyte R, et al. An evaluation of tumor oxygenation and gene expression in patients with early stage non-small cell lung cancers. Clin Cancer Res. 2006;12:1507–14. doi:10.1158/1078-0432.ccr-05-2049.
45. Rube CE, Palm J, Erren M, Fleckenstein J, KÃnig J, Remberger K, et al. Cytokine plasma levels: reliable predictors for radiation pneumonitis? PLoS One. 2008;3:e2898.
46. Oh JH, Craft J, Al Lozi R, Vaidya M, Meng Y, Deasy JO, et al. A Bayesian network approach for modeling local failure in lung cancer. Phys Med Biol. 2011;56:1635–51. doi:10.1088/0031-9155/56/6/008. S0031-9155(11)60164-4 [pii].

Index

A
Acute haematologic toxicity, 289, 290
Akaike information criteria (AIC), 18
Anomaly detection, radiotherapy
   decision function, 239
   kernel function, 239
   probability density function, 238
   QA, 239–241
   quadratic programming, 239
   SVM, 239
Artificial neural networks (ANNs)
   area under the curve (AUC), 293, 295
   beam/tumor alignment, 204
   breathing pattern, 204
   chemotherapy and dosimetry, 292
   correlation over time, 204–206
   dose-surface histogram, 294, 295
   dosimetric information, 293, 296
   EPID, 257
   EUD, 296
   FEV1, 294
   linear accelerator (LINAC), 204
   LKB model, 293
   lung injury, 292
   lung tumors, 203
   machine learning algorithm, 221
   multivariate logistic regression, 292
   PCA, 228
   pneumonitis, 293, 297
   prostate radiotherapy, 294
   rectal bleeding, 295, 296
   regression model, 122
   ROC analysis, 292
   sigmoidal activation function, 292
   SVM, 228
   treatment targets, 203
   xerostomia, 294
   X-ray fluoroscopic imaging., 204

B
Backward pruning algorithm, 163–165
Bayesian Information Criterion (BIC) score, 18, 19, 25, 35, 246, 249
Bayesian networks (BNs)
   conditional probability values, 34
   DAG (see Directed acyclic graph (DAG))
   dose-volume, 300
   graph modeling technique, 8
   maximum likelihood estimate or MAP, 34
   probabilistic approach, 35
   radiation pneumonitis, 268, 305
   for rectal cancer, 94
   SVM, 35
   techniques, 13
Best matching unit (BMU), 25
BIC. See Bayesian information criteria (BIC) score
Big data resource, radiotherapy
   cancer diagnostics, 266
   DVHs, 266
   exogenous biomarkers, 267
   genetic biomarkers, 267
   genomics and proteomics, 266
   GWAS studies, 267
   NLP techniques, 266
   radiomics, 266–267
   SNPs, 267
Bioinformatics
   data-driven outcomes, 265
   dose-response modeling, 269
   dosimetric radiation pneumonitis, 268
   DREES, 269
   multicenter communication systems, 271
   multivariate modeling, 269
   NLP techniques, 271
   optimal treatment planning, 264
   protein expression assay, 271

Bioinformatics (cont.)
   QUANTEC consortium, 270, 271
   radiation treatment, 264
   radiotherapy outcomes, 264
   rapid learning, 270
   RTOG, 271
   systems radiobiology, 268–269
   TCP/NTCP models, 264, 268
   TPS, 268
   web resources, 270
Biomedicine, 8, 9
BMU. *See* Best matching unit (BMU)
Breathing motion, ANN
   closed-loop beam, 206, 207
   correlator/predictor, 208
   discrete measurements, 208
   open control loop, 206, 207
   respiration and tumor position, 208
   signal amplitude, 208, 209
   spatial correlation, 209
   target signal, 208
Brier score, 92

C
CADe. *See* Computer-aided detection (CADe)
CADx. *See* Computer-aided diagnosis (CADx)
Cancer
   machine learning algorithms, 8
   radiotherapy, 5, 57–58, 66
   SVM, 31, 35
Caradigm Intelligence Platform, 64
CBIR. *See* Content-based image retrieval (CBIR)
Cell kill equivalent uniform dose (cEUD), 316, 318
Centralized machine learning, 88
Chest radiographs (CXRs)
   convolution neural network, 118
   difference-image technique, 117–118
   FDA-approved product, 120
   FP reduction technique, 119
   imaging examination, 117
   medium-resolution image, 118
   multiresolution composition technique, 118
   nodule detection, 120
   overlying bones, 118
   suppression, bones, 119
Classifier function, 134
Clinical data research networks (CDRN), 94
Clinical decision support systems (CDSS), 72
Clinical medical electronic record, 60
Clinical Research Chart (CRC), 93
Clustering
   BMU, 25
   intuitive and succinct representation, 24
   K-means clustering, 25
   optimization method, 25
   radiotherapy toxicity modeling, 26
   SOM/Kohonen map, 25, 26
   vector quantization, 24
Colonic imaging
   colorectal cancer detection, 120–121
   polyps, CTC, 121–122
Colorectal cancer detection, 120–121
Computational Environment for Radiotherapy Research (CERR), 63
Computational learning theory
   information theory, 18–19
   learning capacity/learnability
      definition, 15
      PAC learning, 14, 16
      QA, 15
      training process, 15
      VC dimension, 14, 16–17
   maximum likelihood/Bayesian techniques, 13
   modern computational learning theory, 14
   PCA, 17
   recursive elimination technique, 17
   resampling methods, 19
   statistical learning theory, 17–18
   vs. statistics
      hypothesis generation, 14–15
      hypothesis testing, 14
      QA in radiotherapy, 15
Computed tomography (CT), 58, 60–62
Computer-aided detection (CADe)
   categorization, 101–102
   characterization, 102
   colonic imaging, 120–122
   flowchart, 103
   FPs, 104
   ML, 102, 104–109
   pattern features, 104
   PML, 102–103
   thoracic imaging, 110–120
Computer-aided diagnosis (CADx)
   CBIR, 142–144
   classifier training and performance evaluation, 138–139
   mammography, 134
   microcalcification lesions, 139–141
Computer-aided segmentation, 158
Cone-beam CT (CBCT), 158, 231
Conformal radiation therapy (CRT), 194, 313
Content-based image retrieval (CBIR)
   classification performance, 143, 144
   conventional CADx, 142

features, 143
  Gaussian RBF kernel function, 143
  logistic regression, 142
  tumor classification, 142
Correlation, ANN
  breathing signal inputs, 216–217
  feedforward network, 212–214
  Kalman filter, 214–216
  nonlinear networks, 212
  PCA, 288
  recurrent network, 214
  sigmoid function, 212
  single neuron/linear filter, 209–212
Cost function and MapReduce, 90–91
CT colonography (CTC), polyps
  ANN regression model, 122
  CADe output, 121
  classification method, 121
  detection, 121
  3D MTANN, 121
  ML approaches, 122
CXRs. *See* Chest radiographs (CXRs)

D

DAG. *See* Directed acyclic graph (DAG)
Database/SQL, 65
Data warehousing (DWH), 73–75
Decision function, 134
Decision trees
  boosting, 32–33
  ensemble learning, 32
  ID3 (iterative dichotomizer 3) algorithm, 32
  learning process, 31–32
  NTCP
    AdaBoost algorithm, 301–302
    AUC, 302
    dichotomisation, 301
    Lyman model, 302
    NSCLC, 302
    radiation pneumonitis, 301, 302
    recursive partitioning model, 301
    toxicity, 301
    univariate logistic regression, 302
  radiotherapy, 33
  random forest algorithm, 32
  reduced-error pruning, 32
  structure, 31
Deformable image registration (DIR), 230
Deformable-model-based methods, 183
Dempster–Shafer theory (DST), 196
Dice similarity coefficient (DSC)
  ILSM, 167
  prostate segmentation, 172, 188
  standard deviation, 188

DICOMan, 63
Dicompyler, 62–63
DICOM-RT, 62–63
Digital Imaging and Communications in Medicine (DICOM), 62
Digitally reconstructed radiographs (DRRs), 257
Directed acyclic graph (DAG)
  BIC score, 35
  clinical variables, 33–34
  $K2$ algorithm, 35
  marginal likelihood (Bayesian) score, 35
  MCMC, 35
  probability distributions, 34
  treelike structures (Chow-Liu trees), 35
Discriminant sub-dictionary learning (DSL), 169–170, 172, 173
Displacement vector fields (DVFs), 230, 232
Distributed discriminative dictionary (DDD)
  atoms, 185
  deformable segmentation, 185, 187
  divide-and-conquer strategy, 186
  Fisher-LDA residual integration, 186
  minimal-redundancy, 186
  mRMR, 186
  radiotherapy dose, 184
  sparse representation technique, 184
  SRC, 185
  SSC, 184
Distributed machine learning
  cost function and MapReduce, 90–91
  linear regression implementation
    cost function, 89–90
    MapReduce concept, 88
    training algorithm, 88–89
  MapReduce and multicenter learning, 91–92
  training, testing, and validation, 92
Dose response explorer (DREES), 266, 269
Dose-volume histogram (DVH)
  correlation matrix, 313
  Gaussian distributions, 195
  KBTP, 195
  LKB, 280
  PCA, 286
  TPS, 62
Dosimetric data reduction
  dose-response, 282
  equivalent uniform dose, 281
  FSU, 281, 282
  LKB model, 282
  organ/structure, 281
  relative seriality model, 282

Dosimetric variables, TCP
    cEUD model, 316
    correlation matrix, 313, 314
    CRT, 313
    data exploration, 313
    DVH, 313
    kernel-based modeling, 315–316
    lung cancer, 315–317
    nonlinear prediction model, 316, 317
    NSCLC, 313
    PCA, 313, 314
    Poisson-based, 316, 318
    SVM-RBF, 316–318
Dosimetric verification
    EWMA, 258, 259
    IMRT, 258
    MLC, 258
    out-of-control measurements, 259
    prostate treatments, 258, 259
    SPC, 257–258

E

Electronic medical record (EMR), 73, 81
Electronic portal imaging device (EPID), 256, 257
Entity-relationship (ER) model, 75
Equivalent uniform dose (EUD)
    ANNs, 286
    DVH, 285
    radiation pneumonitis, 300
Error events. *See* Radiotherapy
ETL tooling. *See* Extraction, transformation, and load (ETL) tooling
Euregional Computer-Aided Theragnostics (EuroCAT), 93
Evidence-based medicine (EBM), 72
Exponentially weighted moving average (EWMA), 258, 259
Extended Kalman filter (EKF)
    ANN, 215
    breathing system, 215
    recurrent network, 216
Extraction, transformation, and load (ETL) tooling, 73, 76

F

Failure mode and effect analysis (FMEA), 238
Feature-based machine learning, 104–105
Feature/variable selection
    acute haematological toxicity, 289
    cervical cancer, 289
    DVH, 285
    liver radiotherapy, 286
    logistic regression, 286
    model fitting, 285
    multivariate models, 287
    parotid gland dose, 286, 287
    PCA, 285
    pelvic bone marrow, 289, 290
    prostate radiotherapy, 288
    rectal bleeding, 287
    toxicity/nontoxicity, 290
    varimax rotation, 286
    xerostomia, 286, 287
Feed-forward neural networks (FFNN)
    autocorrelation coefficient, 220
    back propagation, 213
    batch mode or sequential mode, 29
    breathing signal's decay time, 220
    hidden neuron, 218
    LMS algorithm, 213
    lung cancer, 219
    multiple neurons, 218
    neural network architecture, 28
    optimal convergence, 214
    prediction filter, 219
    root-mean square error, 218
    single neuron, 212, 213
    validation signal, 214
Filter learning, 106
Fluorodeoxyglucose (FDG), 316
Forced expiration volume 1 (FEV1), 292, 294
Forward learning algorithm, 164
Friedman's test, 54–56
Functional subunit (FSU), 281, 282

G

Generalized linear discriminant analysis (GLDA), 228
General regression neural networks (GRNN), 29
Genitourinary (GU) toxicity, 31, 288
Genome-wide association studies (GWAS), 267, 284
GIGO principle, 9
Gleason score, 22
Graphic processing unit (GPU), 231
GRNN. *See* General regression neural networks (GRNN)

H

Health Insurance Portability and Accountability Act (HIPPA), 271
HIS. *See* Hospital information system (HIS)
Histogram Analysis in Radiation Therapy (HART), 67
Histogram-of-oriented-gradient (HOG), 176, 179, 181

Hosmer–Lemeshow test, 92
Hospital information system (HIS), 59, 65
Hospital information system (HL7), 65
Human-machine interaction, 7–8

I
I2B2. *See* Informatics for Integrating Biology and the Bedside (I2B2)
ID3 (iterative dichotomizer 3) algorithm, 32
Image biomarker extraction
　communication protocols, 74
　DICOM images, 74
　PACS, 74
　radiomic analysis, 74
Image-guided radiotherapy (IGRT)
　ASM method, 188
　atlas-based segmentation, 159
　automatic segmentations, 184
　CBCT, 158
　deformable segmentation, 160
　Dice ratio, 183
　DSC, 188
　DSL, 173
　elastic net, 173
　hybrid approaches, 159
　inter-operator variation, 158
　organ segmentation, 159
　pelvic bone structures, 181
　PPV, 189
　RBLR, 173
　residue linear regression, 171
　robustness, 178
　SCOTO, 181–182
　segmentation accuracy, 178
　SVR model, 182
　true-positive fraction (TPF), 182, 183
　tumor tissue segmentation, 158
Incremental learning with selective memory (ILSM)
　cascade learning, 163
　patient-specific prostates, 164
　personalized anatomy detector, 165
　population-based landmark, 163
　prostate localization, 167, 168
　pruning and learning, 163
Independent component analysis (ICA), 87
Individualized medicine (IM), 72
InfoMaker, 64
Informatics for Integrating Biology and the Bedside (I2B2), 93
Information theory, 18–19
Institutional infrastructure
　traditional ETL and DWH, 81
　　with RDF store, 81
　　with virtual RDF store, 81–82
　virtual RDF store
　　per institute, 82–83
　　per source and institute, 83–84
Integrating the Healthcare Enterprise-Radiation Oncology (IHE-RO), 59–60
Intensity-modulated radiotherapy (IMRT)
　bladder and rectum, 195, 197
　deformation, 196, 197
　dose delivery process, 258
　dosimetry, 257
　DVHs, 195
　prostate cancer, 195
　treatment delivery validation, 253

J
"Jackknife". *See* "Leave-one-out" cross-validation (LOO-CV) procedure

K
Kalman filter
　breathing signal, 215
　EKF, 215
　error covariance matrix, 216
　plant and measurement noise, 214
　prediction/correction loop, 215
　recurrent network, 215, 216
Kernel-based methods
　decision tree, 31
　dual optimization problem, 30
　radiotherapy, 31
　support vectors, 30
　SVMs, 30, 31
Kernel PCA, 24
Kernel trick, 137
K-fold cross-validation process, 49–50
Knowledge-based treatment planning (KBTP)
　computer-aided process, 193
　dose distributions, 194
　DVHs, 195, 196
　IMRT, 194
　parotid gland, 195
　prostate planning, 198
　quality control, 195
　tumor structures, 193
Kohonen map, 25, 26
KV x-ray computed tomography (kV-CT), 266

L
"Leave-one-out" cross-validation (LOO-CV) procedure, 19, 50–51
Leave-one-out (LOO) procedure, 138, 141, 143, 291, 297

Linear filter, ANN
  breathing signal, 209
  least mean square method, 211
  prediction filter, 210, 211
  sequential training, 211
  signal amplitude, 210
Linear-quadratic (LQ) model, 265, 312
Linked data, 76–80
Local-binary-pattern (LBP), 176, 179, 181
Logistic regression modeling
  acute haematologic toxicity, 289
  artificial intelligence methods, 28
  binomial deviance, 31
  likelihood function, 27
  model parameters, 27
  sigmoidal form, 26–27
  sparsity constraint, 176
  SVM, 141
LOO procedure. *See* Leave-one-out (LOO) procedure
Lung cancer detection, 110
Lung nodules in CT
  axial slice, 113
  CADe outputs, 117
  CADe system, 111
  CXR, 117–120
  3D Gaussian function, 115
  features, 110
  FP reduction, architecture, 114
  FROC curve indication, 116, 117
  ground-glass nodules, 112
  high-contrast image, 113
  k-nearest-neighbor classifier, 111
  lesion enhancement, 113
  MTANNs, 110, 115
  scoring method, 115
  segmentation, 113
  sources, 112
  and suppression, FPs, 116, 117
  thin-slice screening, 111
Lyman–Kutcher–Burman (LKB) model, 280, 282, 293

M
Machine learning (ML). *See also* Bioinformatics; Tumor control probability (TCP)
  advantages, 5
  approaches, 7–8
  Bayesian networks, 300–301
  biomedicine, 8
  characterization, 136
  classification, 136, 138
  confusion matrix, 291
  correlation coefficient, 291
  data mining, 6–7
  decision function, 137
  definition, 6
  discriminant or generative models, 8
  DVH, 285
  EPID, 254, 257
  error detection, 244–245
  evaluation
    application-specific domain, 42
    classifier evaluation framework, 42, 43
    generic algorithm, 42
    generic classifiers, 43
    multiple classifiers, 43
    performance measures, 43–44
  feature-based (segmented-object-based) and classifiers, 104–105
  GIGO principle, 8–9
  IMRT, 258
  input data, 4, 5, 8
  kernel trick, 137
  MDS, 145–148
  medical physics, 8
  multilayer perceptron (MLP), 6
  non-linear model, 285
  optimum principles, 137
  parsimony, 9
  PCA, 227
  perceptron, development of, 6
  PML, 105–109
  radiation delivery process, 254
  radiation oncology, 8
  radiotherapy, 5 (*see also* Motion management)
  reinforcement learning, 8
  respiratory gating, 227
  ROC analysis, 290
  semi-supervised learning, 5, 7
  sensitivity and specificity, 290–291
  supervised learning, 4, 7
  template matching, 227
  training, 4
  transductive and inductive learning, 7–8
  unsupervised learning, 4, 7
Malignant and benign tumors
  CADx schemes, 148–149
  classification framework

Index 331

CADx training and evaluation, 138–139
components, 134
feature extraction, quantification, 135–136
machine learning, 136–138
perception modeling, 135
development, 133
diagnostic accuracy, 148
mammography, 134, 139–144
MDS, visualization tool, 144–148
Mammography
CADx and MCs, 139–141
cancerous/precancerous tumor, 139
CBIR, 142–144
MAP. See Maximum a posteriori (MAP)
MapReduce
cost function, implementation, 90–91
distributed and multicenter learning, 91–92
Marginal likelihood (Bayesian) score, 35
Markov Chain Monte Carlo (MCMC), 35
Massive-training artificial neural networks (MTANNs)
activation functions, 107
ANN, 107
architecture, 107
center voxel, 108
development, 106
single pixels, 108
structure, 108
"teaching" images/volumes, 108
MATLAB, 62
Maximum a posteriori (MAP), 34
Maximum likelihood estimation (MLE), 283
MCMC. See Markov Chain Monte Carlo (MCMC)
McNemar's test, 52–53
MDL. See Minimum description length (MDL)
Microcalcification clusters (MCCs), 299
Microcalcification lesions
CADx techniques, 140
cancer diagnosis, 139
mammogram, 139, 140
MC classification, 141
Microsoft Amalga Unified Intelligence System, 64
Minimum description length (MDL), 25
ML. See Machine learning (ML)
MOSAIQ, 64
Motion management
anterior-posterior direction, 232
dose delivery, 225–226

fiducial markers, 226
real-time tumor, 226
respiratory tumor, 225
tumor localization, 226
x-ray imaging system, 226
MTANNs. See Massive-training artificial neural networks (MTANNs)
Multi-atlas-based image segmentation, 174
Multicenter infrastructure
centralized, 84, 85
distributed, 84–86
Multicenter learning
applications
EuroCAT, 93
I2B2, 93
PCORnet, 94
VATE, 93–94
centralized machine learning, 88
data extraction
biological data, 73
data sources, 73
ETL tooling and data warehousing, 73
image biomarker extraction, 74–75
data representation
ICD-10, 75
National Cancer Institute's Thesaurus (NCIT), 75
relational databases and ontologies, 75–76
semantic interoperability, 75
Semantic Web technologies, 76–80
syntactical interoperability, 75
distributed (see Distributed machine learning)
network infrastructure
institutional infrastructure, 80–84
multicenter infrastructure, 84–86
privacy preservation, 86–87
Multidimensional scaling (MDS)
data embedding technique, 145
MC lesions, 145–148
retrieval framework, 144
Multilayer perceptron (MLP), 6
Multileaf collimator (MLC), 63–64, 204
Multiresolution composition technique, 118

N
Naive Bayes
inaccurate probability, 36
MAP rule, 36
naive independence assumption, 35–36

National Cancer Institute's Thesaurus (NCIT), 75
Natural language processing (NLP), 266
Non-small-cell lung cancer (NSCLC), 298
Normal tissue complication probability (NTCP)
  dose distributions, 278, 280
  dose-volume constraints, 283, 284
  DVH, 279
  fractional irradiation, 280
  LKB, 280
  parameter fitting, 283
  prostate radiotherapy, 283
  QUANTEC report, 280
  quantification, 282–283
  radiotherapy, 277, 278
  rectal toxicity, 284
  robustness, 284
  sigmoidal response, 280
  TCP, 277, 278
  therapeutic ratio, 277
  treatment planning process, 280, 284
NTCP, hybrid models
  acute dysphagia, 304
  ANN, 305
  AUC, 303–304
  Bayesian networks, 305
  dose distribution, 304–305
  prostate radiotherapy, 304
  radiation pneumonitis, 303
  resultant model, 303–304
  sigmoid activation function, 305
  Spearman correlation, 304
  squamous cell histology, 304
  SVM, 305

O

Organs at risk (OARs), 58, 195
Outlier detection, applications
  beam energies, 249
  BIC, 249
  computer clustering, 245
  covariance matrix, 246
  Gaussian distribution, 247
  K-means clustering algorithm, 247
  linear accelerators, 245
  Lloyd's algorithm, 246
  monitor units (MUs), 248, 250
  PCA, 247–249
  probability distribution model, 247
  radiotherapy treatment plans, 245
  sum-squared-error (SSE), 246

P

PAC. See Probably approximately correct (PAC) learning
PACS. See Picture archiving and communication systems (PACS)
Patch-based representation
  LBP, 176
  logistic regression process, 176
  Nesterov's method, 176
  reference voxel, 176
  sparse label propagation, 176
Patch-/pixel-based machine learning (PML)
  convolution neural networks, 106
  and feature-based ML (classifiers), 109
  FP reduction, 106
  medical image processing/analysis, 105
  MTANNs, 106–109
  neural filters, 106
Patient-Centered Outcomes Research Network (PCORnet), 94
Patient management system (PMS), 61, 64
Patient-powered research networks (PPRN), 94
Patient-specific landmark detectors, 166
PCA. See Principal component analysis (PCA)
PCORnet. See Patient-Centered Outcomes Research Network (PCORnet)
Performance metrics
  machine learning evaluation (see Machine learning (ML))
  performance measures
    generic confusion matrix, 44, 45
    matrices, accuracy, 45–46
    overview, 44, 45
    precision and recall, 46
    ROC analysis, 46–48
  significance testing
    Friedman's test, 54–56
    McNemar's test, 52–53
    parametric tests, 52
    resampling, 52
    statistical testing, 52, 53
    Wilcoxon's signed-rank test, 53–54
Physical and biological variables, 320
Picture archiving and communication systems (PACS), 59, 66
Plan validation. See Treatment planning validation
PML. See Patch-/pixel-based machine learning (PML)
PMS. See Patient management system (PMS)
Positive prediction value (PPV), 189
Positron emission tomography (PET), 316

Index 333

Principal component analysis (PCA), 17
   covariance matrix, 289
   DVH, 286
   Kernel, 24
   linear
      esophagitis and pneumonitis, 23
      Gleason score, 22
      kernel methods, 23
      xerostomia, 22–23
      z-scoring, 22
   NTCP, 286
   visualisation, 286
Privacy preservation
   data obfuscation, 87
   data perturbation, 87
   pseudonymization
      bidirectional, 86
      unidirectional, 86
Probably approximately correct (PAC) learning, 14, 16
Prostate localization, 166
Prostate segmentation methods
   cascade learning, 161
   dictionary learning method, 169–170
   DSC, 172
   elastic net, 170–171
   finer resolution, 162
   ILSM, 161
   landmark detection, 161, 162
   multi-atlas segmentation, 163
   optimization, 162
   population-based learning, 163
   real-time face detection, 161
   rigid transform, 162–163
   traditional learning approaches, 166–167

Q

Quality assurance (QA)
   applications, 65
   data collection, 239
   NSCLC, 240
   PCA, 240
   radiation physics, 5, 8
   in radiotherapy, 15
   radiotherapy processes, 241
   RBF, 241
   requirements, 63
   SBRT, 240
   SVM, 239–241
   treatment planning, 239–240
Quality control (QC), 195, 198

Quantitative Analyses of Normal Tissue Effects in the Clinic (QUANTEC), 280

R

Radial basis function (RBF), 241, 315–316
Radiation delivery
   action thresholds, 256
   linear accelerators, 254
   machine learning tools, 254
   QA process, 256
   RBA-5, 255
   standard deviation, 255
Radiation oncology (RO), 8
   computed tomography (CT), 58
   CT scanner, 61–62
   data aggregation and analysis programs, 67
   database/SQL, 65
   DICOM-RT, 62–63, 66
   digital information, 59
   electronic information systems, 59
   EUROCAT project, 67
   guidelines/recommendations, 67
   healthcare process, 58
   high level/external sources, 66–67
   HIS/HL7, 65
   IHE-RO, 59–60
   imaging and radiation dose distribution, 60
   PACS, 59, 66
   patient management systems, 60
   peripheral sources/billing data, 66
   PMS, 64
   QA, 65
   radiotherapy, 59
   R&V system, 63–64
   TPS, 58, 62
   tumor staging, 58
   vendors, 67
Radiation therapy (RT), 57
Radiation therapy oncology group (RTOG), 271
Radiotherapy, 5, 8
   anthropomorphic phantom, 238
   community, 244
   dose delivery, 225–226
   error detection/prediction, 238
   FMEA, 238
   motion management, 226
   outlier detection, 245
   quality assurance, 15
   respiratory tumor motion, 225
   treatment plan, 244

Radiotherapy outcomes, 264, 265
Random projection-based multiplicative perturbation (RPBMP) method, 87
Real-time tumor tracking
  CBCT, 231
  DVFs, 230, 232
  fiducial tracking method, 230
  GPU, 231
  machine learning, 229
  PCA, 228–230
  regression analysis, 228
  ROI, 228, 229
  single x-ray projection, 230–232
  spatiotemporal patterns, 229
  stochastic tracking method, 230
  tumor localization, 231, 232
Receiver-operator curve (ROC) analysis, 292
Record and verify system (R&V), 60, 73
  billing systems, 64
  DICOM-RT Plan, 64
  linear accelerators, 63–64
  multileaf collimator (MLC), 63–64
  tools, 64
Recursive feature elimination (RFE) method, 315
Region of interest (ROI), 228, 229
Reinforcement learning (RL)
  definition, 36
  feedback system, 8
  Markov decision process, 37
Relational Database Management System (RDBMS), 76
Resampling
  bootstrapping, 19, 51
  cross-validation methods, 19
  holdout method, 49
  k-fold cross-validation process, 49–50
  leave-one-out or the jackknife process, 50–51
  multiple resampling, 48
  simple resampling, 48
Residue based linear regression (RBLR), 173
Resource description framework (RDF), 76–77, 94
Respiratory gating
  ANN, 228
  binary classification, 227
  dimensionality reduction, 228
  fluoroscopic images, 226
  GLDA, 228
  PCA, 227
  ROI, 227
  SVM, 227
  template matching, 226
RT_Image, 63
R&V. *See* Record and verify system (R&V)

S

Segmented-object-based machine learning, 104–105
Self-organizing map (SOM), 25, 26
  ANN, 299
  AUCs, 300
  dosimetric variables, 299
  neurons, 300
  PCA, 299
  radiation pneumonitis, 299, 300
Semantic Web technologies
  querying using SPARQL, 78–80
  resource description framework (RDF), 76, 94
  URIs and linked data, 76–78
Semi-supervised learning, 5, 7
Shared Health Research Information Network (SHRINE) tool, 93
Single-nucleotide polymorphisms (SNPs), 267, 284
SPARQL protocol and RDF query language (SPARQL)
  federation, 79
  HTTP protocol, 78
  RDF store, 78
  retrieving patient resources, 78
  Semantic Web technology, 80
  URL locations, 78
Sparse label propagation
  graph weights, 177
  multi-atlas-based labeling, 177
  prostate probability map, 178
  reference voxel, 177
Sparse representation-based classification (SRC)
  automatic organ segmentation, 174
  computer vision, 167
  dictionary learning, 168
  face recognition, 168
  label propagation method, 175
  multi-atlas-based segmentation, 169
  patch-based representation, 175
  probability map, 171, 172
  signal processing, 167

spatial regularization, 171
voxel intensity information, 175
Sparse shape constraint (SSC)
composition method, 187
deformable model, 188
inverse affine transformation, 187
Spatial-constrained multitask
estimation strategy, 180
prostate segmentation, 179
SCOTO, 179
SVR, 179
Spatial-Constrained Transductive Lasso
(SCOTO), 181–182
Statistical learning, 14–15, 52, 53
Statistical process control (SPC),
257–258
Stereotactic body radiotherapy (SBRT), 256
Structured Query Language (SQL), 65
Supervised learning, 4, 7
Bayesian network, 33–35
decision tree, 31–33
FFNN, 28–29
GRNN, 29
input and output samples, 21
Kernel-based methods, 30–31
logistic regression, 26–28
Naive Bayes, 35–36
Support vector machines (SVMs), 30, 31
acute oesophagitis, 298
anomaly detection, 238–239
dosimetric variables, 297
dual optimisation, 296
induction chemotherapy, 297
logistic regression, 298
MCC, 299
non-linear kernel, 296, 297
NSCLC, 298
PCA, 297
pneumonitis, 297
variables, 296
xerostomia, 297–298
Support vector regression (SVR), 179
Surveillance, Epidemiology, and End Results
(SEER) program, 65–66
SVM. See Support vector machines (SVMs)

T
TCP/NTCP, 264, 268, 269
Thin-slice screening, 111
Thoracic imaging
CXR, 117–120

lung cancer detection, 110
lung nodules in CT, 110–117
Translational medicine approach, 93
Treatment Assessment tools, 195
Treatment delivery validation
ANN, 257
breast cancer treatments, 254
DRR, 257
IMRT, 253
liver radiotherapy treatment, 256–257
PCA, 257
QA program, 253
QC, 254
radiotherapy, 254
SBRT, 257
VMAT, 253
Treatment planning system (TPS). See also
Knowledge-based treatment
planning (KBTP)
computed tomography (CT) data sets, 61,
62
computerized, 60
dose-volume histograms (DVH), 62
OARs, 58
PACS, 73
radiotherapy process, 62
Treatment planning validation
cancer radiotherapy, 244
computer-assisted techniques, 244
IMRT, 251
morbidity and mortality, 243
radiation dose, 244
VMAT, 251
Treatment response. See Bioinformatics
Treatment verification
ANN, 257
EPID, 256
SBRT, 257
True-positive fraction (TPF), 182, 183
Tumor classification, 134, 136, 142
Tumor control probability (TCP)
biological markers
BNs, 319
hypoxia and inflammation, 318
NSCLC, 320
therapeutic intervention, 318
clonogenic tumor cells, 312
decision-making, 312
dose distribution, 311
dose–volume metrics, 312
FDG-PET intensity, 316
functional/molecular imaging, 316

Tumor control probability (TCP) (*cont.*)
   locoregional failure probability, 319
   logistic regression, 313, 315
   PCA, 320
   phenomenological models, 312
   radiotherapy outcomes, 311
Tumor response, 312, 316, 318

U
Unique resource identifiers (URIs)
   and linked data, 76–77
   RDF stores, 78, 80
   semantic interoperability, 78
   unique resource locator (URL), 77
   unique resource name (URN), 77
Univariate logistic regression analysis, 288
Unsupervised learning, 4, 7
   clustering, 24–26
   input samples, 21–22
   Kernel PCA, 24
   linear PCA, 22–24
URIs. *See* Unique resource identifiers (URIs)

V
VAlidation of High TEchnology based on large database analysis by learning machine (VATE) project, 93–94
Vapnik-Chervonenkis (VC) dimension, 9, 14, 16–17
VATE. *See* VAlidation of High TEchnology based on large database analysis by learning machine (VATE) project
VC dimension. *See* Vapnik-Chervonenkis (VC) dimension
Volumetric-modulated arc therapy (VMAT), 251

W
Wilcoxon's signed-rank test, 53–54

X
X-ray computed tomography (X-ray CT), 61

MIX
Papier aus verantwortungsvollen Quellen
Paper from responsible sources
FSC® C105338

If you have any concerns about our products,
you can contact us on
ProductSafety@springernature.com

In case Publisher is established outside the EU,
the EU authorized representative is:
**Springer Nature Customer Service Center GmbH
Europaplatz 3, 69115 Heidelberg, Germany**

Printed by Libri Plureos GmbH
in Hamburg, Germany